*Clinical Application
of Neuropsychological
Test Batteries*

Clinical Application of Neuropsychological Test Batteries

Edited by

THERESA INCAGNOLI
Veterans Administration Medical Center
Northport, New York
and School of Medicine
State University of New York
Stony Brook, New York

GERALD GOLDSTEIN
Veterans Administration Medical Center, Highland Drive
and University of Pittsburgh
Pittsburgh, Pennsylvania

and

CHARLES J. GOLDEN
University of Nebraska Medical Center
Omaha, Nebraska

Plenum Press • New York and London

Library of Congress Cataloging in Publication Data

Main entry under title:

Clinical application of neuropsychological test batteries.

Includes bibliographies and index.
1. Neuropsychological tests. 2. Halstead-Reitan Neuropsychological Test Battery. 3. Luria-Nebraska Neuropsychological Test Battery. I. Incagnoli, Theresa. II. Goldstein, Gerald, 1931– . III. Golden, Charles J., 1949– . [DNLM: 1. Neuropsychological Tests. 2. Neuropsychology. WL 103 C6413]
RC386.6.N48C52   1985   152   85-25674
ISBN 0-306-42045-7

© 1986 Plenum Press, New York
A Division of Plenum Publishing Corporation
233 Spring Street, New York, N.Y. 10013

All rights reserved

No part of this book may be reproduced, stored in a retrieval system, or transmitted in any form or by any means, electronic, mechanical, photocopying, microfilming, recording, or otherwise, without written permission from the Publisher

Printed in the United States of America

Contributors

Nelson Butters, *San Diego Veterans Administration Medical Center and Department of Psychiatry, University of California School of Medicine, San Diego, California*

Gordon J. Chelune, *Veterans Administration Medical Center and Department of Psychiatry, University of California, San Diego, California*

Kathleen L. Edwards, *Department of Psychiatry, University of Pittsburgh School of Medicine, Pittsburgh, Pennsylvania*

William Ferguson, *Behavioral Medicine Service, St. Joseph's Hospital, Lancaster, Pennsylvania*

Carol Gainer, *Murfreesboro, Tennessee*

Charles J. Golden, *University of Nebraska Medical Center, Omaha, Nebraska*

Gerald Goldstein, *Veterans Administration Medical Center, Highland Drive, Pittsburgh, and Department of Psychiatry and Psychology, University of Pittsburgh, Pittsburgh, Pennsylvania*

Harold Goodglass, *Boston Veterans Administration Medical Center, and Department of Neurology, Boston University School of Medicine, Boston, Massachusetts*

Theresa Incagnoli, *Veterans Administration Medical Center, Northport, New York, and School of Medicine, State University of New York, Stony Brook, New York*

Robert L. Kane, *Veterans Administration Medical Center, East Orange,*

New Jersey and Department of Neurosciences, University of Medicine and Dentistry of New Jersey, Newark, New Jersey

Andrew Kertesz, *Department of Clinical Neurological Sciences, St. Joseph's Hospital Research Institute, London, Ontario, Canada*

Mark Maruish, *National Computer Systems, Minnetonka, Minnesota*

Kurt Moehle, *Department of Psychology, Purdue University School of Science, Indianapolis, Indiana*

Oscar A. Parsons, *Department of Psychiatry and Behavioral Sciences, University of Oklahoma Health Sciences Center, Oklahoma City, Oklahoma*

Andrew Phay, *Veterans Administration Medical Center, Murfreesboro, Tennessee*

Abigail Sivan, *Child Development Clinic, Department of Pediatrics, University of Iowa, Iowa City, Iowa*

Ralph E. Tarter, *Department of Psychiatry, University of Pittsburgh School of Medicine, Pittsburgh, Pennsylvania*

Nils R. Varney, *Veterans Administration Medical Center, and Department of Psychiatry, University of Iowa College of Medicine, Iowa City, Iowa*

Preface

Clinical neuropsychology has become a field of major prominence during the past several years, as well as a field of great complexity. As a result of the extensive amount of activity that neuropsychology has experienced recently, two major developments have emerged. First, several approaches have evolved regarding comprehensive neuropsychological assessment. There are presently several standard test batteries in common use, as well as an approach to assessment that does not make use of standard batteries, but rather fits the evaluation accomplished to the particular diagnostic problems presented by the individual patient. Second, a great deal of specialization has emerged, with assessment procedures developed for evaluation of specific types of neuropsychological deficit. The purpose of this volume is to review these developments, particularly with reference to their implications for application in clinical settings.

The history of this book's development is of particular interest. Some years ago, the Veterans Administration became concerned with developing an optimal method of neuropsychological assessment for its many health-care related facilities. Initially, the problem was conceptualized in terms of whether the VA should encourage wider use of the long-established Halstead-Reitan Neuropsychological Test Battery (HRB) or promote more extensive use of the recently developed Luria-Nebraska Neuropsychological Test Battery (LNNB). A conference was funded to bring together authorities in clinical neuropsychology to discuss this matter and present a series of papers to invited psychologists from various VA facilities. However, the planners of the conference soon discovered that a simple comparison between these two procedures would not be the most productive route, and that a broader perspective needed to be taken. It was pointed out that many neuropsychologists do not use standard batteries, but have adopted a more flexible approach representing the philosophies of such pioneers in the field as Kurt Goldstein, Heinz Werner, and Alexander R. Luria. It was pointed out further that not all neuropsychological assessment is a matter of administering comprehensive screening batteries of the HRB or LNNB type. Some of it involves detailed assessment of specific areas, notably language, memory, and visual-spatial skills, that may only be briefly evaluated by the standard batteries. Additionally, it became clear that the administration of tests was not the only means of neuropsychological assessment, and that much can be learned from clinical interviews of the patient, from the medical and social history, and from interviews with informants. The conference that was held, therefore, consisted of a much broader array of topics than was originally contemplated, and went far beyond comparisons between the HRB and LNNB. There were also papers on memory evaluation, language testing, flexible and individualized approaches to evaluation, and the use of the interview and history in neuropsychological assessment.

When this book was planned, the editors took note of these developments in the planning of the conference and of the issues and controversies that were raised at the conference itself. They decided that the book should reflect a wide range of assessment strategies and philosophies including contributions by advocates of the two major standard neuropsychological batteries, as well as by advocates of the individualized, flexible battery approach. Furthermore, contributions were invited from individuals with expertise in the use of the interview and history, in the use of standard psychological tests as applied to clinical neuropsychological areas, and in various aspects of specialized testing. The book as a whole therefore reflects a comprehensive overview of the recent status of neuropsychological assessment and of the controversies and issues in the field that are of great current interest. The editors are pleased to note, as students of clinical neuropsychology are no doubt aware, that the list of contributors contains individuals who are in distinct disagreement among themselves regarding major philosophical and practical matters. We would like to note the spirit of friendly controversy that characterized the development of the book, and to compliment the contributors not only on their scholarly work, but on their willingness to participate in a joint effort with others of persuasions different from their own.

Drs. Incagnoli and Goldstein would like to acknowledge the support of the Veterans Administration with regard to preparation of the book and for support of the VA-sponsored research that it contains. We are indebted to Joseph L. Mancusi, Associate Director of Psychology, Veterans Administration, for his encouragement of that conference. Vivian Avery, Nelson Butters, and Andrew Phay were instrumental in planning the conference. Special thanks are also due to Christopher

Carlson, Chief, Psychology Service, VA Medical Center, Northport, New York for his continued support of this project. We would like to acknowledge the assistance of Lorraine Hummel and Adelaide Goertler for assistance in preparation of the manuscript.

THERESA INCAGNOLI
GERALD GOLDSTEIN
CHARLES J. GOLDEN

Contents

CHAPTER 1

Current Directions and Future Trends in Clinical Neuropsychology   1

Theresa Incagnoli

The Evolving Role of the Clinical Neuropsychologist	1
Framework Within Which to Conduct Neuropsychological Evaluation	3
Approaches to Evaluation	7
Current Issues and Future Directions in Clinical Neuropsychology	12
References	32

CHAPTER 2

Clinical Interviewing of the Patient and History in Neuropsychological Assessment   45

Andrew Phay, Carol Gainer and Gerald Goldstein

Purposes of the Interview and History in Neuropsychological Assessment	46
The Role of the Interview and History in Treatment Planning	53
Assessment Methods	56
General Summary and Conclusions	69
References	70

Chapter 3

The Role of Standard Cognitive and Personality Tests in Neuropsychological Assessment   75

Gordon J. Chelune, William Ferguson and Kurt Moehle

Introduction	75
Neuropsychology and the Role of Assessment	76
Neurodiagnosis and the Standard Battery	80
Assessment Versus Screening	107
Future Directions	108
Summary and Conclusions	110
References	111

Chapter 4

The Flexible Battery in Neuropsychological Assessment   121

Harold Goodglass

Introduction	121
Nonfocal Cognitive Deficits and their Evaluation	124
Focal Cognitive Deficits and their Evaluation	128
References	134

Chapter 5

Neuropsychological Batteries   135

Ralph E. Tarter and Kathleen L. Edwards

Introduction	135
Characteristics of Test Batteries	136
Types of Test Batteries	137
Theoretical Bias in Battery Development	138
A Decision Approach to Systematic Assessment	142
The Meaning of Neuropsychological Test Results	146
Dimensions of Neuropsychological Assessment	148
Assessment of Children and Adults: Special Considerations	149
Normative and Criterion Assessment	151
Summary	151
References	152

Chapter 6

Overview of the Halstead-Reitan Battery   155

Oscar A. Parsons

Introduction	155
Development of the Halstead-Reitan Battery	156
Clinical Application of the Halstead-Reitan Battery	163
Examples of Neuropsychological Reports	174
Future Developments of the Halstead-Reitan Battery	184
References	189

Chapter 7

The Luria-Nebraska Neuropsychological Battery   193

Charles J. Golden and Mark Maruish

The Luria-Nebraska Scales	194
Research on the Luria-Nebraska	197
Interpretation of the Luria-Nebraska	204
References	227

Chapter 8

An Overview of Similarities and Differences between the Halstead-Reitan and Luria-Nebraska Neuropsychological Batteries   235

Gerald Goldstein

Introduction	235
Historical Considerations	236
Similarities and Differences	243
Questions and Answers	261
Concluding Remarks	268
References	271

Chapter 9

Comparison of Halstead-Reitan and Luria-Nebraska Neuropsychological Batteries: Research Findings   277

Robert L. Kane

Studies Using Raters	279
Statistical Relationships	281
The HRB, LNNB, and Psychometric Intelligence	294
The LNNB and the Assessment of Memory	295
Concluding Comments	298
References	300

CHAPTER 10

A Comparison of the Halstead-Reitan, Luria-Nebraska, and Flexible Batteries Through Case Presentations 303

Gerald Goldstein and Theresa Incagnoli

Introduction	303
A Case of Huntington's Disease	304
A Case of Schizophrenia	311
A Case of Mnestic Dysfunction in Thalamic Infarction	318
Summary	325
References	326

CHAPTER 11

Assessment of Aphasia   329

Andrew Kertesz

The Development of Aphasia Testing	329
Principles of Aphasia Testing	331
Comprehensive Aphasia Examinations	338
Tests of Communicative Function	351
Short Screening Tests in Aphasia	352
Aphasia Examinations for Polyglots, and in other Languages	354
Modality Specific Tests of Aphasia	354
Conclusion	356
References	356

CHAPTER 12

The Clinical Aspects of Memory Disorders: Contributions from Experimental Studies of Amnesia and Dementia   361

Nelson Butters

Comparisons of the Memory Disorders of Patients with
Alcoholic Korsakoff's Syndrome and Patients with Huntington's
Disease 362
Conclusions 379
References 380

CHAPTER 13

Visual-Spatial Disabilities   383

Nils R. Varney and Abigail B. Sivan

Introduction 383
Constructional Apraxia 384
Recognition and Discrimination of Faces 388
Dressing Apraxia 389
Optic Ataxia 392
Topographical Disorientation 393
Neglect 395
Perception of Direction 398
Conclusion 399
References 400

Index 403

1

Current Directions and Future Trends in Clinical Neuropsychology

THERESA INCAGNOLI

THE EVOLVING ROLE OF THE CLINICAL NEUROPSYCHOLOGIST

The neuropsychological evaluation can be viewed as an extension and quantification of that part of the neurologic examination concerned with higher cortical functions. Luria (1973a) has noted that lesions of the highest zones of the cortex are inaccessible in classical neurological examination. Furthermore, neuropsychological evaluation of cognitive and personality changes may be more sensitive than the neurological exam to early manifestations of brain disease (Reitan, 1976a).

Costa (1983) has noted that clinical neuropsychology is truly a discipline in evolution, as advances in clinical medicine and neuroscience research impact its future direction. The results of basic science research are continually incorporated into assessment techniques and rehabilitation interventions. The work of Benton, Butters, Goodglass, and Kertesz, among others, attests to such a relationship.

The contention that lateralization and localization of the lesion represented one of the two most commonly asked questions in this area would now appear to be outdated in light of recent neuroradiologic

---

THERESA INCAGNOLI • Veterans Administration Medical Center, Northport, NY 11768, and School of Medicine, State University of New York, Stony Brook, NY.

innovations, such as CT and PET scans (Walsh, 1978). Indeed we have already begun to witness the realization of Wedding and Gudeman's (1980) contention that the role of the clinical neuropsychologist would shift from lesion localization to precise specification of deficits for litigation and/or rehabilitation. It is of interest to note the opposing contention (Bigler & Steinman, 1981) that "the increased use of CAT scanners will not supplant a neuropsychological function (i.e., lesion localization), since this function was never widely accepted" (p. 195).

Although these technological advances will alter the referral questions posed to neuropsychologists, differential diagnosis will continue to represent a major role for the clinical neuropsychologist, albeit in a different fashion. For example, as the geriatric segment of our population continues to expand, the clinical neuropsychologist's involvement with diagnostic issues will increase for this group. The neuropsychologist will be asked to assist in the differential diagnosis of depression/dementia and to evaluate patients suspected of Alzheimer's disease, particularly in the incipient stages. Albert (1981) has provided a comprehensive review of the problems associated with such diagnoses and has recommended various approaches to evaluation.

Neuropsychologists will continue to play a role, once again an evolving one, in confirmed cases of neurologic insult. For patients who have suffered acute involvement (cerebrovascular accident, neoplasm, or head trauma), neuropsychologists will continue to delineate strengths and weaknesses, providing a profile of higher cortical functions. Neuropsychological examination will continue to provide quantitative and qualitative information regarding an individual's functioning. However, these results will be increasingly utilized to formulate rehabilitative interventions. Diller and Weinberg (1977) have noted that traditionally, neuropsychologists have operated on the diagnosis and symptom-description phase and have neglected rehabilitation. Neuropsychologists will be asked to make predictions regarding prognosis, to develop appropriate intervention strategies, and to document the rate and degree of recovery through repeated examinations.

For patients with progressive neurologic disease (multiple sclerosis, Alzheimer's disease) neuropsychological examination will be requested to assist in early and/or differential diagnosis, document the degree of change in performance, and provide information regarding the course of cognitive change. For this purpose such patients are often referred for serial examinations.

The evolving role of the neuropsychologist will further expand to encompass the effects of various medical disease entities. The medical disorder/neuropsychology interface is already marked by promising starts in cardiac, pulmonary, renal, hepatic (liver), and behavioral toxicity conditions (Ariel & Strider, 1983). This area is discussed at length in the medical neuropsychology subsection in this chapter.

Although many of the diagnostic issues that neuropsychologists address will change, others will remain constant. Neuropsychologists will continue to perform pre- and postexaminations to assess the effectiveness of an intervention, whether it be neurosurgical (resection of tumor, shunt insertion for normal pressure hydrocephalus) or pharmacologic in nature. The role of the neuropsychologist in the preoperative localization of lesions in focal epilepsy will continue because positive signs are frequently unobserved in neurologic or radiologic tests. Those psychiatric or neurologic cases in which there is a change in behavior/cognition while neurologic tests are normal or equivocal will remain within the diagnostic province of the neuropsychologist.

Other neuropsychologists will continue to function within the purview of pediatric neuropsychology. For a summary of more recent findings the reader is referred to Rourke, Bakker, Fisk, and Strang (1983). Child neuropsychology, however, is beyond the scope of this text.

FRAMEWORK WITHIN WHICH TO CONDUCT NEUROPSYCHOLOGICAL EVALUATION

A framework for neuropsychological evaluation can be conceptualized as consisting of three distinct entities:

1. Current and background information
2. Neuropsychological examination itself
3. Feedback to patient and/or significant others and to referral source

Current and Background Information

The importance of obtaining current and background information cannot be overemphasized; it allows for the generation of hypotheses that will be evaluated and provides a context in which to interpret neuropsychological data (Lezak, 1982; Luria, 1966). An example would be the case of a patient referred for neuropsychological examination due to the recent onset of memory problems. During the clinical interview it becomes apparent that this individual is quite depressed; the patient reports a recent loss of a significant other and queries regarding vegetative signs of depression are answered affirmatively. Based on this information the hypothesis is entertained that such mnestic dysfunction may indeed represent a depressive pseudodementia. Such a hypothesis will alert the examiner to note particular patient characteristics. Is failure of an item characterized by long reaction times and excessive time requirements for completion? If so, the neuropsychologist will feel comfortable in ascertaining that, in all likelihood, we are dealing with a pseudodementia. Here, data obtained from the clinical interview was instrumental in hypothesis generation and in the interpretation of examination data.

Current and background information is obtained from various sources, including medical records and a clinical interview of the patient and/or significant others. Ideally, medical records are surveyed prior to the actual patient interview; this provides direction for inquiry for the examiner as well as the opportunity to compare patient recollection with reported events. The type of information most fruitful includes hospital admission and discharge summaries, consults from neurology, psychiatry, and rehabilitation medicine services, neurosurgical reports and results of neurologic/neuroradiologic procedures, such as EEG, CT, PET scans, and NMR imaging. Information regarding previous psychological evaluations, including dates and tests administered, are also noted.

Although much valuable information is obtained from the clinical interview, this area is a sorely neglected one in clinical neuropsychology, with few exceptions (Filskov & Leli, 1981; Lezak, 1983). An extensive treatment of this area is provided in Chapter 2. The guidelines proposed here are based on those utilized by Dr. Russell L. Adams at the University of Oklahoma Health Sciences Center.

For children or those patients unable to provide a history (due to mnestic or language difficulties), an interview with a significant other is essential. It is very often helpful to interview these individuals in other instances as well. In many cases the patient is unaware of or denies deficits that are conspicuous to those who live with or know the patient.

It is most helpful to begin a clinical interview with a general inquiry regarding presenting complaints, followed by more specific questions regarding frequently noted problem areas. The onset and course of those symptoms are noted. This is followed by detailed medical and psychiatric histories. The latter should include past and current information, with particular attention to depression (Lezak, 1983) and anxiety (King, Hannay, Masek, & Burns, 1978; Mueller, 1979; Wrightsman, 1962) because these can affect examination results. Note should also be made of psychotropic medications and dosage (Heaton & Crowley, 1981). The importance of information regarding alcohol and drug abuse has been noted by Parsons and Farr (1981) and Parsons and Adams (1983). Questions regarding legal involvement should address whether

Table 1. Guide for Clinical Interview

Presenting problem

General—Patient's understanding of problem
Specific—Attention and concentration
        Memory
        Planning and organizing
        Being confused or easily distracted
        Change in personality or behavior
        Driving
        Cognitive skills (arithmetic, reading, writing)
Onset—Acute/chronic
Course—Improvement/deterioration/static

Medical history

Conditions being treated for (CVA, TIA, high blood pressure, encephalitis, meningitis)
Medication and dosage at time of evaluation
Family history of presenting symptoms and medical illness
Epilepsy and convulsions
Auto accidents
Head injury
Unconsciousness from sports, fall or being hit by an object and period of unconsciousness

Psychiatric history

Past
  Previous hospitalizations including reasons for and dates of hospitalization
  ECT
  Family history
Current
  Hallucinations (auditory, visual, tactile, olfactory)
  Paranoia
  Special powers
  Special relationship with God
  Medication and dosage
  Depression
    Vegetative signs of depression including
    Sleep disturbances
    Weight loss/change in appetite
    Change in libido
  Anxiety
    Life stressors

Alcohol and drug abuse

Specific drugs and/or type of alcohol consumed
Length of use

(continued)

Table 1 (Continued)

Alcohol and drug abuse
Quantity consumed
Blackouts, DTs, binges/OD, coma
Accidents related to use
Arrests for DWI (driving while intoxicated)
License revoked

Legal involvement
Problems with the law
Arrests and reason for arrest
Prison sentence and length
Litigation
Disability, pension, workmen's compensation

Education and occupation
Highest grade completed
Type of curriculum
Grades attained
History of learning problems (specify area and type of problem)
History of adjustment problems
Special classes
Repetition of grade
Type and length of employment
Problems on job (cognitive, emotional, interpersonal) and how resolved
Termination of employment (quitting, fired)

Military service
Rank at discharge
Type of discharge
Combat
Injuries sustained
Court marshal

Typical day and plans for future

Sensory and motor problems
Glasses/hearing aid (worn or forgotten)
Paralysis

the patient is currently involved in any litigation because this could affect examination performance. McMahon (1983) cites the possible complicating factors of hysteria and malingering. Information regarding education and occupation allows estimates of premorbid levels of intelligence to be made (Leli & Filskov, 1979; Matarazzo, 1972; Wilson et al., 1978). How patients spend their days and what plans the patients have for their futures provide worthwhile information. Sensory and motor problems should be noted.

Neuropsychological Examination will be discussed in detail in the next section on Approaches to Evaluation. Such an evaluation should address both cognitive and affective components. Russell (1975, 1977) has recommended the sequential use of personality tests with neuropsychological measures.

FEEDBACK

The results of the neuropsychological examination may be reported to the patient and to significant others in a feedback session in which specific recommendations are provided. Depending on the individual case this might consist of cognitive retraining, a form of psychotherapy (individual, group, family), medical consultation, additional diagnostic procedures, institutionalization, or some combination. This feedback should also be provided to the referring professional, and where appropriate the neuropsychologist should meet with the treatment team to formulate an intervention strategy.

APPROACHES TO EVALUATION

Three approaches to neuropsychological evaluation have been identified; these include behavioral neurology, the individual-centered normative approach, and the utilization of neuropsychological batteries (Beaumont, 1983). Each of these will be discussed in turn.

BEHAVIORAL NEUROLOGY

Behavioral neurology, which is rooted in 19th century European neurology, emphasizes a conceptual rather than an operational definition of behavior. American interest was stimulated by Geschwind's (1965) seminal paper on "Disconnexion Syndromes in Animals and Man," in which he reported the European literature. Such an approach has as its goal the in-depth evaluation of an individual, utilizing tests devised specifically to evaluate brain-damaged patients; there is little reliance on standardized tests or normative data. Tests are selectively utilized depending on the patient's responses and constitute a branching examination that relies heavily on the skill of the examiner in deciding how to proceed further. Emphasis is on the manner in which a problem is solved or failed rather than on whether it is solved.

Individual qualitative neuropsychological investigation is exemplified in the works of Luria (1966, 1973b). The task of the neuropsychologist becomes the "qualification of the symptom," which involves the careful analysis of a patient's deficits, whereby one attempts to elucidate a common element underlying the varied symptoms demonstrated in various forms of the patient's behavior. Such syndrome analysis is arrived at through utilization of a sole method of inference, the presence or absence of pathognomonic signs.

Such an investigation consists of four stages (Christensen, 1979b), in which the first stage, the preliminary conversation, is concerned with obtaining information regarding the patient's orientation and level of consciousness, premorbid intelligence, attitude, and symptoms. This is followed by the preliminary examination, in which the individual analyzers (motor, auditory, visual, kinesthetic) are evaluated via standardized tests that are simple and short. The selective examination of stage three focuses in detail on those deficits demonstrated in the preliminary examination. In the fourth, final stage, the formulation of a clinical neuropsychological conclusion occurs based on an analysis of the data.

Many of the test items used by Luria have been compiled by Christensen (1975, 1979a), although they were not meant to be utilized in a fixed battery format. Reitan (1976b) has criticized this investigation because the lack of standardization of procedures and materials precludes cross-validation. The strong emphasis on evaluating the functions of the left (dominant) hemisphere, the disregard for quantification, and the concern with specific deficits and their corresponding lesion locations in view of functional systems are of concern to him. Christensen (1979b; 1984) has provided case illustrations of Luria's methodology. The latter delineates the patient's assets and liabilities and how these were incorporated into a rehabilitation program.

Contemporary proponents of a behavioral neurology approach include Heilman and Valenstein (1979) and Hécean and Albert (1978). These behavioral neurologists are concerned with the delineation of syndromes, syndrome analysis (Luria 1966), rather than the quantitative analysis of psychometric data. In neuropsychology today there is renewed interest in the case-study method and the identification of specific syndromes (Marshall & Newcombe, 1984; Shallice, 1979). The reader is referred to the single case subsection of the Current Issues and Future Directions section in this chapter.

It should be noted that other neuropsychologists (among them Goodglass, Kaplan, Lezak) adhere to many of the above described tenets. However, because of their reliance on norms, their approach to neuropsychological evaluation will be discussed in a separate section, that of the "individual-centered normative" approach.

Individual-Centered Normative Approach

On a continuum with behavioral neurology at one endpoint and fixed neuropsychological batteries (which will be discussed in the next section) at the other, the individual-centered normative one lies at midpoint. Such an approach is called individualized because it emphasizes tailoring the assessment to the particular patient's deficits. In this sense it is akin to the behavioral neurology approach. Unlike that approach, however, it relies on standardized tests and normative data, where available. "Few if any examinations conducted within this framework can be identical in the instruments used, the extent to which limits are tested, or the amount of effort expended in interviewing and history-taking" (Lezak, 1984, p. 30). Such an examination is highly dependent on the examiner's skill where hypotheses must be constantly generated and evaluation instruments chosen. For a review of this approach the reader is referred to Lezak (1984). Other neuropsychologists representative of such an approach include Benton, deS Hamsher, Varney, and Spreen (1983), Goodglass and Kaplan (1979) and Walsh (1978).

Neuropsychological Batteries

Neuropsychological batteries can be either clinically or theoretically based. In a clinical battery, such as the Halstead-Reitan, the tests utilized were originally chosen because they successfully discriminated between brain-damaged individuals and individuals without such damage. Theoretical batteries can be either psychometric, such as the Luria-Nebraska Neuropsychological Battery, or nonpsychometric, as exemplified by Christensen (1979a).

Other ways in which neuropsychological batteries can be classified include the Flexible/Fixed and Qualitative/Quantitative dimensions. We will proceed to discuss each of these two classifications.

Flexible/Fixed

The flexible approach, which Goodglass refers to as "patient-centered testing," is discussed in his chapter in this text. "The examiner should acquaint himself thoroughly with the case history so that he may select his tests appropriately and be alert to interpret his findings in the light of previously known features of the disorder in question" (Goodglass & Kaplan, 1979, p. 4). This approach was exemplified in the case report of a patient who demonstrated a global sequencing processing disorder following head injury (Milberg, Cummings, Goodglass, & Kaplan, 1979).

Christensen's (1979b, 1984) utilization of a flexible battery has been discussed under the behavioral neurology section. Walsh (1978) has provided a list of tests that he cautions does not constitute a "battery". Lezak's (1983) flexible battery consists of a core battery comprising individually administered paper-and-pencil tests that the patient takes alone. In addition, specific tests are utilized to answer the referral question. Russell (1984) considers this to be an example of a "step battery," which will be discussed below. Benton also advocates a flexible battery utilizing many of the tests he and his colleagues have devised. The reader is referred to Benton et al. (1983) for a review of these instruments.

Advantages of such an approach include the lack of redundancy in evaluation and the ability to readily incorporate new research findings and tests that reflect such knowledge. However, because so much is dependent on the skill of the examiner, some deficits may go unnoticed and technicians cannot always be utilized. It is also difficult to compare scores across instruments because they are reported in different formats and are drawn from different normative samples.

The fixed approach to neuropsychological evaluation involves the administration of a standard battery of tests that are routinely administered in the same fashion to all individuals evaluated. For a review of fixed, neuropsychological batteries see Tarter's chapter in this volume. The chief proponent of such an approach is Ralph Reitan (Reitan & Davison, 1974). The reader is referred to the chapters in this text on the Halstead-Reitan Neuropsychological Battery by Oscar A. Parsons and the chapter on the Luria-Nebraska Neuropsychological Battery by its founder Charles J. Golden. An overview of the similarities and differences between these two batteries has been provided by Goldstein (this volume).

The advantages of a standard battery include its comprehensiveness, the suitability for technician administration, and the feasilibity of combining research with clinical objectives. Disadvantages pertain to the redundancy of obtained information, its inappropriateness for particular types of problems, its poor ability to account for failure, and the fact that other areas of functioning may remain unexplained.

Although it appears that the flexible and fixed approaches are mutually exclusive, Russell (1984) notes that they may in fact be complementary. The requirement of flexibility and the advantages of fixed batteries can be attained through what he refers to as a "step battery".

> The step battery is organized into steps, with certain groups of tests designed for specific purposes. For instance, a first step or group of tests would provide information that might indicate the possibility of other problems. If a problem is indicated, then the next set of tests would be administered to verify and elaborate the indications of the first step. Thus, each step acts as a screening battery for the next, more detailed step. (p. 60)

Tarter (this volume) also proposes a decision approach to systematic assessment in which the PINTS (Pittsburgh Initial Neuropsychological Testing System) (Goldstein, Tarter, Shelly, & Hegedus, 1983) is utilized as a screening battery for psychiatric patients.

Qualitative/Quantitative

The qualitative/quantitative dimension refers to the manner in which data are evaluated rather than to the method of administration by which the data are obtained. In a quantitative system, the concern is whether or not a problem is solved correctly, and if so, to what extent. Here performance is compared to a cutoff score that represents the optimal separation between a brain-damaged and a non-brain-damaged group. A qualitative evaluation, on the other hand, addresses itself to how a problem is solved or failed. Werner (1937) cautioned that measuring development via achievement "must be supplanted by an analysis of the mental processes which underlie the achievements themselves" (p. 533). Because defective performance on a task can occur for a variety of reasons, the task becomes that of identifying the underlying reason for failure.

> The process approach informs us of the compensatory strategies, both adaptive and maladaptive, that an individual has developed to cope with his or her deficit(s). . . . [It] provides information concerning the individual's differential response to varying task demands as well as stimulus parameters that may induce a more or less effective response. (Kaplan, 1983, p. 155)

Another important distinction between these two approaches resides in the methods for inferring a psychological deficit. The psychometric methods applicable to quantitative data are reviewed by Walsh (1978). Boll (1978) discusses the methods of clinical inference (level of performance, pattern of performance, pathognomonic signs, comparative performance of the two sides of the body) that have traditionally been applied to the Halstead-Reitan Battery. More recently, Russell (1984) reviewed pattern analysis methods applicable to that battery. In contrast to these approaches, the evaluation of qualitative data is much more dependent on the observations of the examiner.

New instruments, such as the Luria-Nebraska, represent a combination of quantitative performance and qualitative interpretation of results. Golden (this volume) is currently developing a standardized scoring system for qualitative data. Lezak (1983) notes that either qualitative or quantitative data is incomplete without the other; "to do justice to the complexity, variability, and subtleties of patient behavior, the neuropsychologist examiner needs to consider quantitative and qualitative data together" (p. 131).

CURRENT ISSUES AND FUTURE DIRECTIONS IN CLINICAL NEUROPSYCHOLOGY

The Role of Computers

The proliferation of computers that we are currently witnessing has indeed entered neuropsychology. A recent *APA Monitor* article (Turkington, 1984) further attests to the increasing utilization and abuse of computer testing. Concerns were voiced about issues of confidentiality, test and interpretation validity, supervision, copyright infringement, software dissemination, and lack of ethical standards. Swiercinsky (1984) has proposed guidelines regarding ethical and professional standards issues for the utilization of computers in neuropsychology. The author of a program has the responsibility to: (a) follow APA standards in test development and provide a manual that contains logic; (b) stress the unique advantages of computer technology; (c) make provision for clinician input in the processing of data; (d) provide administration standards; (e) provide norms based on computer administration, and; (f) work closely with the test publisher in marketing so that user qualifications are stipulated and advertising and manuals reflect intent and limitations. The responsibility of the publisher includes: (a) translating the author's work into computer code and making provision for compatibility of software designs; (b) writing programs in tamperproof ways; (c) building checks into the hardware and software for reliability; (d) insuring that ranges are being checked; (e) following APA standards in marketing, and; (f) creating fee structures.

Goldstein (1980) has presented a defense of the computer in the areas of diagnosis, service delivery, rehabilitation planning, and theory. Adams (1984) has noted that use of computers for administrative tasks, such as tracking patients in a longitudinal study, and in test administration, interdata analysis (tabulation of data with reference to age, sex, and education) and data analysis. The use of computers in the establishment of data banks (Costa, 1983) has also been suggested. The present discussion will focus on the utilization of computers in neuropsychology in the areas of:

1. Test administration and scoring
2. Analysis and interpretation of data obtained via routine administration of neuropsychological tests
3. Cognitive retraining for deficits in higher cortical functions

Administration and Scoring

Acker (1980) has devised a microcomputer-administered neuropsychological battery to be utilized solely as a screening instrument for chronic alcoholics. This fixed battery consists of measures specifically designed to evaluate areas of functioning in which alcoholics have been reported to be impaired. The characteristic neuropsychological profile consists of intact verbal skills with deficits in memory, particularly visuospatial functions, abstraction tasks (such as the Wisconsin Card Sorting Test or the Category Test from the Halstead-Reitan Battery) and visuospatial skills (Parsons, 1977; Tartar, 1980). The Maudsley Automated Psychological Screening (MAPS) consists of the following seven subtests: Premorbid Intelligence, Perceptual Motor Speed, Visuoperceptual Analysis, Visuospatial Ability, Memory (verbal and visuospatial), and Abstract Problem Solving. Normative data are currently being collected. It is still too early to evaluate the effectiveness of this screening battery. Although designed for a chronic alcoholic population, it may well mark the beginning of an era in which microcomputer-administered screening instruments will be devised for other populations.

In juxtaposition to Acker's (1980) fixed screening battery rests the California Neuropsychological System (Fridlund & Delis, 1983), a series of tests that can be utilized in a flexible fashion depending on the assessment needs of each individual patient. These tests are based on the "process approach" (Kaplan, 1983; Werner, 1937) that has been discussed in the section on approaches to evaluation. For example, after the patient has completed drawing the Rey-Osterrieth Complex Figure on the monitor with a lightpen, the computer will regenerate the patient's drawing at the examiner's request and inquire about process data. One such question is "Was the large rectangle drawn as a unit?" The computer then tabulates and summarizes process data (D. Delis, personal communication, October 19, 1983).

To date, three tests in this system have been completed; these consist of the California Verbal Learning Test, which was modeled after the Rey Auditory Verbal Learning Test (Rey, 1964; Lezak, 1983), the California Proverb Test, and the California Finger Tapping Test. Each of these tests can be administered and scored either manually or via microcomputer (IBM PC and Apple).

Other computer-administered and scoring systems consist of the SAINT-II, System for the Administration and Interpretation of Neuropsychological Tests, which consists of 10 brief neuropsychological tests (Swiercinsky, 1984) in the areas of spatial orientation, motor (finger-tapping), memory (verbal-logical and figural), latent learning (symbol-number coding), vocabulary, abstract sequencing, rhythm discrimination, and number and alternating connection tasks. Many of these tests represent computerized analogues of similar tasks in the Halstead-Reitan Neuropsychological Battery. To prevent confusion, it should be noted that SAINT-II differs from SAINT (Swiercinsky, 1978), which was an actuarial approach for computerized interpretation of the Halstead-Reitan Neuropsychological Battery.

Analysis and Interpretation of Data Obtained via Routine Neuropsychological Test Administration

The first computerized interpretation was the "key approach" (Russell, Neuringer, & Goldstein, 1979) to the Halstead-Reitan Neuropsychological Battery, patterned after the biological key species classification technique. It provides a successive set of decision rules for localization (left, right, or diffuse) and process (acute, static, or congenital) if the criterion for brain damage, an Average Impairment Rating (AIR)[1] of 1.55 or greater, is met.

In the initial validation study, the AIR correctly classified 94% of the brain damaged subjects and 71% of the controls. Predictions regarding localization were correct at 81% for left-hemisphere damage, 80% for right-hemisphere damage, and 62% for diffuse impairment (see Table 2). The process key correctly classified lesions as acute (93%), static (77%), or congenital (88%). The successful delineation regarding the presence or absence of brain damage was upheld in the cross-validation study of Anthony, Heaton, and Lehman (1980), although at lower levels than those originally reported by Russell et al. (1970) (see Table 2). Unfortunately, on cross-validation, the process key in particular did not hold up well, performing at little more than chance level.

The discrepancies between the original Russell et al. (1970) study and the subsequent cross-validation study may well be due to the different neurological criteria employed in the two studies (Anthony et al., 1980). Russell et al. (1970) were well aware of the metholological flaw in their study, in which physical examination and neurological history were the diagnostic criterion. This limitation has been emphasized by Boll and Reitan (1971) and Schwartz (1971) in their reviews of the Russell et al. book. Because lateralization was based on sensory and motor findings from the neurological exam, and because the localization key relied on similar examination procedures, the high degree of concordance between two such measures was not suprising. In the Anthony et al. (1980) study, lateralization was based on neurosurgical and neuroradiological findings. "It is therefore likely that lateralized brain damage resulted in fewer clinically obvious sensory and motor deficits in many of our subjects than in the subjects of the Russell et al. study" (Anthony et al., 1980, p. 323).

---

[1]AIR is based on 11 measures from the Halstead-Reitan battery and one Wechsler Adult Intelligence Scale subtest score.

Table 2. Percentage of Subjects Correctly Classified by Neuropsychological Criteria as Brain Damaged or Not Brain Damaged

Summary index	Brain damaged	Control
Key	94.0% (N=80)	71.0% (N=24)
BRAIN I	96.0% (N=144)	92.0% (N=36)
Cross-validation Key[a]	74.7% (N=150)	89.0% (N=100)
Cross-validation BRAIN I[a]	84.0% (N=150)	71.0% (N=100)

[a]From "An Attempt to Cross-validate Two Actuarial Systems for Neuropsychological Test Interpretation" by W. Z. Anthony, R. K. Heaton, and R. W. Lehman, 1980, Journal of Consulting and Clinical Psychology, 48(3). Copyright 1980 by the American Psychological Association. Adapted by permission.

Comparisons of the key approach with discriminant functions (Swiercinsky & Warnock, 1977) and clinician ratings (Heaton, Grant, Anthony, & Lehman, 1981) have been reported. Because diagnosis in the Swiercinsky and Warnock (1977) study was based on a complete neurological examination, as in the Russell et al. (1970) study, the same methodological shortcomings are applicable. Although the neuropsychological key was more accurate (86.8%) than the discriminant analysis (68.7%) in correctly predicting brain damage, the reverse was true for the non-brain-damaged population. Here, the key correctly predicted only 48.7% of those patients as compared to 71.8% with the discriminant. When the classification predictions for non-brain damage and lateralization (left, right, or diffuse) are considered, both measures performed poorly; the key correctly identified 42.7% of the cases whereas the discriminant function predicted 51.9% of the cases.

The efficacy of the key approach has also been compared with the diagnostic accuracy of two experienced clinicians for the dimensions of presence, laterality, and process of brain damage (Heaton et al., 1981). The clinicians attained a higher overall hit rate than either the Average Impairment Rating or the Localization Key on the presence and laterality criterion. The higher accuracy prediction of the key regarding process (acute vs. chronic) was not statistically significant. Interestingly, neither the clinicians nor the key predicted chronicity any better than the base rate did. The superiority of the clinicians is explained in terms of their flexibility in weighting complex, highly variable combinations of test data. It will be interesting to note whether this current advantage will persist in the light of more sophisticated actuarial techniques.

Adams (1975) subsequently developed another automated program for interpreting input data from the Halstead-Reitan Neuropsychological Battery. The program was validated on 63 cases (21 left-hemisphere damaged, 21 right-hemisphere damaged, 21 diffusely damaged) drawn from the files of Dr. Ward C. Halstead. Although neurological documentation consisting of histological/neurosurgical confirmation was available for the lateralized cases, this was so for only 2 of the 21 patients in the difussely damaged criterion group. Here, classification was on the basis of the neurological examination. The automated report was 75% accurate in determining the presence or absence of brain damage, 66% accurate regarding lateralization, and 44% accurate when both of these criteria were required. These results are less impressive when one considers the lack of a control group.

Gregory (1976) has devised a computer system called Interpretation of Brain lmpairment Tests (IBIT) that accepts as input 41 quantitative scores from the H-R battery (including all the scale scores of the MMPI with the exception of the Si scale) and 31 neurological symptoms that are indicative of either right- or left-hemispheric damage. Output statements deal with location, lateralization, and diagnosis, with the latter qualified by a likelihood score. Target criteria for this computer system consisted of Reitan's personal interpretation of 10 cases for the three categories of output statements. Unfortunately, difficulty in interpreting the meaning of certain cell categories precluded statistical analysis for the site of damage. Concordance was 100% for degree of lateralization, whereas the computer agreed with 18 of 24 of Reitan's diagnostic statements. Aside from the small sample size, the chief limitation of this system (and one of which the author is well aware) rests on its use of case-history information to generate decision rules that were subsequently utilized to classify the same case. The need for a cross-validation study remains.

BRAIN I (Finkelstein, 1977), a Fortran IV program that accepts inputs from the Halstead-Reitan Neuropsychological Test Battery, is more sophisticated neuropsychologically than the key approach. This program utilizes a logical decision-tree model for output diagnostic inferences regarding presence or absence of brain damage, lateralization of the lesion, presence or absence of recent tissue destruction, and the most likely neurological diagnosis. BRAIN I was validated on 144 of Reitan's brain damaged patients and 36 controls; evidence of structural lesion was based on neurosurgical reports, neuroradiological procedures, EEG, autopsy, and neurological examinations. It will be recalled that in the Russell et al. (1970) study, inclusion in the brain-damaged group was based solely on neurological examination.

Preliminary results with BRAIN I were encouraging in that it correctly classified 96% of the brain-damaged and 92% of the controls. Predictions regarding lateralization and recent tissue destruction were correct in 75% and 83% of the cases respectively. BRAIN I correctly predicted one of 11 specific neurological diagnoses in 64% of the cases. In a cross-validation study, such results were, unfortunately, not upheld; BRAIN I correctly classified 84% of the brain-damaged and 71% of the controls. In addition, inferences regarding lateralization and the presence or absence of recent tissue destruction were at chance levels of prediction. Predictions regarding correct neurological diagnoses were accurate in only 30% of the cases identified as brain damaged.

In the cross-validation and comparison study of the key approach and BRAIN I (Anthony et al., 1980), BRAIN I was superior (84% accurate) to the key (74.7% accurate) in classifying brain-damaged individuals. However, for controls, the converse was true, with the key correctly predicting 89% whereas BRAIN I predicted 71% accurately (see Table 2). The poor predictions of the key for diffuse damage and of BRAIN I for all lateralization categories are reported in Table 3.

In summation, although the key and BRAIN I can reliably predict presence or absence of brain damage, neither does so with the same degree of accuracy for either localization or process. "Our results indicate that little confidence can be placed in most of the 'finer' diagnostic distinctions made by these programs regarding the location, chronicity, and etiology of cerebral lesions" (Anthony et al., 1980, p. 322). At the

Table 3. Predictions of Localization by Neuropsychological Criteria

Study	Neurological criterion		
	Right	Left	Diffuse
Original Key[a]	80.0%	81.0%	62.0%
Cross-validation[b]	69.4%	60.6%	32.3%
BRAIN I[c]	86.1%	66.7%	73.6%
Cross-validation[b]	41.0%	31.0%	36.0%

[a]From *Assessment of Brain-damage: A Neuropsychological Key Approach* by E. W. Russell, C. Neuringer, and G. Goldstein. Copyright 1970 by John Wiley and Sons, Limited. Adapted by permission.
[b]From "An Attempt to Cross-validate Two Actuarial Systems for Neuropsychological Test Interpretation" by W. Z. Anthony, R. K. Heaton, and R. W. Lehman, 1980, *Journal of Consulting and Clinical Psychology*, 48(3). Copyright 1980 by the American Psychological Association. Adapted by permission.
[c]From "BRAIN: A Computer Program for Interpretation of the Halstead-Reitan Neuropsychological Test Battery" (Doctoral Dissertation, Columbia University, 1976), by J. N. Finkelstein 1977, in *Dissertation Abstracts International*, 37. Copyright 1977 by University Microfilms International. Adapted by permission. $N=144$

same time, however, those authors, realizing the potential contribution of such programs, suggest ways in which future versions might be improved. These include employment of additional algorithms for age and education, incorporation of other measures of premorbid intellectual ability in lieu of the Full Scale IQ, and utilization of more optimal test score weightings for the lateralization, process, and neurological variables predictions.

Computers are also being utilized to score data from the Luria-Nebraska Neuropsychological Battery (Western Psychological Services, 1984). This program provides results of the clinical and summary, localization and factor scales, as well as relative strengths and weaknesses for each of these scales. In addition, estimated IQ scores are provided for Verbal, Performance, and Full Scale IQs.

Computers and Cognitive Rehabilitation

Computers are also increasingly being utilized in the area of cognitive rehabilitation. The precision with which the computer can present stimuli, time events, and record responses has led Bracy (1983) to view it as perhaps the most valuable tool we have in cognitive retraining. Computers are an integral feature of outpatient retraining programs because they allow establishment of homebound programs. Factors to be considered in choice of a microcomputer include versatility and availability of software, serviceability, modular versus complete unit, speed, color, graphics, cassette versus disk, and memory (Gianutsos, 1981). Software is of two kinds: structured and unstructured. Examples of structured programs include those developed in the areas of perception and memory (Gianutsos & Klitzner, 1981) and those in foundation skills (reorganization of basic processes including attention, initiation/inhibition), visual-spatial, conceptual, memory, and auditory perception skills (Bracy, undated). Unstructured programs consist of games and educational programs that can be utilized or modified for cognitive rehabilitation. Lynch (1982) has discussed the role of video games in the diagnosis and treatment of cognitive deficits.

Computers and the Elderly

Although the computer lends itself to many specialized problems inherent in a geriatric setting, its use remains limited and scarcely explored. Automated systems that have been devised (Gedye & Miller, 1970; Levy & Post, 1975; Perez, Hruska, Stell, & Rivera, 1978) allow evaluation of patients who would be considered untestable utilizing traditional psychometric tests. Because these systems generally operate on a "match-to-sample" procedure, the range of difficulty can be quite wide. Miller (1980) has commented on the sensitivity of automated testing to changes in cognitive functioning at low levels of functioning. The need for brief repeated measures of performance in documenting the variability of cognitive functioning in delirium is well suited by the sophistication and availability of the microcomputer (Albert, 1981).

New Developments in Traditional Neurological Disorders

This section will focus on recent developments within traditional neurological disorders that have been studied by neuropsychologists. The topics of head injury, alcoholism, cerebrovascular disorders, and dementia will each be addressed in turn.

Head Injury

The first comprehensive, critical summary of present knowledge about the behavioral effects of traumatic head injury, particularly closed head injury, is the current text *Neurobehavioral Consequences of Closed Head Injury* (Levin, Benton, & Grossman, 1982). In his review of this book Shore (1982) comments that "Taken as a whole, the book's strengths include breadth of coverage, stringent attention to methodological issues, and objectivity of critical analyses" (p. 376). Emphasis is on research findings rather than impressions accrued during clinical experience. In addition to relevant background material (chapters on pathophysiologic mechanisms, epidemiology) and prognostic indications (outcome and social recovery), Levin *et al.* review the neuropsychologic literature (anterograde and retrograde amnesia, memory function, cognitive effects, language functions, perceptual and motor skills) and psychiatric literature (psychiatric consequences). The concluding chapter offers many fruitful directions as to where future research is needed.

We are witnessing increasing attention to the neuropsychological consequences of mild head injury (Barth *et al.*, 1983; Boll & Barth, 1983; Rimel, Giordani, Barth, Boll, & Jane, 1981), a topic which has only recently been considered worthy of attention. In his presidential address to Division 40 of the APA, Boll (1984) considered the complex interrelationship between head injury and interpersonal adjustment that may have begun prior to the injury to be an area of threshold significance in clinical neuropsychology. There is some controversy as to the persistence of neuropsychological deficits, with some studies (McLean, Temkin, Dikmen, & Wyler, 1983) reporting no significant findings at one month postinjury while other investigations (Gronwall & Wrightson, 1981; Rimel et al., 1981) note significant differences at one and three months postinjury. Such differences "may possibly be explained by a number of factors, including the subject selection criteria and the type of control group used" (McLean et al., 1983, p. 372). Other investigators (Ewing, McCarthy, Gronwall, & Wrightson, 1980) contend that the persisting effects of minor head injury, although subtle, emerge only under the effects of stress.

Alcoholism

Neuropsychological investigations have focused almost entirely on either the effects of chronic alcoholism or on a specific alcohol-related disorder, Korsakoff syndrome. Extensive reviews of chronic alcoholism (Goldstein & Neuringer, 1976; Parsons & Farr, 1981) and Korsakoff syndrome (Butters & Cermak, 1975) have been provided. Although Tarter (1976) believes both of these entities lie on a continuum, Goldstein (1984a) has cautioned that such a distinction may not be a wise one because we may be dealing with two separate entities.

There is a consensus regarding a chronic alcoholic profile in which intact verbal language abilities are accompanied by substantially impaired abstraction and visual-spatial skills (Parsons & Farr, 1981). Most investigations of Korsakoff patients have focused on a selected cognitive function, memory, which has been investigated utilizing an information processing approach (Butters & Cermak, 1980). A temporal gradient has been delineated whereby remote memories are better preserved than those of the last 10 to 15 years (Albert, Butters, & Levin, 1979).

Attention is now being directed toward the ignored topic of the neuropsychological effects of chronic alcohol abuse in women. For example, Fabian, Jenkins, and Parsons (1981) noted the differential effects of chronic alcoholism on tactual-spatial performance (Tactual Performance Test—time) for males and females. Although control males were far superior to alcoholic males ($p < .001$), this was not so for the females. Differences also existed in tactual memory with women remembering more shapes than men. However, the spatial location of such remembered shapes was more impaired for women.

More recently the effects of gender on verbal and visual-spatial learning (Fabian, Parsons, & Sheldon, 1984) have been investigated. Although alcoholic women were unimpaired on either measure, this was not so for the alcoholic men, who demonstrated deficits in visual-spatial learning. Contrary to previous findings, this study suggests that alcoholism may have differential effects on learning for men and women.

There is an increasing emphasis in neuropsychology on understanding the etiology of alcoholism. Evidence for a genetic predisposition has been provided by twin (Kaij, 1960; Partanen, Bruun, & Markkanen, 1966), adoption (Bohman, 1978; Goodwin, Schulsinger, Hermansen, Guze, & Winokur, 1973; Schuckit, Shuskan, Duby, Vega, & Moss, 1982), and familial studies (Amark, 1951; Bleuler, 1955; Pitts & Winokur, 1966). Tarter (1976) has suggested that hyperkinesis and/or minimal brain damage in childhood may predispose one to alcoholism. However, the issue of neuropsychological deficits that antedate the drinking onset has not been systematically investigated in an empirical fashion (Tarter, in press). Only recently (Tarter & Alterman, 1984) has the question been raised as to whether *all* reported deficits are a consequence of the alcoholism.

Cerebrovascular Disease

It is of interest that there is a marked decline in the incidence of strokes (Garraway et al., 1979). Although there was no major change in age of onset, the reduction in rates was more pronounced in the elderly. The reason for this decline in unknown.

Recently investigation has focused on the "aprosodias" (Ross, 1981), which are disorders of affective language and behavior resulting from focal lesions of the right hemisphere. Although the majority of cases in which this dysfunction has been reported (Ross & Rush, 1981) were cerebrovascular in origin, this phenomenon has also been noted to occur in head trauma. These authors believe the aprosodias to be classifiable in a manner similar to the aphasias. Motor, sensory, global, and transcortical sensory aprosodias have been described and are noted to have good anatomical correlation with lesions of the left hemisphere that cause homologous aphasias. For example, in motor aprosodia (Ross, 1981), speech is notable for a flat monotonic voice, devoid of emotion and characterized by the absence of spontaneous gesturing. However, comprehension of the emotional components of language and gesture remains intact whereas repetition of sentences with affective variation is poor.

The relatively new neurosurgical procedures (carotid endarterectomy, vertebral endarterectomy, superficial temporal artery to middle cerebral artery bypass, and posterior fossa bypass) have increased the variety of patients being referred for preoperative neuropsychological evaluation (Dull et al., 1982). The fact that patients with TIAs (transient ischemic attacks; neurologic symptoms remit in 24 hours or less) were the primary candidates for this surgery may well explain why most preoperative neuropyschological investigations focused on this patient group. It was noted, for example, that the occurrence of TIA was associated with significant impairments in the more complex functions of abstraction, new learning and perceptual-motor integration that extended well beyond the 24 hour definitional limit (Delaney, Wallace, & Egelko, 1980).

More recently, however, patients with longer episodes, such as those with RINDS (reversible ischemic neurological deficits) and mild to moderate strokes, have been considered potential candidates for such procedures (Lee et al., 1979). Preoperative neurobehavioral impairments of these diverse patient groups has been provided by Dull et al. (1982).

Dementia

Although there is a general consensus that Alzheimer-type degenerative changes exist in over 50% of the dementias of old age, the reported prevalence of ischemic lesions ranges from 10% (Marsden & Harrison, 1972) to 30%–40% (Busse & Blazer, 1980). In addition, cerebral vascular disease in and of itself is probably not a sufficient cause of dementia unless widespread multiple areas of infarction occur (Obrist, 1980). Hence, the term "multi-infarct dementia" (Hachinski, Lassen, & Marshall, 1974) was proposed to distinguish such pathology from the more ambiguously defined and erroneously inferenced "cerebral arteriosclerosis."

Patterns of neuropsychological impairment that would discriminate between the subtypes of dementia are being investigated. Discriminant-function analyses have been utilized to discriminate among Alzheimer's disease, vertebrobasilar insufficiency and multi-infarct dementia utilizing both the Wechsler Adult Intelligence Scale (WAIS) (Perez, Gay, & Taylor, 1975) and the Wechsler Memory Scale (WMS) (Perez, Gay, Taylor, & Rivera, 1975). Investigations have also focused on the differences among Alzheimer, multi-infarct dementia, normal pressure hydrocephalus, Parkinson and Huntington disease patients, and depressives (Gainotti, Caltagirone, Masulli, & Miceli, 1980).

Whereas these studies have attempted to discriminate between dementia subgroups, other investigations are concerned with the more specific delineation of higher cortical functions within each of these dementia subtypes. For example, in Alzheimer's Disease, recognition memory (Wilson, Bacon, Kramer, Fox, & Kaszniak, 1983), primary and secondary memory (Wilson, Bacon, Fox, & Kaszniak, 1983), and verbal fluency (Rosen, 1980) have been investigated.

Investigations concerned with the relationship between cognitive findings and CT scan measures of brain atrophy have yielded inconsistent results. Although some studies (such as Brinkman, Sarwar, Levin,

& Morris, 1981; Kaszniak, Garron, Fox, Bergen, & Huckman, 1979; Soininen, Puranen, & Riekkinen, 1982) report significant relationships, others (such as Berg et al., 1984; Earnest, Heaton, Wilkinson, & Manke, 1979) note CT findings to be poor predictors of dementia. Such inconsistencies may well be due to methodological limitations. These deficiencies in measurement include comparing subjects on the basis of absolute values (whether linear or area measurements are utilized), which does not take into account the wide variation in brain and head size from individual to individual (Damasio et al., 1979). Such reported inconsistencies are attributable to unitary and incomplete measures. For example, the ventricular measurement in which the largest width of the body of the lateral ventricle is utilized is unidimensional and may not relate to overall ventricular size, shape, or volume or with the presence of sulci enlargement (Bird, 1982).

Another methodologic problem that may have contributed to the reported inconsistencies in the literature concerns the choice of controls. These subjects were either volunteers who were not carefully screened for the presence of CNS disease, or they were patients whose CT scans were obtained for diagnosis and then judged normal by visual inspection (Damasio et al., 1983). The lack of adequate control groups becomes further highlighted when one considers Haug's (1977) findings of the relation of age and sex to normal ventricular size.

Several investigators (Bigler, Hubler, Cullum, & Turkheimer, in submission; Damasio et al., 1983; deLeon & George, 1983; Jernigan, Zatz, Feinberg, & Fein, 1980; Storandt, Botwinick, Danziger, Berg, & Hughes, 1983) have begun utilizing volumetric measurement techniques in dementia. For example, Damasio et al. (1983) correctly identified 84% (n = 46) of their dementia patients utilizing ventricle and interhemispheric fissure brain ratios. In Alzheimer patients, it was found that despite ventricular volumes in excess of 60% larger than normal, there was no significant correlation between ventricular size and WAIS or WMS performance.

Another approach that is being utilized for CT scan evaluation is density CT numbers (CT value, delta number or Hounsfield unit), which are related to the coefficient of attenuation of the cerebral cortex in a given site. Significant differences in CT numbers between atrophied and normal brains have been reported (Pullan, Fawcitt, & Isherwood, 1978). Nondemented individuals had higher CT numbers (41 or greater) than demented patients (CT numbers of 40 and below) (Naeser, Gebhardt, & Levine, 1980). CT numbers have been utilized to arrive at a Brain Volume Index that correlated ($p < .01$) with a dementia rating scale (Ito, Hatazawa, Yamaura, & Matsuzawa, 1981).

One of the most recent techniques for quantifying cortical atrophy proposes measurement of the surface area of the cortex (Turkheimer et al., 1984). This is in contrast to most previous methods, which involve the estimation of the volume of enlarged sulci in the cortex.

Cerebral blood flow (CBF) and cerebral metabolism investigations have been conducted to study dementia and to differentiate degenerative changes produced by the various dementias. Although it is well known that average cerebral blood flow and metabolism are reduced in demented individuals (see review by Yamaguchi, Meyer, Yamamoto, Sakai, & Shaw, 1980), there is controversy regarding regional cerebral blood flow abnormalities in the dementias (Alzheimer's disease, multi-infarct dementia, or both) that affect the elderly. Some investigations (Hachinski et al., 1975; O'Brien & Mallett, 1970) found significantly lower CBFs in cerebrovascular disease than in primary neuronal degeneration whereas other studies (Simard, Olesen, Paulson, Lassen, & Shinhoj, 1971) found the converse. Still other investigations (Ingvar & Gustafson, 1970; Obrist, Thompson, Wang, & Wilkinson, 1975) were unable to distinguish AD from a vascular etiology. The majority of studies, however, do not differentiate AD from MID and do not employ age-matched controls. Studies (such as Yamaguchi et al., 1980) that considered these methodological variables noted reduced bilateral and symmetrical reduction of F_1 (gray matter but not white matter) in AD, which correlated with atrophy estimated by CT. In MID bilateral hemispheric F_1 was patchily reduced. More recent studies utilize programs that provide rapid computer derivation for values of local cerebral blood flow (LCBF) and local tissue: blood partition coefficients using inhaled xenon as an indicator (Amano, Meyer, Okabe, Shaw, & Mortel, 1982). Here, cortical and thalamic gray matter LCBF values were significantly reduced when compared with age-matched controls, whereas there was no change in local tissue (blood partition coefficients).

In Alzheimer's disease there is a dramatic loss of choline acetyltransferase (Davies, 1979), which has been correlated with both histopathological changes in the brain characteristics of the disease and the degree of the dementia (Perry et al., 1978). A recent update on the neurochemistry of this disease has been provided by Davies (1983). Because choline, a precursor of acetylcholine, can increase acetylcholine concentrations in the brain, numerous studies have investigated its effect on memory in Alzheimer's disease. Unfortunately, the effects of choline on memory have been ineffective (see review by Greenwald and Davis, 1983). Other ways to enhance cholinergic activity include utilization of cholinesterase inhibitors and cholinergic agonists. Physostigmine, a short-acting cholinesterase inhibitor, has been noted to modestly enhance memory in Alzheimer's disease (Christie, Shering, Ferguson, & Glen, 1981; Davis, Mohs, & Tinklenberg, 1979; Davis & Mohs, 1982; Kaye et al., 1982; Muramoto, Sugishita,

Sugita, & Toyokura, 1979; Peters & Levin, 1979; Smith & Swash, 1979; Sullivan, Shedlack, Corkin, & Growdon, 1981; Summers, Viesselman, Marsh, & Candelora, 1981). It has been suggested that cholinergic therapy can only be maximally effective when neuropeptide cofactors that may regulate sensitivity to ACh are also administered (Johns, Greenwald, Mohs, & Davis, 1983).

MEDICAL NEUROPSYCHOLOGY

Neuropsychology is becoming increasingly concerned with the effects of various medical disease entities. Adams, Sawyer, and Kvale (1980) advocate a new focus in clinical neuropsychology on studies of neurobehavioral dysfunction with lowered cerebral oxygenation that results from medically reversible conditions. A comprehensive overview of the medical disorder/neuropsychology interface has been provided by Ariel and Strider (1984).

In the cardiovascular area, the neuropsychological sequelae of open heart surgery, hypertension, and diabetes are promising new areas for investigation. To date, most neuropsychological studies have focused on the effect of carotid endarterectomy on cognitive functioning. Although a literature review of that area is beyond the scope of this chapter, it should be noted that discrepant findings regarding changes in mental status following surgery have been reported, although the majority of studies interpreted increments in neuropsychological test scores as indicative of improved cognitive functioning. Matarazzo, Matarazzo, Gallo, and Wiens (1979) have cautioned that such retest changes "are most likely the result of a test–retest practice effect" (p. 115). In addition to controlling for practice effects, future studies in this area will no doubt incorporate other methodological considerations, such as appropriate control groups and follow-up intervals, insurance of optimal motivation for all subjects (Asken & Hobson, 1977; Matarazzo et al., 1979), and control for examiner bias (Parsons & Prigatano, 1978).

In the pulmonary realm, initial studies (such as Krop, Block, & Cohen, 1973) focused on patients with chronic obstructive pulmonary disease (COPD) and hypoxemia. More recently, in the only reported study to date, attention has been focused on mildly hypoxemic COPD patients (Prigatano, Parsons, Wright, Levin, & Hawryluk, 1983). Opportunities for future research in this area include the reversibility of neuropsychological deficit following oxygen therapy (Adams et al., 1980). Whether the "premature aging" hypothesis (reduced oxygen availability and abnormal aging, Adams et al., 1980) is operative, whether lowered levels of blood oxygenation lead to inefficiencies in neural functioning and subsequently to impairment or finally, whether COPD enhances vascular disease and leads to reduced cerebral blood flow and oxygen consumption remains to be investigated (Prigatano et al., 1983).

An experimental paradigm that is beginning to be utilized in the area of urinary disorders compares chronic hemodialysis patients with nondialyzed patients and control subjects. Representative of studies employing such an approach are investigations of Ryan, Souheaver, and DeWolfe (1981) and Hart, Pederson, Czerwinski, and Adams (1983).

The study of neuropsychological effects of environmental toxins is in its infancy and represents an area of increasing concern, which was reflected in the recent 1984 INS symposium on "Neurobehavioral Toxicology and Teratology". The new multiple-effects model of environmental toxins stresses subtle behavioral alteration as an early indicator of toxicity and as evidence that a particular chemical agent can produce long-term dysfunction in susceptible individuals (Fein, Schwartz, Jacobson, & Jacobson, 1983). The purview of this new specialty encompasses the entire spectrum of environmental chemicals, including heavy metals, solvents, fuels, pesticides, air pollutants, and food additives (Weiss, 1983). Reviews of neuropsychological effects of toxins in general (Ariel & Strider, 1983) and more specifically of toxic gases (Adams et al., 1980) are noted. The need for future studies on the relation between toxins and behavioral development (Fein et al., 1983) and on the effects of long-term exposure to lower levels of oxygen-displacing gas in environmental and industrial environments (Adams et al., 1980) is noted.

This review of the medical disorder/neuropsychology interface would be incomplete without mention of the investigations dealing with hepatic (liver) functioning. Representative of such studies are those concerned with cirrhotic patients with and without alcoholism (Rehnström, Simert, Hansson, Johnson, & Vang, 1977).

SINGLE CASE INVESTIGATIONS/EXPERIMENTAL DESIGNS

Intensive investigation of the single case, that is, "the systematic testing of explanations about certain aspects of a patient's psychological disorder" (p. 663), has been discussed by Shapiro (1970). Although Shapiro was aware of a few studies in the literature that could be considered investigations of this kind, he noted the limited development of this line of investigation. It now appears that we are beginning to witness a burgeoning of experimental investigation of the single case in neuropsychology. Shallice (1979) has concluded that the neuropsychological case study approach is the most valuable neuropsychological technique for the establishment of the functional organization of cognition subsystems. This is so despite potential problems with "resource artifacts,"[2] the nature of the lengthy clinical experimental procedure used, statistical selection artifacts, reorganization of function, atypical lateralization, and the existence of associated deficits. Such objections have been refuted by Marshall and Newcombe (1984), who conclude that in neuropsychology a group has no significance over and above the individual members it contains. The potential of single-case experimental designs, in contrast to case studies, for controlled assessment of rehabilitation interventions in neuropsychology, has also been noted (Gianutsos & Gianutsos, 1979, 1980).

GERIATRIC NEUROPSYCHOLOGY

As the elderly segment of the population continues to expand, so too will geriatric neuropsychology. Comprehensive reviews of this area have been provided by Albert (1981) and Botwinick (1981). Problems in the evaluation of the elderly are numerous; these include the absence of adequate age norms, derivation of norms from different populations, reporting of norms in different statistical forms which precludes comparison for profile analysis, inappropriateness of tasks in terms of difficulty levels and physical demands (hearing, vision, manipulability of items), decrement in speed of performance in the elderly, and length of the neuropsychological measures (Faibush, Thornby, Auerbach, & Burgum, 1981). In addition, concerns have been raised about the nature and course of normal aging, the significance of diversive interactive effects and the influence of disease and affective processes (Benton & Sivan, 1984). Goldstein (1983) has stressed the role of health in relation to intellectual decline.

Attempts to ameliorate these deficiencies are noted. More appropriate norms for the Halstead-Reitan (Harley, Leuthold, Matthews, & Bergs, 1980) and Benton tests (Benton, Eslinger, & Damasio, 1981) have been provided. Abbreviated neuropsychological examinations have been constructed (Folstein, Folstein, & McHugh, 1975; Mattis, 1976) and tests with ecological validity (Crook, Ferris, & McCarthy, 1979) are beginning to appear.

The delineation of deficits in dementia, and more specifically in certain types of dementia, is occurring. As an illustration, witness the

---

[2]The problem where different tasks may not only involve different subsystems but may also make different resource requirements of a subsystem.

evaluation of primary and secondary memory in Alzheimer patients (Wilson, Bacon, Fox, & Kaszniak, 1983) and the occurrence of marked dissociation in visual-perceptual performance in dementia (Eslinger & Benton, 1983). What are needed are overall general profiles to distinguish one disease entity from the other.

FORENSIC NEUROPSYCHOLOGY

Neuropsychologists are being called on with increasing frequency by attorneys and the judiciary to evaluate clients in both criminal and civil litigation (Matarazzo, 1972; Strub & Black, 1977), although by far the majority of cases are in the civil sphere, particularly in the personal injury realm. In criminal law neuropsychologists have been asked to render opinions regarding an individual's competency to stand trial, sanity or insanity at the time of the offense, the degree of intent with which an act was committed, and the presence or absence of mitigating or aggravating circumstances with regard to the death penalty (McMahon & Satz, 1981). McMahon (1983) has characterized the role of the neuropsychologist in the forensic setting as specifying the extent of the cortical dysfunction and making inferences regarding prognosis and/or rehabilitation.

Of particular concern to forensic neuropsychologists are "intentional malingerers." Heaton, Smith, Lehman, and Vogt's (1978) landmark study investigated this topic in nonlitigating head-trauma patients and volunteers. Although the results of this study support the contention that naive individuals cannot easily fake brain damage, the lack of a control group and the spurious inflation of the degree of classification accuracy due to the number of predictor variables (37) exceeding the number of subjects (32), caution against such interpretation (Goebel, 1983). Besides inclusion of a control group, a larger sample size ($N = 254$), and a heterogeneous group of neurologically impaired patients, the issue of whether lateralized deficits were any more difficult to fake than diffuse patterns was investigated (Goebel, 1983).

Neuropsychologists who wish to pursue working in this area would do well to become familiar with the law. The issues of reports, privileged communication, advocacy, consultation, and courtroom testimony assume special significance in this area (McMahon, 1983).

REHABILITATION

Traditionally, neuropsychologists have operated on the diagnosis and symptom description phase and have sorely neglected rehabilitation (Diller & Weinburg, 1977). Although at present the primary emphasis continues at this level, neuropsychologists are becoming increasingly involved in the rehabilitation of the neurologically impaired. We are indeed witnessing the actualization of Wedding and Gudeman's (1980) contention that the role of the neuropsychologist would shift from site localization of brain damage to precise delineation of functional deficits for rehabilitation counseling.

A comprehensive evaluation is the prerequisite to a successful rehabilitation program. Diagnosis must take into account both emotional and cognitive factors. The specific evaluation instruments chosen, whether they be standard or flexible batteries of neuropsychological tests, will depend on both the training and the orientation of the neuropsychologist. To evaluate affective factors, clinical interviews of both the patient and significant others is essential. This can be supplemented with other objective and projective techniques that are deemed appropriate.

In addition to providing a baseline of current level of functioning, the neuropsychological evaluation lends itself to the establishment of a profile of an individual's strengths and deficits. Such an assessment will assist the neuropsychologist in determining whether the patient can profit from rehabilitation and then whether the patient will favorably participate in such a program. Only then should consideration be given to those functions that can be most effectively compensated for or resolved, and those which will present maximum benefit for the patient's restoration of an adaptive lifestyle.

Based on the conclusions drawn from such an evaluation, the neuropsychologist is in the position to offer recommendations. These include a follow-up session with both the patient and his family (Parsons & Prigatano, 1978) and formulation of a treatment plan. To be effective such programs must be highly individualized and deficit-specific (Gudeman, Golden, & Crane, 1978). Formulation alone is not enough because it is necessary for the neuropsychologist to interact closely with the treatment team and the family. Diller and Gordon (1981) note that rehabilitation is a comprehensive process of which cognitive retraining is only a part; it also includes matters of a psychosocial and practical nature related to a patient's capacity to adapt to a natural environment.

In devising a treatment plan, the neuropsychologist is expected to specify the deficit, the level of dysfunction, the specific techniques to be utilized in rehabilitation, and how progress will be evaluated. To prevent undue frustration and to encourage the patient's motivation, rehabilitation efforts must begin at a level where the patient experiences mastery. In patients with several deficits, it will be necessary to prioritize the order in which areas of dysfunction will be treated. In some instances, the development of such a hierarchy is clearly delineated; this would indeed be the case for patients with attentional or motivational deficits. For other patients, priority may be given to the deficit that is identified by the patient as being most distressing. Once a training program has been initiated the patient should be provided with constant feedback.

BIOLOGICAL SUBSTRATES OF MENTAL ILLNESS

Depression

The hypothesis has been advanced that schizophrenic individuals have a lateralized impairment of the left hemisphere, whereas depressives have a right hemisphere defect (Flor-Henry & Yeudall, 1979). Evidence supporting a right hemisphere deficit in depression has been amassed from diverse sources. Neuroradiological studies utilizing both cerebral blood flow and PET scanning (Buchsbaum, 1982; Golden, 1982) report a right frontal lobe deficit in affective disorders. Neuropsychological test performance of depressed patients notes greater evidence of right hemisphere dysfunction (Flor-Henry, 1976; Flor-Henry & Yeudall, 1979; Goldstein, Filskov, Weaver, & Ives, 1977; Kronfol, Hamsher, Digre, & Warizi, 1978; Taylor, Greenspan, & Abrams, 1979). Evidence from central auditory processing of dichotic click stimuli (Bruder & Yozawitz, 1979; Yozawitz et al., 1979) is consistent with the right hemisphere dysfunction hypothesis. Electrophysiologic studies (Abrams & Taylor, 1979; D'Elia & Perris, 1973; Flor-Henry, 1976; Monakhov, Perris, Botskarev, von Knorring, & Nikiforov, 1979; Perris & Monakhov, 1979) and lateralized skin conductance abnormalities (Gruzelier & Venables, 1974; Myslobodsky & Horesh, 1978) support the nondominant hemisphere dysfunction hypothesis. Additional documentation was provided by Zenhausern and Parisi (1981), who noted that on a lateralized discriminative reaction time task, psychotic depressives had a significant 100 microsecond faster reaction time in the right as opposed to the left hemisphere.

In their review of 10 cases (Heaton, Baade, & Johnson, 1978) in which neuropsychological tests attempted to discriminate mixed affective disorders from organic patients, a 77% correct classification was reported. Interestingly, only 2 of these 10 studies concerned depressives (Heaton & Crowley, 1981). The latter authors also report an additional study by Watson, Davis, and Gasser (1978) in which there was an 83% correct classification of 25 depressed and 40 organic patients.

No relation has been found between either psychiatric or self-ratings of depression and an extensive battery of ability tests (Friedman, 1964), or between self-ratings of depression and memory test performance (Squire & Chace, 1975). Findings relating depression and learning and memory tasks are inconsistent, with Miller (1975) and Sternberg and Jarvik (1976) noting the effect of depression on such tasks. Perhaps such controversy will be resolved when subtypes of affective disorders, as noted in the *Diagnostic and Statistical Manual* of the American Psychiatric Association (1980), are investigated systematically, as such subtypes are known to differ in both course and etiology.

Schizophrenia

A traditional concern of clinical neuropsychology has been the differential diagnosis between schizophrenia and brain damage. When chronic or process schizophrenics are excluded from psychiatric diagnostic categories, a median hit rate of 75% was noted for 84 classification attempts. The 34 studies concerned with the chronic/process schizophrenic from organic differential achieved a median hit rate of only 54%, which barely exceeds chance prediction (Heaton, Baade, & Johnson, 1978).

Such attempts have failed in part due to methodological considerations. Goldstein (1978) has reviewed the persistent methodological problems of diagnostic inaccuracy, inadequate sampling, and difficulty interpreting subject's performance. Malec (1978) reported that few studies of the previous 10 years utilized the most valid criteria in defining brain-damaged and schizophrenic samples. In addition, few studies controlled for IQ, age, education, length of hospitalization or drug effects. All but 7 of 94 studies published between 1960–1975 dealing with the accuracy of neuropsychological diagnoses in psychiatric settings contained three or more of the above methodological flaws (Heaton *et al.*, 1978). This was also true for 24 studies published between 1975 and July, 1976, and reviewed by Heaton and Crowley (1981).

In addition to these methodological problems, there are substantive issues that must be considered in this differential diagnostic problem (Goldstein, 1984b). Foremost among these is the increasing body of evidence documenting brain damage within a subgroup of the schizophrenic population. CT data suggest subgroups of schizophrenia characterized by either the lack of morphological brain abnormality, cerebral atrophy with enlarged ventricles, or lack of atrophy with atypical asymmetry (see the review by Seidman, 1983). Regional cerebral blood flow (rCBF) studies demonstrate significant low frontal activation in schizophrenics. Although most studies (Ariel *et al.*, 1983; Franzén & Ingvar, 1975a, 1975b; Ingvar & Franzén, 1974) note that such reduction occurs in the left frontal area, one study has reported significantly lower right frontal activation (Mathew et al., 1981). Evidence of left frontal deficit has also come from positron emission tomography (PET) studies (Buchsbaum, 1982; Farkas et al., 1980).

The differential diagnostic question that has been posed may indeed have been an inappropriate one. We are witnessing much needed changes in the manner in which neuropsychological instruments are being utilized. One need only review the recent studies in which neuropsychological measures are being correlated with CT abnormalities. Such correlations have been positive for both the Halstead-Rietan (Donnelly, Weinberger, Waldman, & Wyatt, 1980) and Luria-Nebraska Batteries (Golden, Graber, Moses, & Zatz, 1980; Golden et al., 1982).

ACKNOWLEDGMENT

I wish to thank Dr. Gerald Goldstein for his careful review of this manuscript.

REFERENCES

Abrams, R., & Taylor, M. A. (1979). Differential EEG patterns in affective disorders and schizophrenia. *Archives of General Psychiatry, 36,* 1355–1358.

Acker, W. (1980). A microcomputer administered neuropsychological assessment system for use with chronic alcoholics. *Substance and Alcohol Actions/Misuse, 1*(5–6), 545–550.

Adams, K. M. (1975). Automated clinical interpretation of the neuropsychological battery: An ability based approach (Doctoral dissertation, Wayne State University, 1974). *Dissertation Abstracts International, 35,* 6085B. (University Microfilms No. 75-13, 289).

Adams, K. M. (1984, February). The use of computers in neuropsychological assessment: A state of the art review. In A. J. McSweeny (Chair), *Professional and ethical issues in the use of computers in neuropsychological assessment.* Symposium conducted at the meeting of the International Neuropsychological Society, Houston.

Adams, K. M., Sawyer, J. D., & Kvale, P. A. (1980). Cerebral oxygenation and neuropsychological adaptation. *Journal of Clinical Neuropsychology, 2*(3), 189–208.

Albert, M. (1981). Geriatric neuropsychology. *Journal of Consulting and Clinical Psychology, 49,* 835–850.

Albert, M. S., Butters, N., & Levin, J. (1979). Temporal gradients in the retrograde amnesia of patients with alcoholic Korsakoff's disease. *Archives of Neurology, 36,* 211–216.

Amano, T., Meyer, J. S., Okabe, T. Shaw, T., & Mortel, K. F. (1982). Stable Xenon CT cerebral blood flow measurements computed by a single compartment-double integration model in normal aging and dementia. *Journal of Computer Assisted Tomography, 6*(5), 923–932.

Amark, C. (1951). A study in alcoholism: Clinical, social-psychiatric and genetic investigations. *Acta Psychiatrica et Neurologica Scandinavica,* (Suppl. 70), 1–283.

Anthony, W. Z., Heaton, R. K., & Lehman, R. W. (1980). An attempt to cross-validate two actuarial systems for neuropsychological test interpretation. *Journal of Consulting and Clinical Psychology, 48*(3), 317–326.
Ariel, R. N., & Strider, M. A. (1983). Neuropsychological effects of general medical disorders. In C. J. Golden & P. J. Vicente (Eds.), *Foundations of clinical neuropsychology*. New York: Plenum Press.
Ariel, R. N., Golden, C. J., Berg, R. A., Quaife, M. A., Dirksen, J. W., Forsell, T., Wilson, J., & Graber, B. (1983). Regional cerebral blood flow in schizophrenia. Tests using the Zenon Xe^{133} inhalation method. *Archives of General Psychiatry, 40*(3), 258–263.
Asken, M., & Hobson, R. (1977). Intellectual change and carotid endarterectomy, subjective speculation or objective reality: A review. *Journal of Surgical Research, 23*, 367–375.
Barth, J. T., Macciocchi, S., Giordani, B., Rimel, R., Jane, J., & Boll, T. J. (1983). Neuropsychological sequelae of minor head injury. *Neurosurgery, 13*, 529–532.
Beaumont, J. G. (1983). *Introduction to neuropsychology*. New York: Guilford.
Benton, A. L., deS Hamsher, K., Varney, N. R., & Spreen, O. (1983). *Contributions to neuropsychological assessment*. New York: Oxford University Press.
Benton, A. L., & Sivan, A. B. (1984). Problems and conceptual issues in neuropsychological research in aging and dementia. *Journal of Clinical Neuropsychology, 6*(1), 57–63.
Benton, A. L., Eslinger, P. J., & Damasio, A. T. (1981). Normative observations on neuropsychological test performances in old age. *Journal of Clinical Neuropsychology, 3*(1), 33–42.
Berg, L., Danziger, W. L., Storandt, M., Choen, L. A., Gado, M., Hughes, C. P., Kneservich, J. W., & Botwinick, J. (1984). Predictive features in mild senile dementia of the Alzheimer type. *Neurology, 34*, 563–569.
Bigler, E. D., & Steinman, D. R. (1981). Neuropsychology and computerized axial tomography: Further comments. *Professional Psychology, 12*, 195–197.
Bigler, E. D., Hubler, D. W., Cullum, C. M., & Turkheimer, E. (in submission). Intellectual and memory impairment in Alzheimer's disease: CT volume correlations.
Bird, J. M. (1982). Computerized tomography, atrophy and dementia: A review. *Progress in Neurobiology, 19*, 91–115.
Bleuler, M. (1955). Familial and personal background of chronic alcoholics. In O. Diethelm (Ed.), *Etiology of chronic alcoholism*. Springfield, IL: Charles C Thomas.
Bohman, M. (1978). Some genetic aspects of alcoholism and criminality: A population of adoptees. *Archives of General Psychiatry, 35*, 269–276.
Boll, T. J. (1978). Diagnosing brain impairment. In B. B. Wolman (Ed.), *Clinical diagnosis of mental disorders* (pp. 601–675). New York: Plenum Press.
Boll, T. J. (1984, August). *Developing issues in clinical neuropsychology*. Paper presented at the meeting of the American Psychological Association, Ontario, Canada.
Boll, T. J., & Barth, J. (1983). Mild head injury. *Psychiatric Developments, 3*, 263–275.
Boll, T. J., & Reitan, R. M. (1971). Review of "Assessment of brain damage: A neuropsychological key approach" by E. W. Russell, C. Neuringer, & G. Goldstein. *Professional Psychology, 2*, 410, 412–413.
Botwinick, J. (1981). Neuropsychology of aging. In S. B. Filskov & T. J. Boll (Eds.), *Handbook of clinical neuropsychology*. New York: Wiley.
Bracy, O., L. (1983). Computer based cognitive rehabilitation. *Cognitive Rehabilitation, 1*(1), 7–8.
Bracy, O. L. (undated). *Foundation Skills Package. Visuospatial Skills Package. Conceptual Skills Package. Memory Skills Package. Auditory Perception Skills Package*. [computer program]. Indianapolis, IN: Psychological Software Services.

Brinkman, M. A., Sarwar, M., Levin, H. S., & Morris, H. H. (1981). Quantitative indexes of computed tomography in dementia and normal aging. *Radiology, 138,* 89–92.

Bruder, G. E., & Yozawitz, A. (1979). Central auditory processing and lateralization in psychiatric patients. In J. Gruzelier & P. Flor-Henry (Eds.), *Hemisphere asymmetries of function in psychopathology* (pp. 561–580). Amsterdam: Elsevier/North Holland Biomedical Press.

Buchsbaum, M. (1982). *Positron emission tomography.* Paper presented at the Laterality Conference, Banff.

Busse, E. W., & Blazer, D. G. (1980). The theories and processes of aging. In E. W. Busse & D. G. Blazer (Eds.), *Handbook of geriatric psychiatry* (pp. 3–27). New York: Van Nostrand Reinhold.

Butters, N., & Cermak, L. S. (1975). Some analyses of amnesia syndromes in brain-damaged patients. In R. L. Isaacson & K. H. Pribram (Eds.), *The Hippocompus* (Vol. 2). New York: Plenum Press.

Butters, N., & Cermak, L. S. (1980). *Alcoholic Korsakoff's Syndrome: An information-processing approach to amnesia.* New York: Academic Press.

Christensen, A.-L. (1975). *Luria's neuropsychological investigation.* New York: Spectrum.

Christensen, A.-L. (1979a). *Luria's neuropsychological investigation* (2nd ed.). Copenhagen: Munksgaard.

Christensen, A.-L. (1979b). A practical application of the Luria methodology. *Journal of Clinical Neuropsychology, 1*(3), 240–247.

Christensen, A.-L. (1984). The Luria method of examination of the brain-imparied patient. In P. E. Logue & J. M. Schear (Eds.), *Clinical neuropsychology. A multidisciplinary approach* (pp. 5–28). Springfield, IL: Charles C Thomas.

Christie, J. E., Shering, A., Ferguson, J., & Glen, A. I. M. (1981). Physostigmine and arecoline: Effects of intravenous infusions in Alzheimer presenile dementia. *British Journal of Psychiatry, 138,* 46–50.

Costa, L. (1983). Clinical neuropsychology: A discipline in evolution. *Journal of Clinical Neuropsychology, 5*(1), 1–11.

Crook, T., Ferris, S., & McCarthy, M. (1979). The misplaced object task: A brief test for memory dysfunction in the aged. *Journal of the American Geriatrics Society, 27,* 284–287.

Damasio, H., Eslinger, P., Damasio, A. R., Rizzo, M., Huang, H. K., & Demeter, S. (1983). Quantitative computed tomography analysis in the diagnosis of dementia. *Archives of Neurology, 40,* 715–719.

Davies, P. (1979). Biochemical changes in Alzheimer's disease-senile dementia. In R. Katzman (Ed.), *Congenital and acquired cognitive disorders.* New York: Raven Press.

Davies, P. (1983). An update on the neurochemistry of Alzheimer's disease. In R. Mayeux & W. G. Rosen (Eds.), *The dementias.* New York: Raven Press.

Davis, K. L., & Mohs, R. C. (1982). Enhancement of memory processes in Alzheimer's disease with multiple dose intravenous physostigmine. *American Journal of Psychiatry, 139,* 1421–1424.

Davis, K. L., Mohs, R. C., & Tinklenberg, J. R., Pfefferbaum, A., Hollister, L. E., & Kopell, B. S. (1978). Physostigmine: Improvement of long-term memory processes in normal subjects. *Science, 201,* 272–274.

Delaney, R. C., Wallace, J. D., & Egelko, S. (1980). Transient cerebral ischemic attacks and nueropsychological deficit. *Journal of Clinical Neuropsychology, 2,* 107–114.

deLeon, M. J., & George, A. E. (1983). Computed tomography in aging and senile dementia of the Alzheimer's type. In R. Mayeux & W. G. Rosen (Eds.), *The dementias.* New York: Raven Press.

D'Elia, G., & Perris, C. (1973). Cerebral functional dominance and depression: An analysis of EEG amplitude in depressed patients. *Acta Psychiatrica Scandinavica, 49,* 191–197.

Delis, D. C., Kramer, J., Ober, B. A., & Kaplan, E. (1983). *The California Verbal Learning Test: Administration and implementation.* Preliminary manual.

Diller, L., & Gordon, W. (1981). Interventions for cognitive deficits in brain injured adults. *Journal of Consulting and Clinical Psychology, 49,* 822–834.

Diller, L., & Weinberg, J. (1977). Hemi-attention in rehabilitation: The evolution of a rational remediation program. In E. Weinstein & R. Friedland (Eds.), *Advances in neurology* (Vol. 18). New York: Raven Press.

Donnelly, E. F., Weinberger, D. R., Waldman, I. N., & Wyatt, R. J. (1980). Cognitive impairment associated with morphological brain abnormalities on computed tomography in chronic schizophrenic patients. *Journal of Nervous and Mental Disease, 168*(15), 305–308.

Dull, R. A., Brown, G. G., Adams, K. A., Shatz, M. W., Diaz, F. G., & Ausman, J. I. (1982). Preoperative neurobehavioral impairment in cerebral revascularization candidates. *Journal of Clinical Neuropsychology, 4*(2), 151–165.

Earnest, M. P., Heaton, R. K., Wilkonsin, W. E., & Manke, W. F. (1979). Cerebral atrophy, ventricular enlargement and intellectual impairment in the aged. *Neurology, 29,* 1138–1143.

Eslinger, P. J., & Benton, A. L. (1983). Visuoperceptual performances in aging and dementia: Clinical and theoretical implications. *Journal of Clinical Neuropsychology, 5*(3), 213–220.

Ewing, R., McCarthy, D., Gronwall, D., & Wrightson, P. (1980). Persisting effects of minor head injury observable during hypoxic stress. *Journal of Clinical Neuropsychology, 2*(2), 147–155.

Fabian, M. S., Jenkins, R. L., & Parsons, O. A. (1981). Gender, alcoholism, and neuropsychological functioning. *Journal of Consulting and Clinical Psychology, 49*(1), 138–140.

Fabian, M. S., Parsons, O. A., & Shelton, M. D. (1984). Effects of gender and alcoholism on verbal and visual-spatial learning. *Journal of Nervous and Mental Disease, 172,* 16–20.

Faibush, G. M., Thornby, J. I., Averbach, V. S., & Burgum, D. L. (1981). Developing a test battery for the neuropsychological assessment of older adults. In *Perspectives in Veterans Administration Neuropsychology and Rehabilitation: Proceedings of Mental Health and Behavioral Sciences Service Conference* (pp. 119–129). Veteran's Administration.

Farkas, T., Reivich, M., Alavi, A., Greenberg, J. H., Fowler, J. S., MacGregor, R. R., Christman, D. R., & Wolf, A. P. (1980). The application of ^{18}F-deoxy-2-flouro-D-glucose and positron emission tomography in the study of psychiatric conditions. In J. V. Passonneau, R. A. Hawkins, W. D. Lust, & F. A. Welsh (Eds.), *Cerebral metabolism and neural function.* Baltimore: Williams & Wilkins.

Fein, G. G., Schwartz, P. M., Jacobson, S. W., & Jacobson, J. L. (1983). Environmental toxins and behavioral development. A new role for psychological research. *American Psychologist, 11,* 1188–1197.

Filskov, S. B., & Leli, D. A. (1981). Assessment of the individual in neuropsychological practice. In S. B. Filskov & T. J. Boll (Eds.), *Handbook of clinical neuropsychology.* New York: Wiley.

Finkelstein, J. N. (1977). BRAIN: A computer program for interpretation of the Halstead-Reitan Neuropsychological Test Battery (Doctoral dissertation, Columbia University, 1976). *Dissertation Abstracts International, 37,* 5349B. (University Microfilms No. 77-8, 8864)

Flor-Henry, P. (1976). Lateralized temporal-limbic dysfunction and psychopathology. *Annals of the New York Academy of Sciences, 280*, 777–797.

Flor-Henry, P., & Yeudall, L. T. (1979). Neuropsychological investigation of schizophrenia and manic depressive psychosis. In J. Gruzelier & P. Flor-Henry (Eds.), *Hemisphere asymmetries of function in psychopathology.* Amsterdam: Elsevier/North Holland Biomedical Press.

Folstein, M. F., Folstein, S. E., & McHugh, R. R. (1975). Mini-mental state. *Journal of Psychiatric Research, 12,* 189–198.

Franzén, G., & Ingvar, D. H. (1975a). Absence of activation in frontal structures during psychological testing of chronic schizophrenia. *Journal of Neurology, Neurosurgery, and Psychiatry, 38,* 1027–1032.

Franzén, G., & Ingvar, D. H. (1975b). Abnormal distribution of cerebral activity in chronic schizophrenia. *Journal of Psychiatric Research, 12*(3), 199–214.

Fridlund, A. J., & Delis, D. C. (1983b). *The California Neuropsychological System.* Bayport, NY: Life Science Associates.

Friedman, A. S. (1964). Minimal effects of severe depression in cognitive functioning. *Journal of Abnormal and Social Psychology, 69,* 237–243.

Gainotti, G., Caltagirone, G., Masullo, C., & Miceli, C. (1980). Patterns of neuropsychologic impairment in various diagnostic groups of dementia. In L. Amaducci, A. Davison, & P. Antuono (Eds.), *Aging of the brain and dementia* (pp. 245–250). New York: Raven Press.

Garraway, W. M., Whisnant, J. P., Furlan, A. J., Phillips, L. H., Kurland, L. T., & O'Fallan, W. M. (1979). The declining incidence of stroke. *New England Journal of Medicine. 300,* 449–452.

Gedye, J. L., & Miller, E. (1970). Developments in automated testing systems. In P. J. Mittler (Ed.), *The psychological assessment of mental and physical handicap* (pp. 735–760). London: Methuen.

Geschwind, N. (1965). Disconnexion syndrome in animals and man. *Brain, 88,* 237–294, 585–644.

Gianutsos, R. (1981, August). *Using microcomputers for cognitive rehabilitation.* Paper presented at the meeting of the American Psychological Association, Los Angeles.

Gianutsos, R., & Gianutsos, J. (1979). Rehabilitating the verbal recall of brain injured patients by mnemonic training: An experimental demonstration using single-case methodology. *Journal of Clinical Neuropsychology, 1,* 117–135.

Gianutsos, R., & Gianutsos, J. (1980). *Single-case and group-oriented experimental approaches to the assessment of interventions in neuropsychology.* Paper presented at the meeting of the International Neuropsychological Society, San Francisco.

Gianutsos, R., & Klitzner, C. (1981). *Handbook computer programs for cognitive rehabilitation* [Computer program]. Bayport, NY: Life Science Associates.

Goebel, R. A. (1983). Detection of faking on the Halstead-Reitan neuropsychological test battery. *Journal of Clinical Psychology, 39*(5), 731–742.

Golden, C. J. (1981). A standardized version of Luria's neuropsychological tests: A quantitative and qualitative approach to neuropsychological evaluation. In S. B. Filskov & T. J. Boll (Eds.), *Handbook of clinical neuropsychology* (pp. 608–642). New York: Wiley.

Golden, C. J., (1982). *Cerebral blood flow.* Paper presented at the Laterality Conference, Banff.

Golden, C. J., Graber, B., Moses, J. A., & Zatz, L. M. (1980). Differentiation of chronic schizophrenia with and without ventricular enlargement by the Luria-Nebraska Neuropsychological Battery. *International Journal of Neuroscience, 11*(2), 131–138.

Golden, C. J., MacInnes, W. D., Ariel, R. N., Reudrich, S. L., Chu, C. C., Coffman, J. A.,

Graber, B., & Bloch, S. (1982). Cross-validation of the ability of the Luria-Nebraska Neuropsychological Battery to differentiate chronic schizophrenics with and without ventricular enlargement. *Journal of Consulting and Clinical Psychology, 50*(1), 87–95.
Goldstein, G. (1978). Cognitive and perceptual differences between schizophrenics and organics. *Schizophrenia Bulletin, 4*(2), 161–185.
Goldstein, G. (1980, September). In defense of the computer: Remarks on objective neuropsychological interpretation. *Quantitative and qualitative approaches to clinical neuropsychology*. Symposium conducted at the meeting of the American Psychological Association, Montreal.
Goldstein, G. (1983). Normal aging and the concept of dementia. In C. J. Golden & P. J. Vicente (Eds.), *Foundations of clinical neuropsychology* (pp. 249–271). New York: Plenum Press.
Goldstein, G. (1984a). Contributions of clinical neuropsychology to psychiatary. In P. E. Logue & J. M. Schear (Eds.), *Clinical neuropsychology. A multidisciplinary approach*. Springfield, IL: Charles C Thomas.
Goldstein, G. (1984b). Neuropsychological assessment of psychiatric patients. In G. Goldstein (Ed.), *Advances in clinical neuropsychology* (pp. 55–87). New York: Plenum Press.
Goldstein, G., & Neuringer, C. (1976). *Empirical studies of alcoholism*. Cambridge, MA: Ballinger.
Goldstein, G., Tarter, R., Shelly, C., & Hegedus, A. (1983). The Pittsburgh Initial Neuropsychological Testing System (PINTS): A neuropsychological screening battery for psychiatric patients. *Journal of Behavioral Assessment, 5*, 227–238.
Goldstein, S. G., Filskov, S. B., Weaver, L. A., & Ives, J. O. (1977). Neuropsychological effects of electroconvulsive therapy. *Journal of Clinical Psychology, 33*, 798–806.
Goodglass, H., & Kaplan, E. (1979). Assessment of cognitive deficit in the brain injured patient. In M. S. Gazzaniga (Ed.), *Neuropsychology*. New York: Plenum Press.
Goodwin, D., Schulsinger, F., Hermansen, L., Guze, S., & Winokur, G. (1973). Alcohol problems in adoptees raised apart from alcoholic biological parents. *Archives of General Psychiatry, 28*, 238–243.
Greenwald, B. S., & Davis, K. L. (1983). Experimental pharmacology of Alzheimer disease. In R. Mayeux & W. G. Rosen (Eds.), *The dementias* (pp. 87–102). New York: Raven Press.
Gregory, R. J. (1976). Computerized interpretation of brain impairment tests: Preliminary results. Abstracted in the JSAS *Catalog of Selected Documents in Psychology, 6*(1), 17.
Gronwall, D., & Wrightson, P. (1981). Memory and information processing capacity after closed head injury. *Journal of Neurology, Neurosurgery, & Psychiatry, 44*, 889–895.
Gruzelier, J., & Venables, P. (1974). Bimodality and lateral asymmetry of skin conductance orienting activity in schizophrenics: Replication and evidence of lateral asymmetry in patients with depression and disorders of personality. *Biological Psychiatry, 8*, 55–73.
Gudeman, H., Golden, C., & Craine, J. (1978). The role of neuropsychological evaluation in the rehabilitation of the brain-injured patient: A program in neurotraining. *JSAS Catalog of Selected Documents in Psychology, 8*, 44. (Ms. No. 1693).
Hachinski, V. C., Iliff, L. D., Zilhka, E., DuBoulay, G. M., McAllister, V. L., Marshall, J., Russell, R. W. R., & Symon, L. (1975). Cerebral blood flow in dementia. *Archives of Neurology, 32*, 632–637.
Hachinski, V. C., Lassen, N. A., & Marshall, J. (1974). Multi-infarct dementia: A cause of mental deterioration in the elderly. *Lancet, 2*, 207–209.

Harley, J. P., Leuthold, C. A., Matthews, C. G. & Bergs, L. E. (1980). *Wisconsin neuropsychological test battery T-score norms for older VAMC patients* (mimeograph). Madison, WI: Author.

Hart, R., Pederson, J. A., Czerwinski, A. W., & Adams, R. L. (1983). Chronic renal failure, dialysis, and neuropsychological function. *Journal of Clinical Neuropsychology*, 5(3), 301–312.

Haug, G. (1977). Age and sex dependence of the size of the normal ventricles on computed tomography. *Neuroradiology*, 14, 201–204.

Heaton, R. K., Baade, L. E., & Johnson, K. L. (1978). Neuropsychological test results associated with psychiatric disorders in adults. *Psychological Bulletin*, 85, 141–162.

Heaton, R. K., & Crowley, J. J. (1981). Effects of psychiatric disorders and their somatic treatments on neuropsychological test results. In S. B. Filskov & T. J. Boll (Eds.), *Handbook of clinical neuropsychology* (pp. 481–525). New York: Wiley.

Heaton, R. K., Grant, I., Anthony, W. Z., & Lehman, R. A. W. (1981). A comparison of clinical and automated interpretation of the Halstead-Reitan Battery. *Journal of Clinical Neuropsychology*, 3(2), 121–141.

Heaton, R. K., Smith, H. H., Lehman, R. A. W., & Vogt, A. T. (1978). Prospects for faking believable deficits on neuropsychological testing. *Journal of Consulting and Clinical Psychology*, 46(5), 892–900.

Hécaen, H., & Albert, M. L. (1978). *Human neuropsychology*. New York: Wiley-Interscience.

Heilman, K. M., & Valenstein, E. (1979). *Clinical neuropsychology*. New York: Oxford University Press.

Incagnoli, T., & Newman, B. (1982, October). *Cognitive and behavioral rehabilitation interventions*. Paper presented at the meeting of the National Academy of Neuropsychologists, Atlanta.

Ingvar, D. H., & Franzén, G. (1974). Abnormalities of cerebral blood flow distribution in patients with chronic schizophrenia. *Acta Psychiatrica Scandinavica*, 50, 425–462.

Ingvar, D. H., & Gustafson, L. (1970). Regional cerebral blood flow in organic dementia with early onset. *Acta Neurologica Scandinavica*, 46(Suppl. 43), 42–73.

Ito, M., Hatazawa J., Yamaura, H., & Matsuzawa, T. (1981). Age-related brain atrophy and mental deterioration—a study with computed tomography. *British Journal of Radiology*, 54, 384–390.

Jernigan, T. L., Zatz, C. M., Feinberg, I., & Fein, G. (1980). Measurement of cerebral atrophy in the aged by computed tomography. In C. W. Poon (Ed.), *Aging in the 1980's: Psychological issues*. Washington, DC: American Psychological Association.

Johns, C. A., Greenwald, B. S., Mohs, R. C., & Davis, K. L. (1983). The cholinergic treatment strategy in aging and senile dementia. *Psychopharmacology Bulletin*, 19(2), 185–197.

Kaij, L. (1960). *Studies on the etiology and sequels of abuse of alcohol*. Unpublished manuscript, University of Lund, Department of Psychiatry, Lund, Sweden.

Kaplan, E. (1983). Process and achievement revisited. In S. Wapner & B. Kaplan (Eds.), *Toward a holistic developmental psychology*. Hillsboro, NJ: Lawrence Erlbaum.

Kaszniak, A. W., Garron, D. C., Fox, J. H., Bergen, D., & Huckman, M. (1979). Cerebral atrophy, EEG slowing, age, education, and cognitive functioning in suspected dementia. *Neurology*, 29, 1273–1279.

Kaye, W. H., Sitaram, N., Weingartner, H., Ebert, M. H., Smallberg, S., & Gillan, J. (1982). Modest facilitation of memory in dementia with combined lecithin and anticholinesterase treatment. *Biological Psychiatry*, 17, 275–280.

King, G. D., Hannay, H. J., Masek, B. J., & Burns, J. W. (1978). Effects of anxiety and sex on neuropsychological tests. *Journal of Consulting and Clinical Psychology*, 46, 375–376.

Kronfol, Z., Hamsher, K., Digre, K., & Warizi, R. (1978). Depression and hemispheric functions: Changes associated with unilateral ECT. *British Journal of Psychiatry*, *132*, 560–567.

Krop, H. D., Block, A. J., & Cohen, E. (1973). Neuropsychological effects of continuous oxygen therapy in chronic obstructive pulmonary disease. *Chest*, *64*, 317–322.

Lee, M. C., Ausman, J. I., Geiger, J. D., Latchaw, R. E., Klassen, A. C., Chou, S. N., & Resch, J. A., Superficial temporal to middle cerebral artery anastomosis: Clinical outcome in patients with ischemia or infarction in the internal carotid artery distribution. *Archives of Neurology*, *36*, 1–4.

Leli, D. A., & Filskov, S. B. (1979). Relationship of intelligence to education and occupation as signs of intellectual deterioration. *Journal of Consulting and Clinical Psychology*, *47*, 702–707.

Levin, H. S., Benton, A. L., & Grossman, R. G. (1982). *Neurobehavioral consequences of closed head injury*. New York: Oxford University Press.

Levy, R., & Post, F. (1975). The use of an interactive computer terminal in the assessment of cognitive functions in elderly psychiatric patients. *Age and Aging*, *4*, 110–115.

Lezak, M. D. (1982). Coping with head injury in the family. In G. A. Brol & R. L. Tate (Eds.), *Brain impairment. Proceedings of the Fifth Annual Brain Impairment Conference*. Sydney, Australia: Postgraduate committee in Medicine of the University of Sydney.

Lezak, M. D. (1983). *Neuropsychological assessment* (2nd ed.). New York: Oxford University Press.

Lezak, M. D. (1984). An individualized approach to neuropsychological assessment. In P. E. Logue & J. M. Schear (Eds.), *Clinical neuropsychology. A multidisciplinary approach* (pp. 29–49). Springfield, IL: Charles C Thomas.

Luria, A. R. (1966). *Higher cortical functions in man* (B. Haigh, Trans.). New York: Basic Books and Plenum Press. London: Tavistock. (Original work published 1962).

Luria, A. R. (1973a). Neuropsychological studies in the U.S.S.R. A review (Part 1). *Proceedings of the National Academy of Science U.S.A.*, *70*(3), 959–964.

Luria, A. R. (1973b). *The working brain. An introduction to neuropsychology* (B. Haigh, Trans.) New York: Basic Books.

Lynch, W. J. (1982). The use of electronic games in cognitive rehabilitation. In L. Trexler (Ed.), *Cognitive rehabilitation. Conceptualization and intervention*. New York: Plenum Press.

Malec, J. (1978). Neuropsychological assessment of schizophrenia versus brain damage: A review. *The Journal of Nervous and Mental Disease*, *166*(7), 507–516.

Marsden, C. D., & Harrison, M. J. (1979). Outcome of investigation of patients with presenile dementia. *British Medical Journal*, *2*, 249–252.

Marshall, J. C., & Newcombe, F. (1984). Putative problems and pure progress in neuropsychological single case studies. *Journal of Clinical Neuropsychology*, *6*(1), 65–70.

Matarazzo, J. D. (1972). *Wechsler's measurement and appraisal of adult intelligence* (5th ed.). Baltimore: Williams & Wilkins.

Matarazzo, R. G., Matarazzo, J. D., Gallo, A. E., Jr., & Wiens, A. N. (1979). IQ and neuropsychological changes following carotid endarterectomy. *Journal of Clinical Neuropsychology*, *1*(2), 97–116.

Mathew, R. J., Meyer, J. S., Francis, D. J., Schoolar, J. C., Weinman, M., & Mortel, K. F. (1981). Regional cerebral blood flow in schizophrenia: A preliminary report. *American Journal of Psychiatry*, *138*(1), 112–113.

Mattis, S. (1976). Mental status examination for organic mental syndrome in the elderly patient. In L. Bellak & T. Karasu (Eds.), *Geriatric Psychiatry: A handbook for psychiatrists and primary care physicians* (pp. 77–121). New York: Grune & Stratton.

McLean, Jr., A., Temkin, N. R., Dikmen, S., & Wyler, A. R. (1983). The behavioral sequelae of head injury. *Journal of Clinical Neuropsychology, 5*(4), 361–376.

McMahon, E. A. (1983). Forensic issues in clinical neuropsychology. In C. J. Golden & P. J. Vicente (Eds.), *Foundations of clinical neuropsychology* (pp. 401–427). New York: Plenum Press.

McMahon, E. A., & Satz, P. (1981). Clinical neuropsychology. Some forensic applications. In S. B. Filskov & T. J. Boll (Eds.), *Handbook of clinical neuropsychology* (pp. 686–701). New York: Wiley.

Milberg, W., Cummings, J., Goodglass, H., & Kaplan, E. (1979). Case report: A global sequential processing disorder following head injury: A possible role for the right hemisphere in serial order behavior. *Journal of Clinical Neuropsychology, 1*(3), 213–225.

Miller, E. (1980). Cognitive assessment of the older adult. In J. E. Birren & R. B. Sloane (Eds.), *Handbook of mental health and aging.* New Jersey: Prentice-Hall.

Miller, W. R. (1975). Psychological deficit in depression. *Psychological Bulletin, 82,* 238–260.

Monakhov, K., Perris, C., Botskarev, V. K., von Korring, L., & Nikiforov, A. I. (1979). Functional interhemispheric differences in relation to various psychopathological components of the depressive syndromes: A pilot international study. *Neuropsychobiology, 5,* 143–155.

Mueller, J. E. (1979). Test anxiety and the encoding and retrieval of information. In I. G. Sarason (Ed.), *Test anxiety: Theory, research and applications.* Hillsdale, NJ: Lawrence Erlbaum.

Muramoto, O., Sugishita, M., Sugita, H., & Toyokura, Y. (1979). Effect of physostigmine on constructional and memory tasks in Alzheimer's disease. *Archives of Neurology, 36,* 501–503.

Myslobodsky, M. S., & Horesh, N. (1978). Bilateral electrodermal activity in depressive patients. *Biological Psychology, 6,* 111–120.

Naeser, M. A., Gebhardt, C., & Levine, H. L. (1980). Decreased computerized tomography numbers in patients with presenile dementia: Detection in patients with otherwise normal scans. *Archives of Neurology, 37,* 401–409.

O'Brien, M. D., & Mallett, B. L. (1970). Cerebral cortex perfusion rates in dementia. *Journal of Neurology, Neurosurgery and Psychiatry, 33,* 497–500.

Obrist, W. D. (1980). Cerebral blood flow and EEG changes associated with aging and dementia. In E. W. Busse & D. G. Blazer (Eds.), *Handbook of geriatric psychiatry* (pp. 83–101). New York: Van Nostrand Reinhold.

Obrist, W. D., Thompson, H. K., Jr., Wang, H. S., & Wilkinson, W. E. (1975). Regional cerebral blood flow estimated by ^{133}Xenon inhalation. *Stroke, 6,* 245–256.

Parsons, O. A. (1977). Neuropsychological deficts in alcoholics: Facts and fancies. *Alcoholism: Clinical and Experimental Research, 1,* 51–56.

Parsons, O. A., & Adams, R. L. (1983). The neuropsychological examination of alcohol and drug abuse patients. In C. J. Golden & P. J. Vicente (Eds.), *Foundations of clinical neuropsychology* (pp. 215–248). New York: Plenum Press.

Parsons, O. A., & Farr, S. P. (1981). The neuropsychology of alcohol and drug abuse. In S. B. Filskov & T. J. Boll (Eds.), *Handbook of clinical neuropsychology* (pp. 320–365). New York: Wiley.

Parsons, O. A., & Prigatano, G. (1978). Methodological considerations in clinical neuropsychological research. *Journal of Consulting and Clinical Psychology, 46,* 608–619.

Partanen, J., Bruun, K., & Markkanen, T. (1966). *Inheritance of drinking behavior: A study on intelligence, personality, and use of alcohol of adult twins.* Helsinki, Finland: Finnish Foundation for Alcohol Studies.

Perez, F. I., Gay, F. R., & Taylor, R. L. (1975). WAIS performance of neurologically impaired aged. *Psychological Reports, 37,* 1043–1047.

Perez, F. I., Gay, F. R., Taylor, R. L., & Rivera, V. M. (1975). Patterns of memory performance in the neurologically impaired aged. *Canadian Journal of Neurological Sciences, 2,* 347–355.

Perez, F. I., Hruska, N. A., Stell, R. I., & Rivera, V. M. (1978). Computerized assessment of memory performance in dementia. *Canadian Journal of Neurological Science, 5,* 307–312.

Perris, C., & Monakhov, K. (1979). Depressive symptomatology and systemic structural analysis of the EEG. In J. Gruzelier & P. Flor-Henry (Eds.), *Hemisphere asymmetries of function in psychopathology.* Amsterdam: Elsevier/North Holland.

Perry, E. K., Tomlinson, B. E., Blessed, G., Bergman, K., Gibson, P. H., & Perry, R. H. (1978). Correlation of cholinergic abnormalities with senile plaques and mental test scores in senile dementia. *British Medical Journal, 2,* 1457–1459.

Peters, B. H., & Levin, H. S. (1979). Effects of physostigmine and lecithin on memory in Alzheimer's disease. *Annals of Neurology, 6,* 219–221.

Pitts, F., & Winokur, G. (1966). Affective disorder, VII. Alcoholism and affective disorder. *Journal of Psychiatric Research, 4,* 37–50.

Prigatano, G. P., Parsons, O., Wright, E., Levin, D. C., & Hawryluk, G. (1983). Neuropsychological test performance in mildly hypoxemic patients with chronic obstructive pulmonary disease. *Journal of Consulting and Clinical Psychology 51*(1), 108–116.

Pullan, B. R., Fawcitt, R. A., & Isherwood, I. (1978). Tissue characterization by analysis of the distribution of attenuation values in computed tomography scans: A preliminary report. *Journal of Computer Assisted Tomography, 2,* 49–54.

Rehnström, S., Simert, G., Hansson, J. A., Johnson, G., & Vang, J. (1977). Chronic hepatic encephalopathy: A psychometrical study. *Scandinavian Journal of Gastroenterology, 12,* 305–311.

Reitan, R. M. (1976a). Neurological and physiological bases of psychopathology. *Annual Review of Psychology, 27,* 189–216.

Reitan, R. M., (1976b). Neuropsychology: The vulgarization Luria always wanted. *Contemporary Psychology, 21*(10), 737–738.

Reitan, R. M., & Davison, L. A. (1974). *Clinical neuropsychology: Current status and applications.* Washington, DC: V. H. Winston.

Rey, A. (1964). *L'Examen clinique en psychologie.* Paris: Presses Universitaires de France.

Rimel, R. W., Giordani, D., Barth, J. T., Boll, T. J., & Jane, T. (1981). Disability caused by minor head injury. *Journal of Neurosurgery, 9,* 221–228.

Rosen, W. G. (1980). Verbal fluency in aging and dementia. *Journal of Clinical Neuropsychology, 2*(2), 135–146.

Ross, E. D. (1981). The aprosodias: Functional-anatomic organization of the affective components of language in the right hemisphere. *Archives of Neurology, 38,* 561–569.

Ross, E. D. & Rush, J. (1981). Diagnosis and neuroanatomical correlates of depression in brain-damaged patients. *Archives of General Psychiatry, 38,* 1344–1354.

Rourke, B. P., Bakker, D., Fisk, J., & Strang, J. (1983). *Child neuropsychology: An introduction to theory, research, and clinical practice.* New York: Guilford.

Russell, E. W. (1975). Validation of a brain-damage versus schizophrenia MMPI. *Journal of Clinical Psychology, 31,* 659–661.

Russell, E. W. (1977). MMPI profiles of brain-damaged and schizophrenic subjects. *Journal of Clinical Psychology, 33,* 190–193.

Russell, E. W. (1984). Theory and development of pattern analysis methods related to the Halstead-Reitan Battery. In P. E. Logue & J. M. Schear (Eds.). *Clinical neuropsychology. A multidisciplinary approach.* Springfield, IL: Charles C Thomas.

Russell, E. W., Neuringer, C., & Goldstein, G. (1970). *Assessment of brain-damage: A neuropsychological key approach.* New York: Wiley.

Ryan, J. J., Souheaven, G. T., & DeWolff, A. S. (1981). Halstead-Reitan test results in chronic hemodialysis. *Journal of Nervous and Mental Disease, 169,* 311–314.

Schuckit, M., Shuskan, E., Duby, J., Vega, R., & Moss, M. (1982). Platelet monoamine oxidase activity in relatives of alcoholics. *Archives of General Psychiatry, 39,* 137–140.

Schwartz, M. (1971). Review of "Assessment of brain-damage: A neuropsychological key approach" by E. W. Russell, C. Neuringer, & G. Goldstein *Psychophysiology, 8,* 417.

Seidman, L. J. (1983). Schizophrenia and brain dysfunction: An integration of recent neurodiagnostic findings. *Psychological Bulletin, 94*(2), 195–238.

Shallice, T. (1979). Case study approach in neuropsychological research. *Journal of Clinical Neuropsychology, 1*(3), 183–211.

Shapiro, M. B. (1970). Intensive assessment of the single case: An inductive-deductive approach. In P. E. Mittler (Ed.), *The psychological assessment of mental and physical handicaps.* London: Tavistock Publications.

Shore, D. (1982). Closed head injury. [Review of Neurobehavioral consequences of closed head injury.] *Journal of Clinical Neuropsychology, 4*(4), 373–376.

Simard, D., Olesen, J., Paulson, O. B., Lassen, N. A., & Shinhoj, E. (1971). Regional cerebral blood flow and its regulation in dementia. *Brain, 94*(2), 273–288.

Smith, C. M., & Swash, M. (1979). Physostigmine in Alzheimer's disease. *Lancet, 1,* 42.

Soinenin, H., Puranen, M., & Reikkinen, P. J. (1982). Computed tomography findings in senile dementia and normal aging. *Journal of Neurology, Neurosurgery and Psychiatry, 45,* 50–54.

Squire, L. R., & Chace, P. M. (1975). Memory functions six to nine months after electroconvulsive therapy. *Archives of General Psychiatry, 32,* 1557–1564.

Sternberg, D. E., & Jarvik, M. E. (1976). Memory functions in depression. *Archives of General Psychiatry, 33,* 219–224.

Storandt, M., Botwinick, J., Danziger, W. L., Berg, L., & Hughes, C. P. (1983). Psychometric differentiation of mild senile dementia of the Alzheimer's type. *Archives of Neurology, 41,* 497–499.

Strub, R. L., & Black, F. W. (1977). *The mental status examination in neuropsychology.* Philadelphia: F. A. Davis.

Sullivan, E. V., Shedlack, K. J., Corkin, S., & Growdon, J. H. (1982). Physostigmine and lecithin in Alzheimer's disease. In S. Corkin, K. L. Davis, J. H. Growdon, E. Usdin, & R. J. Wurtman (Eds.), *Alzheimer's disease: A report of progress in research* (Vol. 19, pp. 361–368). New York: Raven Press.

Summers, W. K., Viesselman, J. O., Marsh, G. M., & Candelora, K. (1981). Use of THA in treatment of Alzheimer-like dementia: Pilot study in twelve patients. *Biological Psychiatry, 16,* 145–153.

Swiercinsky, D. P. (1978, September). *Computerized SAINT: System for analysis and interpretation of neuropsychological tests.* Paper presented at the meeting of the American Psychological Association, Toronto.

Swiercinsky, D. P. (1984, February). Computerized neuropsychological assessment: Commercial and ethical challenges. In A. J. McSweeney (Chair), *Professional and ethical issues in the use of computers in neuropsychological assessment.* Symposium conducted at the meeting of the International Neuropsychological Society, Houston.

Swiercinsky, D. P., & Warnock, J. K. (1977). Comparison of neuropsychological key and discriminant analysis approaches in predicting cerebral damage and localization. *Journal of Consulting and Clinical Psychology, 45*, 808–814.
Tarter, R. E. (1976). Neuropsychological investigations of alcoholism. In G. Goldstein & C. Neuringer (Eds.), *Empirical studies of alcoholism.* Cambridge, MA: Ballinger.
Tarter, R. E. (1980). Brain damage in chronic alcoholics: A review of the psychological evidence. In D. Richter (Ed.), *Addiction and brain damage.* London: Croom Helm.
Tarter, R. E., & Alterman, A. I. (1984). Neuropsychological deficits in alcoholics: Etiological considerations. *Journal of Studies on Alcohol, 45*, 1–9.
Tarter, R. E., Alterman, A. I., & Edwards, K. L. (in press). Neurobehavioral theory of alcoholism etiology. In C. Chaudron & D. A. Wilkinson (Eds.), *Theories of alcoholism.* Toronto: Addiction Research Foundation.
Taylor, M. A., Greenspan, B., & Abrams, R. (1979). Lateralized neuropsychological dysfunction in affective disorder and schizophrenia. *American Journal of Psychiatry, 136*, 1031–1034.
Turkington, C. (1984, January). The growing use and abuse of computer testing. *American Psychological Association Monitor*, pp. 7, 26.
Turkheimer, E., Cullum, C. M., Hubler, D. W., Paver, S. W., Yeo, R. A., & Bigler, E. D. (1984). Quantifying cortical atrophy. *Journal of Neurology, Neurosurgery and Psychiatry, 47*, 1314–1318.
Walsh, K. W. (1978). *Neuropsychology: A clinical approach.* Edinburgh: Chruchill-Livingstone.
Watson, C. G., Davis, W. E., & Gasser, B. (1978). The separation of organics from depressives with ability and personality based tests. *Journal of Clinical Psychology, 34*, 393–397.
Wedding, D., & Gudeman, H. (1980). Implications of computerized axial tomography for clinical neuropsychology. *Professional Psychology, 11*(1), 31–35.
Weiss, B. (1983). Behavioral toxicology and environmental health science: Opportunity and challenge for psychology. *American Psychologist, 11*, 1174–1187.
Werner, H. (1937). Process and achievement. A basic problem of education and developmental psychology. *Harvard Educational Review, 7*(3), 353–368.
Western Psychological Services. (1984). WPS test report microcomputer diskette: LNNB, forms I and II [computer program]. Los Angeles, CA: Western Psychological Services. (Catalog No. W-1002).
Wilson, R. S., Rosenbaum, G., Brown, G., Rourke, D., Whitman, D., & Grissel, J. (1978). An index of premorbid intelligence. *Journal of Consulting and Clinical Psychology, 46*, 1554–1555.
Wilson, R. S., Bacon, L. D., Fox, J. H., & Kaszniak, A. W. (1983). Primary memory and secondary memory in dementia of the Alzheimer type. *Journal of Clinical Neuropsychology, 5*(4), 337–344.
Wilson, R. S., Bacon, L. D., Kramer, R. L., Fox, J. H., & Kaszniak, A. W. (1983). Word frequency effect and recognition memory in dementia of the Alzheimer type. *Journal of Clinical Neuropsychology, 5*(2), 97–104.
Wrightsman, L. S. (1962). The effects of anxiety, achievement motivation, and task importance on intelligence test performance. *Journal of Educational Psychology, 53*, 150–156.
Yamaguchi, F., Meyer, J. S., Yamamoto, M., Sakai, F., & Shaw, T. (1980). Noninvasive regional cerebral blood flow measurements in dementia. *Archives of Neurology, 37*, 410–418.
Yozawitz, A., Bruder, G., Sutton, S., Sharpe, L., Gurland, B., Fleiss, J., & Costa, L. (1979).

Dichotic perception: Evidence for right hemisphere dysfunction in affective psychosis. *British Journal of Psychiatry, 135,* 224–237.

Zenhausern, R., & Parisi, I. (1981, February). *Brain disintegration in schizophrenia and depression.* Paper presented at the meeting of the International Neuropsychological Society, Atlanta.

2

Clinical Interviewing of the Patient and History in Neuropsychological Assessment

ANDREW PHAY, CAROL GAINER, and GERALD GOLDSTEIN

Although most of this volume is concerned with the application of formal tests to the assessment of brain-damaged patients, few would deny the importance of some form of clinical assessment as a component of a comprehensive evaluation. Users of standard test batteries may not use clinical data in selecting tests to be administered, but may make extensive use of such material in their interpretations and report preparations. Neuropsychologists who use flexible test batteries may also use clinical data in interpretation, but the actual selection of tests may be based in whole or part on the initial interview and the review of historical material. The major exception to the practice of conducting an initial interview and taking a history prior to test administration is in the case of those clinicians who prefer to do blind interpretation, and who do not want to review the history or even see the patient prior to making an interpretation based exclusively on test data. The practice of

---

This chapter was written while Andrew Phay was affiliated with the Brooklyn Veterans Administration Medical Center.

ANDREW PHAY • Veterans Administration Medical Center, Murfreesboro, TN 37130. CAROL GAINER • 1510 Huntington Drive, Apartment Q4, Murfreesboro, TN 37130. GERALD GOLDSTEIN • Veterans Administration Medical Center, Highland Drive, Pittsburgh, PA 15206.

blind interpretation is a controversial matter, with compelling justifications provided by those for and those opposed to it. It seems likely that the wisdom of using blind interpretation or not using it is a complex matter involving one's approach to neuropsychological assessment, the context in which one practices, the relative significance of research and direct patient care goals in one's practice, and the degree of independent clinical responsibility of the neuropsychologist. For example, blind interpretation may be quite reasonable in a setting in which comprehensive history taking and interviewing is routinely accomplished by members of the clinical team other than the neuropsychologist, but may not be as appropriate in situations in which the neuropsychologist has major clinical responsibility for the assessment and disposition of patients, without support from other clinicians. It will therefore not be our purpose here to encourage or discourage the practice of blind interpretation, but rather to provide some pertinent considerations and materials for those clinicians who wish to use interviewing and history taking as part of their evaluations. It should also be emphasized in that regard that it is not the intent of this chapter to take the stance that diagnostic error can be reduced through utilization of historical and interview data in neuropsychological interpretation. However, we hope to show that such data may often be quite helpful as part of a comprehensive evaluation.

PURPOSES OF THE INTERVIEW AND HISTORY IN NEUROPSYCHOLOGCAL ASSESSMENT

The purposes of the interview and history in neuropsychological assessment are quite straightforward. First, the interview is generally used to obtain the patient's complaint or information about the presenting problem. It is also possible at this point to evaluate the patient's potential for further assessment. Often, the initial contact is helpful with regard to determining whether the patient can cooperate for standard neuropsychological testing. In Luria's (Christensen, 1975, 1984) formulation of the preliminary conversation, emphasis is placed on the patient's level of consciousness. Some brain-damaged patients cannot sustain a state of normal wakefulness, and may appear half asleep or drowsy much of the time. Other patients are too feeble or insufficiently intellectually intact to cooperate for standard testing.

Some clinicians use the interview and history to provide an estimate of premorbid level of ability. Although the availability of such data as intelligence or achievement test scores obtained prior to the illness is a real advantage, such information is seldom readily available and it is usually necessary to make estimates based on such considerations as educational level and reports of the patient and informants about school grades, occupation, and socioeconomic status. Obviously, one crucial function of the interview and history is that of obtaining diagnostic information. Within the purview of neuropsychology, there are generally two important diagnostic issues; whether or not the patient has a medical history consistent with impairment of brain function, and whether or not the patient, in addition to whatever neurological dysfunction the patient has, also has a psychiatric illness other than an organic mental disorder. Although extensive efforts are being undertaken to discover biological and other objective markers of the major psychiatric disorders, it is generally granted that few if any of these markers have achieved definitive status as yet, and the psychiatric interview remains the state-of-the-art instrument for establishing psychiatric diagnoses. The important point for both the general medical and psychiatric interview is that the patient or informant may not report on important information unless specifically asked. In the case of psychiatric interviewing, it is generally not only a matter of asking the right questions, but of conducting the interview in a skilled and organized manner.

It is important to assess the patient's attitude toward the illness. As is well known, attitudes toward general medical or psychiatric illness may range from substantial denial and/or lack of awareness to severe reactive anxiety and depression, sometimes reaching states of panic. Knowledge of this dimension can be of crucial importance for treatment planning. If the patient is particularly distressed about the illness, it is often necessary to treat that distress behaviorally, pharmacologically, or in a combined program before rehabilitation efforts can be initiated. In the case of the patient who denies or is unaware of the illness, strategies must be developed to deal with these matters in some way.

The initial interview can also provide an estimate of the severity of the illness, either from the clinical phenomenology or through a comparison of the patient's present state with what is known of the premorbid level. The patient's level of functioning at the time of initial evaluation can also serve as a baseline against which changes occurring later on can be evaluated. Those neuropsychologists who use flexible or individualized test batteries may also use the initial interview to aid in determining which tests to use. Indeed, Hamsher (1984) has argued that this procedure is a highly reasonable one for neuropsychological assessment because it may, on the one hand, reduce the length of the testing procedure, and on the other hand, assure that the tests administered are appropriate for assessment of the diagnostic problems stated in the referral and presented by the patient.

In summary, the neuropsychologically oriented interview and history is generally directed toward (a) determining the complaint or presenting problem; (b) determining the patient's level of consciousness and capacity to cooperate for formal testing; (c) estimating the premorbid level of ability; (d) obtaining pertinent general medical, neurological, and psychiatric diagnostic data; (e) assessing the patient's attitude toward the illness; (f) estimating the severity of impairment; and (g) in some cases, obtaining information pertinent to test selection. We will very briefly elaborate on these points in the following.

THE COMPLAINT AND PRESENTING PROBLEM

In general clinical practice, the patient's complaint or a problem presented by an informant usually provides the initial focus of the evaluation. In the case of brain-damaged or psychiatric patients, the complaint may sometimes not be clearly expressed, but often the patient can provide significant information. For example, the complaint might involve progressive loss of memory, lack of energy, difficulty finding words or related matters. It is often quite helpful to obtain information about the presenting problem from an informant, particularly when it is difficult for the patient to articulate the nature of the difficulty. The complaint or presenting problem generally provides important leads with which to initiate the evaluation. In neuropsychology, critical complaints or presenting problems generally involve memory, speech, capacity to perform routine activities of daily living, perceptual or motor difficulties, and difficulties with various cognitive abilities, such as calculation, solving work related problems, and maintaining attention.

LEVEL OF CONSCIOUSNESS AND CAPACITY TO COOPERATE

One important consideration for neuropsychological assessment is the testability of the patient. In some settings, notably outpatient and general medical inpatient practice, most patients referred are testable. However, in psychiatric inpatient and geriatric settings, testability is often an issue for numerous reasons. In some cases, the patient may be partially testable, and relatively simple, brief tests can sometimes be substituted for lengthier, more complex procedures. For example, a rating scale can be used instead of a standard intelligence test, or a short form of some procedure can be used when the patient cannot cooperate for the full test. The patient's level of consciousness is both diagnostic in itself, and its assessment can be useful in planning further evaluation. Sometimes the patient is sleepy or inattentive temporarily, and testing can be postponed until the patient is more alert. If it is a permanent condition, that knowledge can be used in organizing the evaluation, and interpreting what testing is accomplished.

ESTIMATING PREMORBID INTELLIGENCE

Ideally, the level of premorbid ability is determined on the basis of objective data contained in the history. Intelligence or achievement test scores obtained prior to illness are of course highly desirable. With the increasing use of objective tests in educational and occupational settings, this information should become available for increasing numbers of people who become ill, but at present it is generally necessary for the clinician to resort to some form of estimate. There are basically two ways of generating these estimates, one based on psychometric and the other on demographic data. The psychometric method is a part of the formal evaluation itself, and takes the form of utilizing such techniques as Wechsler's (1944) Deterioration Index. The demographic methods provide estimates on the basis of such considerations as educational level, occupational history, socioeconomic status, race, sex, and reports of informants. Wilson, Rosenbaum, and Brown (1979) have combined several of these indexes into an equation that predicts premorbid IQ.

It may be noted that the research literature has not been favorable with regard to the accuracy of the Deterioration Index and related measures (Matarazzo, 1972), and it is not clear that the demographic methods are that much better, particularly in atypical cases; for example, the exceptionally bright individual with little formal education. At present, a combination of psychometric indexes and the use of demographic information and historical search is probably the optimal way of assuring accuracy of the premorbid ability estimate.

OBTAINING DIAGNOSTIC DATA

We will defer our discussion of obtaining diagnostic data to the section of this chapter that deals with assessment methods. Probably most neuropsychologists, particularly those in independent practice, utilize some form of brief medical history that emphasizes disorders that have direct or indirect implications for brain function. Some use more general procedures, such as the Cornell Medical Index. We would only point out here that for neuropsychology, it is particularly important to obtain, in addition to the usual history of past illnesses, as detailed a family history as possible, a thorough assessment of early life illnesses, such as birth injury, high fever, severe infection, or seizures, and an assessment of prenatal difficulties. Reported histories of head trauma should be examined in detail, with specific inquiries about the nature of the trauma, length of time unconscious, length of post-traumatic amnesia and sequelae, including such matters as perceived change in function following the injury and history of seizures. The family history is particularly important because certain disorders, notably Down's syndrome and Huntington's disease, are established genetic diseases, whereas others, such as Alzheimer's disease, may have some hereditary or familial component. It is now thought that some functional psychiatric disorders, such as alcoholism and certain forms of affective disorder, have a genetic component. It is sometimes necessary for various reasons to do a psychiatric evaluation in addition to an interview directed solely toward obtaining a general medical history. Some clinicians use objective tests for this purpose, notably the MMPI, whereas others prefer to use formal, structured, or more informal interview procedures. We will discuss some of these procedures in the section on assessment methods.

THE PATIENT'S ATTITUDE

In the realms of psychiatric and neurological illness, the expression "attitude toward illness" has somewhat different implications than is generally the case for physical illness. There is, first of all, an extremely wide range of capacity among neurological patients with regard to comprehension of the nature of their illnesses. Significantly demented patients may not appreciate the fact that they are ill, whereas there are certain disorders in which denial of the illness is a component of the neurological syndrome itself (Weinstein & Kahn, 1955). Psychiatric patients, particularly psychotic individuals, may not fully appreciate the nature of their disorders. In these cases, the situation is quite different from what is generally the case for physically ill individuals. Physically ill patients generally understand the nature of their illnesses, but may vary with regard to how they react to them. There may be anything ranging from denial to panic, but there is generally awareness that the illness is present. In the case of the brain-damaged patient, the problem of attitude generally represents a combination of the capacity of the patient to assess the nature and consequences of his or her illness, and the patient's affective reaction to it. For example, in the case of stroke, there are some patients who can tell you that they had a stroke, how long ago they had it and what its consequences were (e.g. "I was paralyzed on my right side and had difficulties in talking").

Many of these patients will make extremely strenuous efforts to rehabilitate themselves through sometimes arduous physical and speech therapy. There are other stroke patients who completely deny that they had a stroke, and may even persist in that denial even when their hemiplegia is demonstrated to them. Finally, there are some patients who are aware that they have had strokes, but who develop such severe anxiety or affective disorders that they cannot cooperate for rehabilitative efforts. Thus, the task of the interviewer is frequently that of assessing the level of awareness the patient has of his or her illness, as well as the affective reaction to it.

The assessment of attitude should include some evaluation of whether the patient's initial affective state has significant implications for the outcome of treatment or rehabilitation. Quite frequently, secondary depression associated with the neurological illness or severe anxiety states may influence prognosis, and it becomes necessary to diagnose and treat these conditions. Whereas it may not be necessary to do a complete psychiatric evaluation on every patient undergoing neuropsychological assessment, it is generally helpful to do some form of brief screening so that patients with what seem to be significant affective or motivational difficulties can be evaluated further. This aspect of the initial interview process may aid significantly in providing information about how the personal aspects of the patient's life in areas other than the specific illness impact on ultimate outcome and the course of the illness. It provides an opportunity for the patient and informants to express feelings and give information that might not otherwise be made available. Perhaps the clearest example of the significance of this process is to be found in the assessment of the so-called pseudodementia of the elderly (Wells, 1979). If it becomes apparent on interview that an elderly individual presenting with symptoms of dementia is also significantly depressed, the treatment strategy may be quite different from what would be the case if signs of depression were absent. Generally, if depression is present it is the first treatment focus with the hope that the dementia is in fact a pseudodementia, and that the patient's mental status will improve following successful treatment of the depression. Wells (1979) has indicated that formal neuropsychological tests may not be as accurate with regard to distinguishing between actual dementia and pseudodementia as are clinical signs.

Estimating Severity of Impairment

Severity can be said to have three components: the level of deficit in an absolute sense, the degree of impairment relative to premorbid level and the morbidity–mortality characteristics of the underlying disease entity. It is generally important to estimate the absolute level of deficit early on in the assessment process, because the patient's level of functioning may play some role in determining the nature of the formal neuropsychological assessment that will be made. For example, some patients may not be able to cooperate for a lengthy standard battery, and it may only be possible to do ratings and observations, and to obtain information from informants. It is important to know about severity of impairment relative to premorbid level, because such information may aid in reducing the diagnostic errors of overestimating the amount of illness-related impairment in individuals of limited premorbid ability and doing the reverse in the case of individuals with high pre-illness levels. Thus, for example, an obtained Wechsler Adult Intelligence Scale (WAIS) IQ of 90 may reflect devastatingly severe impairment in an individual functioning at a superior intellectual level prior to acquisition of the illness, whereas it may reflect no impairment at all in a person who has always functioned at a low average level of intelligence. The aspect of the assessment having to do with morbidity and mortality considerations has particularly crucial implications for treatment, intervention, and rehabilitation planning. For example, when we first assess a patient with Huntington's disease, the indications of cognitive or motor deficit may be quite minimal. However, if the diagnosis is correct, it is certain that over the next decade the patient will get progressively worse, and will probably die at the end of that time period or shortly thereafter. The patient with multiple sclerosis with early visual changes may become progressively blind over the years. In cases of nonprogressive brain damage, the prognosis is far more benign, and treatment planning may be of quite a different nature from what would be the case for the patient with a progressive illness. Thus, the patients that come to the attention of neuropsychologists may demonstrate a broad range of severities of impairment and of the underlying disease entities that produce the impairment. These dimensions of severity have serious implications for the kinds of assessment done, for formulation of a prognosis, and for treatment planning.

OBTAINING INFORMATION FOR TEST SELECTION

Advocates of the flexible approach to neuropsychological assessment, notably Luria and his followers (Christensen, 1975, 1984; Luria, 1973), feel that the preliminary conversation or interview with the patient is in essence the beginning of the formal neuropsychological assessment, and is of great value in determining what tests should be used in what follows. Those who use standard batteries, but who interview the patient and take a history prior to testing, may use the interview data they obtain in a somewhat different way. If at all possible, the patient gets the standard battery regardless of the results of the interview, but those results may be used in suggesting supplemental procedures and in formulation of the interpretation of the test results. With regard to the first point, for example, on the basis of the clinical interview, the neuropsychologist may wish to obtain a formal psychiatric evaluation, or to administer additional specialized tests that the standard battery does not include. With regard to the latter point, the neuropsychologist may wish to discuss all or some of the various issues reviewed above when writing the report. The neuropsychologist who does blind interpretations may wish to reconcile interview and historical data, obtained by some other person, with the initial interpretation. At that point, a supplemental report may be prepared that contains an analysis of the reconciliation.

THE ROLE OF THE INTERVIEW AND HISTORY IN TREATMENT PLANNING

In general clinical practice, it is customary to use the initial examination, history, and various laboratory procedures as indicated to formulate a treatment plan. Ideally, the development of the treatment plan is an interdisciplinary activity, and is based on as comprehensive an evaluation as is necessary and feasible. In this context, neuropsychologists are being increasingly asked to make recommendations in the form of contributions to the treatment plan. This increase may well be associated with a decreased interest by neuropsychologists and those who use their services with regard to topographical localization of brain pathology, and an increase in interest in detailed analyses of functional deficits. This functional analysis is thought to have major implications for the design of rehabilitation strategies (Diller & Gordon, 1981; Goldstein & Ruthven, 1983). The issue raised here relates to how the interview and history are used by the neuropsychologist, in combination with the formal tests, to assist in formulation of the treatment plan. In general, the formal testing is used in this regard to analyze the pattern of the patient's preserved and impaired cognitive abilities, and to design a program that deals as specifically as possible with restoration of impaired functions. For example, Diller and Weinberg (1971) have devised a procedure to correct for the neglect of the left-visual field commonly found in patients who sustained strokes to their right cerebral hemispheres. Speech therapists generally design their treatment programs on the basis of the pattern of preserved and impaired language abilities found in the individual patient (Albert, Goodglass,

Helm, Rubens, & Alexander, 1981). It would be our view that the interview and history are generally more related to aspects of treatment other than those specifically involving remediation of the patient's cognitive deficits. These aspects can be conveniently divided into patient, family, and environmental considerations.

The Patient

The initial interview and history often aid in identifying treatment needs in addition to those that are directly associated with the patient's neuropsychological or medical disorder. Previously we alluded to the not uncommon need of brain-damaged patients for treatment of affective or anxiety disorders. It is also not uncommon to elicit histories of alcoholism or other substance abuse during the initial interview, which if still active, may require treatment. Sometimes these disorders work in the direction of making the immediate medical problem worse, as, for example, in the case of alcoholism in the diabetic or hypertensive patient. Sometimes patients require rather straightforward counseling and education concerning their illnesses. There are sometimes situations in which more or less formal psychotherapy is indicated, and the initial interview and history can fulfill their traditional roles of providing the therapists with diagnostic information concerning such areas as the patient's conflicts, interpersonal relationship patterns, character structure, and related matters. It is also not uncommon for a physician to prescribe psychopharmacological treatments when the patient or an informant reports appropriate symptoms; for example, treatment with neuroleptics for nocturnal agitation in Alzheimer disease patients.

The Family

Often, family members serve to a greater or lesser extent as caretakers for brain-damaged patients. Sometimes it is necessary to provide the family with instructions for managing patients who essentially require total care, but often it is only necessary to provide the family with information about the patient's medication, habits and diet, and about signs of recurrence of the illness. In epileptic patients, it is often helpful to instruct the family regarding what to do when the patient has a seizure. During the initial assessment process, the clinician has the opportunity to evaluate the family's capacity and motivation to care for the patient. The particular intervention taken may be based to a large extent on this evaluation. Sometimes families simply lack the resources to care for the patient, and it is necessary to arrange for visiting health care providers, or to consider institutionalization. Sometimes the family can provide excellent care, and only requires the needed information and periodic monitoring.

The tragic nature of many brain disorders often necessitates direct treatment of the family members themselves. Living with a spouse who is deteriorating, or who, owing to the consequences of brain damage, has significant behavioral difficulties, is obviously stressful. The whole area of living with brain-damaged patients (Logue, 1975) has been examined in recent years, and there is a popular literature available that may be of help to family members. Sometimes formal family therapy or marriage counseling may be indicated. Thus, the clinician, as part of the initial assessment, might do well to assess the family situation, and institute appropriate interventions.

The Environment

Sometimes, it may not be possible to alter the patient's personal situation, but changes may be made in the world in which the patient lives. Generally, the major consideration here involves providing the patient with sufficient support to maintain good health and an optimal quality of life. This area of intervention can be divided into the physical and the behavioral environment. It is often quite important to determine whether or not the patient, particularly the patient who cannot work, has adequate financial resources and is living in a sufficiently sheltered and supportive setting to assure maintenance of health. Typically, a social worker is involved in this aspect of the assessment. It may sometimes be necessary to assist the patient in applying for benefits, or to work with the family to determine the extent to which it is willing and able to support the patient. Sometimes patients can live at home if the appropriate engineering is accomplished in the house and in vehicles used by the patient. The installation of railings, widening of doors, and construction of ramps to allow wheelchair access and related modifications can sometimes permit a patient to live at home and forestall institutionalization.

The behavioral environment, as we use the term here, has largely to do with how people respond to the patient. The manner of response to brain-damaged patients may range from extreme neglect to suffocating overindulgence. During the initial assessment period, the clinician may wish to evaluate how people in daily contact with the patient deal with the problems generated by the illness. How does a husband deal with his wife who recently had a stroke? How does a mother treat her seriously head-injured child? These observations can often be used in interventions with family members and other associates of the patient that aim toward producing an optimal behavioral environment. In this regard, the role of behavioral assessment and the application of treatments based on contingency management principles could be quite helpful (Goldstein, 1979).

ASSESSMENT METHODS

Thus, far, we have been considering the role of the history and interview as a component of a comprehensive neuropsychological assessment, but have not dealt specifically with the methods that could be employed in performing these assessments. In this section, we will discuss some of these methods, primarily through borrowing a great deal from those procedures that are commonly used with psychiatric patients. There are several reasons for taking this approach. First, it is not at all uncommon to have diagnostic problems related to differentiating between the organic mental disorders and other psychiatric conditions for which there are often no clear organic bases. Second, neuropsychologists, as psychologists, are largely interested in behavior, and it is rare for a psychologist to take a complete medical history. Indeed, it may be inappropriate to do so without the necessary medical training. However, psychologists and other mental health practitioners can and do perform assessments of psychiatric patients using interviewing and history-taking methods, and may also perform such assessments with medical patients when there is some question of a behavioral problem. Finally, neuropsychological deficits are primarily mental deficits involving some aspect of perception or cognition, and so it would appear to be quite appropriate to utilize methods designed to assess mental disorders; that is, psychiatric conditions. However, the clinical neuropsychologist should have the capability of assessing certain limited aspects of the medical history, particularly with regard to those historical factors that may have implications for central nervous system function. We will therefore discuss some instruments that address themselves to those matters.

There are two major forms of psychiatrically oriented assessment methods, excluding psychological tests and physiological measurements. They are the interview and the clinical rating scale. Within these two major divisions, numerous subclassifications may be made. Interviews can be structured or unstructured. They can be general, or targeted to some specific problem area, such as the presence of an affective disorder. Rating scales may be of the self-rating type or the scale may be completed by a physician, nurse, relative, or other individual who has had the opportunity to observe the patient. In each of these subclassifications, there are several different instruments available. There are at least four general structured psychiatric interviews and many general rating scales in current use. It may also be noted that several of these instruments are designed for specific populations, notably children, adults, and elderly adults. For example, there are scales designed specifically for assessment of depression in the elderly. There are two very useful books that provide reviews of many of these methods. One by Crook, Ferris, and Bartus (1983) is concerned with assessment of geriatric patients; the other, by Burdock, Sudilovsky, and Gershon (1982), reviews methods appropriate for psychiatric patients in general.

THE INTERVIEW

General Interviews

There has been a revolution in psychiatric interviewing over the past decade, although its impact has not completely permeated the field as yet. The revolution was provoked by a number of studies in which it was shown that when more than one psychiatrist interviewed a patient, there was a distressingly high level of disagreement concerning the diagnosis. In other words, psychiatric diagnosis was found to be unreliable (Hines & Williams, 1975; Zubin, 1967), and steps had to be taken to correct that situation. One of these steps was the development of a new diagnostic system, DSM-III (APA, 1980), in which highly specific criteria have to be met to make a particular diagnosis, and the other step was the development of various structured psychiatric interviews. Great attention was paid to the matter of reliability in the development of these procedures, to the extent that we now have available several highly reliable, structured, general psychiatric interviews. Two groups of investigators were primarily responsible for the development of these procedures, one located at Washington University in St. Louis (Helzer, Robins, Croughan, & Welner, 1981) and the other at the New York State Psychiatric Institute in New York City (Endicott & Spitzer, 1978). Another group in England (Wing, Cooper, & Sartorius, 1974) also developed a structured interview, the Present State Examination, but it appears to be used more in Europe than in the United States. These interviews are closely associated with several objective classificatory systems that were ultimately incorporated, at least to some extent, into DSM-III. The most well known systems are the Research Diagnostic Criteria (Spitzer, Endicott, & Robins, 1978) and the so-called Feighner criteria (Feighner et al., 1972).

The structured general interviews are outlines that guide the clinician through the criteria needed to establish or rule out some particular diagnosis or diagnoses. It may be pointed out that these interviews do not contain specific questions asked of the patient in a mechanical manner, but rather consist of points about which information must be obtained. The interviewer must formulate the specific questions needed to gather this information, and so these procedures can generally not be conducted by untrained individuals. It is generally felt that one needs general training as a mental health practitioner and specific training in the interview procedure itself. There are three structured general interviews that are commonly used in the United States; The Schedule for Affective Disorders and Schizophrenia (SADS) (Endicott & Spitzer, 1978), the Renard Diagnostic Interview (Helzer et al., 1981) and the Diagnostic lnterview Schedule (DIS) (Robins, Helzer, Croughan, & Ratcliff, 1981). As indicated previously, the Present State Examination (PSE) (Wing et al., 1974) is commonly used in Europe, and to some extent in the United States. There are differences among these interviews, and some controversy about their appropriate applications, but suffice it to say that they all provide psychiatric diagnoses, generally compatible with DSM-III criteria, of well established reliability.

Rather than going on to describe these instruments in greater detail, we will focus on their potential role in clinical neuropsychological assessment. We use the word potential advisedly, because to the best of our knowledge, these procedures have had little use as a component of neuropsychological assessment, with evaluation of psychopathology generally accomplished with some objective test, notably the MMPI. Although the MMPI is a convenient, well standardized and researched test, there may be some advantages obtained through use of the structured interviews. First, a structured clinical interview typically takes into account conditions that might moderate and greatly influence interpretation of particular behaviors. First, for example, observation of a patient in a toxic state may greatly affect one's inclination to assign the diagnosis of schizophrenia. Second, although the structured interviews can rather unequivocally determine a diagnosis, and a DSM-III compatible diagnosis in particular, current evidence appears to suggest that it is quite difficult to extract a DSM-III diagnosis from tests of the MMPI type. Third, investigators such as Zubin (1984) have suggested that because the generally accepted validity criteria for objective tests of psychopathology consist of psychiatric diagnoses, and because such diagnoses are almost always based solely on interview and historical data, there seems little point in using a test when one can just as well conduct an interview. It used to be thought that objective tests were to be preferred because interviews tended to be unreliable, but now we have interviews with established reliability.

The question remains as to whether it is useful to administer these structured general interviews as part of neuropsychological assessment, particularly because most patients referred for such assessments have known or suspected brain damage. In DSM-III terms, one generally attempts to identify an organic mental disorder with neuropsychological tests, and the assessment of organic mental disorders does not appear to be one of the strong points of the structured interviews. Indeed, the SADS barely deals with the matter at all. Psychiatrists interested in the diagnosis of dementia have developed special interviewing techniques, perhaps the most extensive one being the Comprehensive Assessment and Referral Evaluation (CARE) (Gurland et al., 1977). Our view of the matter would be that the general structured techniques could be useful in those situations in which there is some suspicion of a psychiatric illness other than an organic mental disorder. Indeed, there might be some hope that the application of these interviews can provide some components of a solution to the venerable problem of discriminating between brain-damaged and schizophrenic patients, and between elderly patients with dementia and depressive pseudodementia (Wells, 1979). In general, then, the application of structured psychiatric interviews to neuropsychological assessment is something for the future, but one might become quite curious as to the value of doing them, and about how the data obtained might influence interpretation of neuropsychological test results.

Specific Interviews

What we are describing as specific interviews are structured procedures targeted toward obtaining information about one or a small number of disorders. The most well known specific interviews are for depression, and the most well known of those is the Hamilton Depression Scale (Hamilton, 1960). Although it is described as a rating scale, it actually consists of a brief structured interview that yields a score reflecting severity of depression. It consists of 21 items, which are rated on the basis of a semi-structured interview generally lasting 30 to 60 minutes. A modified version of the Hamilton scale has been developed for elderly patients by Yesavage, Brink, Rose, and Adey (1983). There are also several structured interviews or interview-like techniques for alcoholism and drug abuse. Generally, the aim here is to obtain a reliable history from the patient or an informant concerning duration, intensity, and pattern of drug abuse. Perhaps the most popular of these procedures is the Michigan Alcoholism Screening Test (MAST) (Moore, 1972), but other procedures have been developed by Marlatt (1976), Khavari and Farber (1978), and McLellan, Luborsky, Woody, and O'Brien (1980).

The mental status examination is actually a form of structured interview, and is commonly used as a standard part of the psychiatric evaluation. However, in recent years, we have seen the development of a number of structured mental-status examinations targeted toward the assessment of dementia. Perhaps the most well known of these instruments are the Dementia Scale of Blessed, Tomlinson, and Roth (1968), the Mini-Mental State examination of Folstein, Folstein, and McHugh (1975), and the Mattis Dementia Rating (Coblentz et al., 1973). These instruments, although sometimes described as rating scales, are actually brief interviews with patients or informants that generally cover such areas as orientation, memory, attention, and the capacity to perform everyday activities (ADL status). At a more basic level, an instrument called the Glasgow Coma Scale (Jennett & Teasdale, 1981) is frequently used to assess comatose patients on the basis of eye opening, motor, and verbal responses. A thorough review of the various dementia scales is contained in Mohs, Rosen, Greenwald, & Davis (1983).

THE CLINICAL RATING SCALES

General Scales

Clearly, the most widely used of the rating scales of psychopathology is Overall and Gorham's Brief Psychiatric Rating Scale (BPRS) (1962). The BPRS is very commonly used in pharmacological or other treatment-related studies in which change in clinical status is the major consideration. It assesses a broad range of psychopathology with 18 items, each of which reflects a factor analytically based symptom construct. Some of the less commonly used general scales are the Inpatient Multidimensional Psychiatric Scale (IMPS) (Lorr & Klett, 1966) and various global assessment techniques, such as the Global Assessment Scale (GAS) of Spitzer, Endicott, and Robins (1978), and the Physician's Global Assessment Scale (Bech, Gram, Dein, Jacobsen, Vitger, & Bolwig, 1975).

There are several rating scales based on direct observation on the ward rather than through use of an interview. The most widely used of these techniques are the Nurses' Observation Scale for Inpatient Evaluation (NOSIE) (Honigfeld & Klett, 1965) and the Ward Behavior Inventory (Burdock, Hardesty, Hakerem, Zubin, & Beck, 1968). There are numerous related techniques. A review of them may be found in Raskin (1982). Generally, these procedures are accomplished by nurses or psychiatric aides who must rate the patient on a number of items of observable behavior (e.g., "Complains about the staff's behavior"). There is also a very large number of general self-rating scales, but these instruments tend to blend in with standardized psychological tests. For example, some would characterize the MMPI as a general self-rating instrument. Because we are trying to focus on nonpsychometric metholodologies here, we will not attempt to review the voluminous amount of material in this area.

Specific Scales

Specific clinical rating scales are also extremely numerous and involve a variety of areas. For example, in the areas of assessment for depression, such instruments as the Beck Depression Inventory (Beck, 1967) or the Zung Depression Scale (Zung, 1965) are the rating-scale equivalents of interview-based techniques of the Hamilton type. Self-rating scales are also available in the areas of alcoholism and drug abuse, anxiety disorders, social adjustment, antisocial behavior, and many other areas of personality and psychopathology. In neuropsychology, perhaps the self-rating scales of greatest interest are those in which patients are asked to rate their own functioning in a number of areas. Often, the clinician has the benefit of comparing these self-assessments with information provided by informants, and with formal neuropsychological and other psychometric test data. The most well known of these procedures is the Patient Assessment of Own Functioning Inventory (PAOFI), reported on in Heaton and Pendleton (1981). This instrument is a 32 item questionnaire used to obtain patients' self-ratings in four areas; cognition, memory, language communication, and sensorimotor function. An example of an item would be, "How often do you lose things or have trouble remembering where they are?" Recently, Parsons (1984) used the PAOFI with a sample of alcoholics, finding that alcoholics do provide realistic estimates of their impairments.

Functional Assessment

Functional assessment, broadly speaking, is a means of evaluating how and how well the impaired individual utilizes the residual functions in everyday life. It deals with practical considerations, such as how effectively the individual can communicate, ambulate, and generally live independently. Functional assessment is highly related to the concept of ADL (Activities of Daily Living), which orginally developed in the fields of physiatry and occupational therapy. Therefore, ADL evaluations are traditionally done by occupational therapists, whereas physiatrists assess the functional status of the various motor and sensory systems. Physical therapists are also sometimes involved in this process. As in the case of the interview and rating scales, there are general and specific forms of functional assessment, and functional assessments may be accomplished by means of interview, questionnaire, or performance tests. In psychiatry and psychology, probably the most widely used functional assessments are contained in the Comprehensive Assessment and Referral Evaluation (CARE) (Gurland, et al., 1977) and the OARS Multidimensional Functional Assessment Questionnaire (MFAQ) (Duke University Center for the Study of Aging, 1978). have also prepared an extensive manual for assessment of numerous skills needed to function effectively independently. A typical ADL item might be, "Do you have any difficulties or problems doing your shopping?" or "Can you climb stairs without taking frequent rests?" In general, the areas assessed are personal hygiene, dressing, household activities, social activities, and ambulation.

Whereas physiatrists, occupational therapists, and physical therapists have developed highly specific forms of functional assessment (Hirschberg, Lewis, & Vaughan, 1976), probably the most well known procedures to psychologists are in the area of language and communication. The two major procedures are the Functional Communication Profile (FCP) (Sarno, 1969) and the Communicative Activities of Daily Living (CADL) (Holland, 1980). The purpose of these procedures, as opposed to the conventional aphasia tests, is that of determining how, and how well, the aphasic patient utilizes residual language capacities to communicate in everyday life situations. For example, a CADL question is, "How would you let somebody know that you're cold?" There is also a recently developed interest in practical memory. Sunderland, Harros, and Gleve (1984) have validated a questionnaire to measure aspects of everyday memory of patients with histories of head injury. The questionnaire must be completed by an informant who has daily contact with the patient. A sample item is, "Forgetting when it was that something happened; for example, whether it was yesterday or last week." The items are rated on a scale of frequency of occurrence, ranging from "not at all" to "more than once a day." Recently, Wilson, Baddeley, and Hutchins (1984) have published what they describe as a behavioral memory test, which actually is a combination of a conventional or formal memory test and a practical memory test. An example item from the practical memory section involves borrowing of a possession of the subject and asking that he or she ask for it back at the end of the test session, and recall where it was placed. It seems apparent that there is a growing interest in functional assessment in clinical neuropsychology, particularly with reference to the implications formal neuropsychological test results have for the capacity of the patient to perform in everyday life situations.

NEUROPSYCHOLOGICAL HISTORIES

As in most clinical specialties, clinical neuropsychologists are concerned with the problem of taking a thorough but pertinent history. Elsewhere in this chapter the importance of the history in neuropsychological assessment is considered, and here we will address ourselves to the topic of the efforts made to develop structured history protocols that are relevant to neuropsychology. To the best of our knowledge, the earliest attempt to develop such an instrument eventuated in Halstead's unpublished "Head Injury Questionnaire." The name is somewhat misleading, as the questionnaire covers much more than the area of traumatic head injury. It has been revised extensively over the years, but is still traditionally given as part of the Halstead-Reitan Battery. Clinicians who believe in blind interpretation do not look at the Head Injury Questionnaire prior to preparation of their reports, but whether or not one wishes to do so is a matter of personal preference. The Head Injury Questionnaire, despite its informal format and its age, contains many crucial items. The patient's level of education is recorded as is handedness and the handedness of both parents. Use of tobacco, alcohol, street drugs, sedatives, and hypnotics are also inquired about, as is the patient's current medication status. The patient is also asked to provide his or her chief complaint, or reason for hospitalization. Current diagnoses are also recorded. A large section of the questionnaire consists of a list of diseases and symptoms. The patient is simply told to indicate whether he or she was ever informed by a doctor that the condition was present (e.g., tuberculosis, high blood pressure). The patient is also asked about seizures, and about a number of conditions that could be associated with brain damage or injury, such as automobile and motorcycle accidents, exposure to high voltages, exposure to toxins, such as paint and exhaust fumes, and episodes of partial drowning. The history taker is encouraged to have the patient elaborate upon any positive responses. The Head Injury Questionnaire, at least in its early forms, is somewhat out of date. For example, it lists high-altitude flying (over 10,000 feet) as a possible risk factor for brain damage, whereas it does not include disease entities that have recently been found to be associated with brain dysfunction, such as chronic obstructive pulmonary disease and renal disease. However, it is readily updated, and there have been numerous attempts made to do so.

An effort at developing a more structured approach has been provided by Melendez (1978) who has developed child and adult forms of what is called a Neuropsychological Questionnaire. The adult version consists of 54 questions, which may be answered yes or no. Examples of questions are, "Have you been in an accident?" or "Does any part of your body feel numb?" Space is available on the questionnaire form for recording detailed information when there is a positive response. The Psychological Assessment Resources Corporation has prepared a comprehensive Neuropsychological Status Examination (Schinka, 1984), which is, in essence, a kit for complete neuropsychological assessment. The kit contains material to record patient and referral data, information concerning premorbid status, and the patient's physical, emotional, and cognitive condition. Results of the neuropsychological testing, diagnostic comments, and recommendations are also entered into the kit forms. However, for our purpose here, the most interesting section is an instrument called the Neuropsychological Symptom Checklist (NSC). This 93-item instrument may be filled out by the patient, a relative, a friend, or a doctor. The items cover various physical symptoms (e.g., loss of vision, balance problems), some aspects of the medical history (e.g., history of kidney problems or cancer) and mental symptoms (e.g., problems with concentration or depression). Inquiry is also made about the family history. Swiercinsky (1978) has provided an additional outline for a structured clinical neuropsychological interview with sections for personal and family data, assessment of premorbid level of functioning, history of drug and alcohol abuse, orientation and state of consciousness, current diagnoses, complaints, treatments used, reasons for current hospitalization, history of previous hospitalizations, current and recently used medication, patient's statement concerning reason for hospitalization, and current or previous illnesses. Additional questions concerning a list of specific pieces of information may be answered through talking to our observation of the patient, or through reviewing the medical records. Such matters as appearance, cooperativeness, and anxiety level are assessed here. Schear (1984) has also proposed an interview outline that covers handedness, sensorium, subjective impressions concerning changes in mental function, family, social and medical history, survey of physical and mental health, past psychiatric or general medical treatments, and drug use.

There is another instrument called the Neuropsychological Impairment Scale (O'Donnell & Reynolds, 1983), which provides a screening for neuropsychological dysfunction. It consists of eight subscales, as follows; Global Measure of Impairment, Total Items Checked, Symptom-Intensity Measure, a Lie scale, a General scale, a Pathognomic scale, a Learning-Verbal scale, and a Frustration scale. The items are answered by the respondent on a 4-point scale ranging from 0 to 3. The symptom-intensity measure is the average rating, the Lie scale evaluates test-taking attitude, the General scale evaluates mental efficiency and alertness, the Pathognomic scale is sensitive to differences between brain-damaged and non-brain-damaged patients, the Learning-Verbal scale assesses communicative and learning abilities, and the Frustration scale evaluates affective and motivation problems. Thus, the instrument, as a whole, is a multidimensional self-rating measure of perceived level of function, and thus resembles the PAOFI to some extent.

MEDICAL RECORD DOCUMENTATION FOR NEUROPSYCHOLOGY

Thus far we have been dealing with interviewing and rating techniques, but it is clear that information about the patient can also be obtained from the medical records. However, it is generally necessary to perform some degree of data reduction in order to abstract pertinent material from the records. It is therefore often useful to have some systematic plan of search, and some standard document with which to record findings in an organized manner. The Neuropsychological Status Examination, discussed previously, provides such a document, but most neuropsychologists have tended to develop documentation forms of their own that meet the specific needs of their research or practice. There are several ways of going about the process of developing a form. For example, one can develop a procedure-oriented or diagnosis-oriented document. The procedure-oriented approach would codify the results of various tests and examinations. In neuropsychology, that would largely amount to the results of neurological examination, the EEG, the CT scan, and other radiological techniques, and a relatively limited number of clinical laboratory findings, such as evidence for infection found on examination of cerebral spinal fluid. Diagnostic oriented systems would codify conclusions of the assessment rather than the procedures used. For example, the presence of an infarct in the left-posterior parietal lobe would be coded, rather than a codification of the radiological or electroencephalographic evidence that established that diagnosis. In the diagnostic approach one might wish to emphasize general medical, psychiatric, or neurological considerations, depending on the setting in which one works.

Figure 1 contains a documentation form that we have worked with for many years. It is clearly an example of the diagnosis-oriented approach, with a strong emphasis on neurological diagnosis. The form is filled out through an examination of the available medical records, and is done without knowledge of the neuropsychological test results, in order to minimize bias.

In addition to biomedical and diagnostic data, records may contain

Unless otherwise noted: 1 = yes, 0 = no	Col.
PSYCHIATRIC DIAGNOSIS: SCHIZ=4 1 = yes, 0 = no, 2 = ?	1
PRESENCE OF BRAIN DAMAGE 1 = yes, 0 = no, 2 = ?	2

Name:_____

Date of Neurological_____

Date of Neuropsychological_____

Localization

	Col.
Left hemisphere	3
Right hemisphere	4
Diffuse	5
(localization) Multiple diagnosis	6
Strength of localization 1 = strong, 0 = weak	7

Specific localization

	Col.
Frontal	8
Temporal	9
Parietal	10
Occipital	11
Multiple focal	12
Subcortical	13
Peripheral nerve damage	14

Process

	Col.
Acute	18
Static	19
Congenital	20
(process) Multiple diagnosis	21

Specific Process

	Col.
Brain malformation	22
Neoplasm	23
Trauma	24
Vascular	25
MS & other demyelinating diseases	26
Neuronal degenerative	27
Toxic (non-alcohol)	28

Figure 1. Neurological documentation sheet.

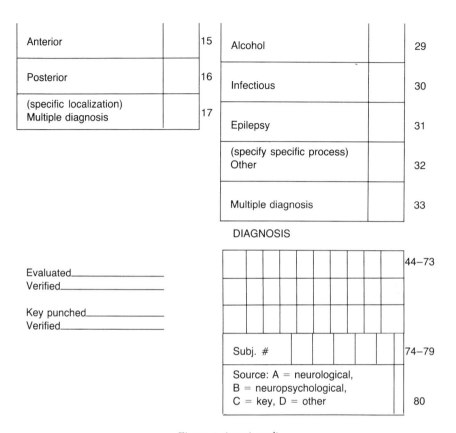

Figure 1. (continued).

other items of information that may help in interpretation of the neuropsychological assessment. For example, they may provide family history data, something that might be crucial with regard to the establishment of a diagnosis. There may be something found in the remote history that is helpful in understanding the patient's current functional status, but that is not a part of the presenting illness or problem. For example, within neuropsychology, a remote history of head injury may often be of great significance, although the patient may be under evaluation for something entirely different at the time of the neuropsychological assessment.

Although there are no commonly used published instruments for documentation of records from the point of view of neuropsychology, most neuropsychologists utilize a decision making system of the type illustrated in Figure 1. First, a decision is made concerning presence or absence of a lesion. If present, lateralization and localization are noted in as much detail as possible. There is then some consideration of the nature of the lesion. It may be rated as recently acquired and active or rapidly progressive, or acquired at some time in the past and static. Lesions acquired during the perinatal period tend to have different implications from what is the case later in life, and so the presence of congenital, or early life, brain damage is also noted. If the specific disease process is identified, that is also coded in as much detail as possible. That is, the rater codes whether the lesion is a result of trauma, vascular disease, neoplastic disease, etc. The document contained in Figure 1 does not go into a great deal of detail in that regard, because further branching is often possible. For example, under neoplasm, it may be important to note whether the neoplasm is intrinsic, extrinsic, or metastatic, because neuropsychological consequences may be quite different. Under trauma, it may be important to note whether the patient had a closed- or open-head injury, etc.

Because of recent neuropsychological findings concerning cognitive and perceptual impairment accompanying a number of general medical disorders (e.g., Ryan, Vega, Longstreet, & Drash, 1984), it is becoming increasingly important to document general systemic illness. Such symptoms and diseases as hypertension, chronic obstructive pulmonary disease, thyroid disease, and diabetes are particularly important to note. Individuals with these disorders may have generally negative findings for the central nervous system, but may have neuropsychological deficits that may be explainable on the basis of the implications these pathological processes may have for CNS function.

With the implementation of DSM-III and the structured interviews described previously, documentation of psychiatric illness is becoming increasingly straightforward. If available, the multiaxial DSM-III diagnosis or diagnoses should be noted, as well as the results of any behavioral or laboratory procedures (e.g., dexamethasone suppression test, sleep studies) that could have implications for psychiatric status. If the patient is taking any medication for his or her psychiatric disorder, that is extremely important to note, because some of these compounds may have significant implications for neuropsychological test results (Heaton & Crowley, 1981). Perhaps one of the most frequently asked clinical questions about neuropsychological test results has to do in some way with whether at least some component of those results could be explained on the basis of the patient's medication. Although it is generally not feasible or even necessarily desirable to withdraw patients from medication prior to assessment, the clinician should at least know and document what medication the patient was taking at the time of the assessment, and should also have some knowledge of the effects the particular medication taken could have on the test results.

GENERAL SUMMARY AND CONCLUSIONS

It would be difficult to imagine conducting almost any form of clinical practice without interviewing the patient and taking a pertinent history. In acute medicine, and to some extent in psychiatry, the most pertinent information may be obtained from an informant or from a review of existing records, but in any event, it is often difficult, if not dangerous, to proceed with active treatment until crucial diagnostic data are obtained in some way. The purpose of this chapter was that of reviewing those aspects of the interview and history that are important for neuropsychological assessment; that is, evaluation of patients with known or suspected brain dysfunction. It was noted that some clinical neuropsychologists prefer to interpret their test material without knowledge of the patient's diagnosis or history, and others may conduct an interview prior to formal testing but use a standard battery and do not use interview data for test selection. However, it was suggested that at some point in the evaluation it becomes necessary to integrate interview, historical, and objective psychometric and laboratory data in order to formulate an optimal treatment plan. Thus, whether the neuropsychologist or some other clinician does the interview and history, it generally has to be done and integrated into the diagnostic data base.

The objectives of doing an interview and taking a history as part of a neuropsychological assessment are in many respects the same as in the case of general clinical practice. Such matters as obtaining the patient's complaint or the presenting problem and diagnostic information are routine components of essentially all health related evaluations. However, in the case of neuropsychological assessment, it is also particularly important to estimate the patient's premorbid functional capacity, to determine the level of consciousness or wakefulness, and to assess the patient's attitude toward and awareness of the illness. Furthermore, there are some particular difficulties with obtaining interview and historical data from a substantial portion of those patients that form the clientele for many neuropsychologists. That is, many brain-damaged and some psychiatric patients may not be reliable informants for numerous reasons, notably impairment of consciousness, intellectual deterioration, or psychotic level disorganization. It is therefore often necessary to rely on informants and to have available instruments, such as the Glasgow Coma Scale or brief mental status examinations, that can be used efficiently to obtain the kinds of data that can be provided by the patient.

In the last section of this chapter, we have attempted to provide the reader with some information concerning the various procedures that are now available for interviewing and history taking. The procedures covered were the general and specific structured psychiatric interviews, general and specific clinical rating scales, scales developed specifically for neuropsychology, and some proposed outlines for systematically searching for and codifying information obtained from medical records. The application of primarily structured methods was stressed because of the frequently noted unreliability of psychiatric diagnosis based on the use of more informal, unstructured interviewing and rating procedures. With regard to the search of medical records, the availability of some formal instrument aids in imposing some degree of order on what amounts in many cases to be many pounds of hospital chart, not necessarily organized for the convenience of the individual seeking specific items of information. In clinical assessment situations, these procedures are not viewed as a replacement for the neuropsychological tests discussed elsewhere in this volume, but rather as a set of methods that may aid significantly in the interpretation and enrichment of the neuropsychological test findings.

REFERENCES

Albert, M. L., Goodglass, H., Helm, N. A., Rubens, A. B., & Alexander, M. P. (1981). *Clinical aspects of dysphasia*. New York: Springer-Verlag/Wein.

American Psychiatric Association. (1980). *Diagnostic and statistical manual of mental disorders*. Washington, DC: Author.

Bech, P., Gram, L. F., Dein, E., Jacobsen, O., Vitger, J., & Bolwig, T. G. (1975). Quantitative rating of depressive states. *Acta Psychiatrica Scandinavica, 51*, 161–170.

Beck, A. T. (1967). Measure of depression: The depression inventory. In A. T. Beck (Ed.), *Depression: Clinical, experimental and theoretical aspects*. New York: Harper & Row.

Blessed, G., Tomlinson, B. E., & Roth, M. (1968). The association between quantitative measures of dementia and of senile change in the cerebral grey matter of elderly subjects. *British Journal of Psychiatry, 114*, 797–811.

Burdock, E. I., Hardesty, A. S., Hakerem, G., Zubin, J., & Beck, Y. M. (1968). *Ward behavior inventory*. New York: Springer.

Burdock, E. I., Sudilovsky, A., & Gershon, S. (1982). *The behavior of psychiatric patients*. New York: Marcel Dekker.

Christensen, A.-L. (1975). *Luria's neuropsychological investigation*. New York: Spectrum.

Christensen, A.-L. (1984). The Luria method of examination of the brain-impaired patient. In P. E. Logue & J. M. Schear (Eds.), *Clinical neuropsychology: A multidisciplinary approach*. Springfield, IL: Charles C Thomas.

Coblentz, J. M., Mattis, S., Zingesser, L., Kasoff, S., Wisniewski, H., & Katzman, R. (1973). Presenile dementia: Clinical aspects and evaluation of cerebrospinal fluid dynamics. *Archives of Neurology, 29*, 302.
Crook, T., Ferris, S., & Bartus, R. (1983). *Assessment in geriatric psychopharmacology.* New Canaan, CT: Mark Powley Associates.
Diller, L., & Gordon, W. A. (1981). Rehabilitation and clinical neuropsychology. In S. B. Filskov & T. J. Boll (Eds.), *Handbook of clinical neuropsychology.* New York: Wiley.
Diller, L., & Weinberg, J. (1971). Studies in scanning behavior in hemiplegia. In L. Diller & J. Weinberg (Eds.), *Studies in cognition and rehabilitation in hemiplegia.* New York: New York Institute of Rehabilitation Medicine, New York University.
Duke University Center for the Study of the Aging. (1978). *Multidimensional functional assessment: The OARS methodology.* Durham, NC: Duke University Press.
Endicott, J., & Spitzer, R. L. (1978). A diagnostic interview: The schedule for affective disorders and schizophrenia. *Archives of General Psychiatry, 35*, 837–844.
Feighner, J., Robins, E., Guze, S., Woodruff, R., Winokur, G., & Munoz, R. (1972). Diagnostic criteria for use in psychiatric research. *Archives of General Psychiatry, 26*, 57–63.
Folstein, M. F., Folstein, S. E., & McHugh, P. R. (1975). Mini-mental state. A practical method for grading the cognitive state of patients for the clinician. *Journal of Psychiatric Research, 12*, 189–198.
Goldstein, G. (1979). Methodological and theoretical issues in neuropsychological assessment. *Journal of Behavioral Assessment, 1*, 23–41.
Goldstein, G., & Ruthven, L. (1983). *Rehabilitation of the brain damaged adult.* New York: Plenum Press.
Gurland, B. J., Kuriansky, J., Sharpe, L., Simon, R., Stiller, P., & Birkett P. (1977). The comprehensive assessment and referral evaluation (CARE)—rationale, development and reliability. *International Journal of Aging and Human Development, 8*, 9–42.
Hamilton, M. (1960). A rating scale for depression. *Journal of Neurology, Neurosurgery, and Psychiatry, 23*, 56–61.
Hamsher, K. deS. (1984). Specialized neuropsychological assessment methods. In G. Goldstein & M. Hersen (Eds.), *Handbook of psychological assessment.* New York: Pergamon Press.
Heaton, R. K., & Pendleton, M. G. (1981). Use of neuropsychological tests to predict adult patients' everyday functioning. *Journal of Consulting and Clinical Psychology, 49*, 807–821.
Helzer, J., Robins, L., Croughan, J., & Welner, A. (1981). Renard diagnostic interview. *Archives of General Psychiatry, 38*, 393–398.
Hines, F. R., & Williams, R. B. (1975). Dimensional diagnosis and the medical student's grasp of psychiatry. *Archives of General Psychiatry, 32*, 525–528.
Hirschberg, G. G., Lewis, L., & Vaughan, P. (1976). *Rehabilitation: A manual for the care of the disabled and elderly* (2nd ed.). Philadelphia, PA: Lippincott.
Holland, A. L. (1980). *CADL Communicative abilities in daily living. A test of functional communication for aphasic adults.* Baltimore: University Park Press.
Honigfeld, G., & Klett, C. (1965). The Nurse's Observation Scale for Inpatient Evaluation (NOSIE): A new scale for measuring improvement in schizophrenia. *Journal of Clinical Psychology, 21*, 65–71.
Jennett, B., & Teasdale, G. (1981). *Management of head injuries.* Philadelphia, PA: F. A. Davis Company.
Khavari, K., & Farber, P. (1978). A profile instrument for the quantification and assessment of alcohol consumption. The Khavari Alcohol Test. *Journal of Studies on Alcohol, 39*, 1525–1539.

Logue, P. E. (1975). *Understanding and living with brain damage*. Springfield, IL: Charles C Thomas.

Lorr, M., & Klett, C. J. (1966). *Inpatient multidimensional psychiatric scale, revised*. Palo Alto, CA: Consulting Psychologists Press.

Luria, A. R. (1973). *The working brain*. New York: Basic Books.

Marlatt, G. A. (1976). The drinking profile: A questionnaire for the behavioral assessment of alcoholism. In E. J. Mash & L. G. Terdal (Eds.), *Behavior therapy assessment: Diagnosis, design and evaluation*. New York: Springer.

Matarazzo, J. D. (1972). *Wechsler's measurement and appraisal of adult intelligence*. Baltimore, MD: Williams & Wilkins.

McLellan, A., Luborsky, L., Woody, G., & O'Brien, C. (1980). An improved diagnostic evaluation instrument for substance abuse patients: The Addiction Severity Index. *Journal of Nervous and Mental Disease, 168*, 26–33.

Mohs, R. C., Rosen, W. G., Greenwald, B. S., & Davis, K. L. (1983). Neuropathologically validated scales for Alzheimer's disease. In T. Crook, S. Ferris, & R. Bartus (Eds.), *Assessment in geriatric psychopharmacology*, New Canaan, CT: Mark Powley Associates.

Moore, R. A. (1972). The diagnosis of alcoholism in a psychiatric hospital *American Journal of Psychiatry, 128*, 1565–1569.

O'Donnell, W. E., & Reynolds, D. Mc. Q. (1983). *Neuropsychological Impairment Scale*. Annapolis, MD: Anapolis Psychological Services.

Overall, J. E., & Gorham, J. R. (1962). The brief psychiatric rating scale. *Psychological Reports, 10*, 799–812.

Parsons, O. A. (1984, May). *Neuropsychological consequences of alcohol abuse: Many questions—some answers*. Paper presented at NIAAA Conference on Clinical Implications of Recent Neuropsychological Findings, Boston, MA.

Raskin, A. (1982). Assessment of psychopathology by the nurse or psychiatric aide. In E. I. Burdock, A. Sudilovsky, & S. Gershon (Eds.), *The behavior of psychiatric patients: Quantitative techniques for evaluation*. New York: Marcel Dekker.

Robins, S. L., Helzer, J., Croughan, N. A., & Ratcliff, K. (1981). National institute of mental health diagnostic interview schedule. *Archives of General Psychiatry, 18*, 381–389.

Ryan, C., Vega, A., Longstreet, E., & Drash, A. (1984). Neuropsychological changes in adolescents with insulin-dependent diabetes. *Journal of Consulting and Clinical Psychology, 52*, 335–342.

Sarno, M. T. (1969). *The Functional Communication Profile: Manual of directions*. New York: New York Institute of Rehabilitation Medicine, New York University Medical Center.

Schear, J. M. (1984). Neuropsychological assessment of the elderly in clinical practice. In P. E. Logue & J. M. Schear (Eds.), *Clinical neuropsychology. A multidisciplinary approach*. Springfield, IL: Charles C Thomas.

Schinka, J. A. (1984). *Neuropsychological Status Examination*. Odessa, FL: Psychological Assessment Resources.

Spitzer, R. L., Endicott, H., & Robbins, E. (1978). Research diagnostic criteria rationale and reliability. *Archives of General Psychiatry, 35*, 773–782.

Sunderland, A., Harros, J. E., & Gleave, J. (1984). Memory failures in everyday life following severe head injury. *Journal of Clinical Neuropsychology, 6*, 127–142.

Swiercinsky, D. (1978). *Manual for the Adult Neuropsychological Evaluation*. Springfield, Il: Charles C Thomas.

Wechsler, D. (1944). *The Measurement of Adult Intelligence*, 3rd Ed. Baltimore: Williams & Wilkins.

Weinstein, E. A., & Kahn, R. L. (1955). *Denial of illness: Symbolic and physiological aspects.* Springfield, IL: Charles C Thomas.

Wells, C. E. (1979). Pseudodementia. *American Journal of Psychiatry, 136,* 895–900.

Wilson, B., Baddeley, A., & Hutchins, H. (1984). *The Rivermead Behavioral Memory Test: A preliminary report.* Oxford: Rivermead Rehabilitation Center.

Wilson, R. S., Rosenbaum, G., & Brown, G. (1979). The problem of premorbid intelligence in neuropsychological assessment. *Journal of Clinical Neuropsychology, 1,* 49–53.

Wing, J. K., Cooper, J. E., & Sartorius, N. (1974). *The measurement and classification of psychiatric symptoms.* Cambridge, MA: Cambridge University Press.

Yesavage, J. A., Brink, T. L., Rose, T. L., & Adey, M. (1983). The geriatric depression rating scale: Comparison with other self-report and psychiatric rating scales. In T. Crook, S. Ferris, & R. Bartus (Eds.), *Assessment in geriatric psychopharmacology.* New Canaan, CT: Mark Powley Associates.

Zubin, J. (1967). Classification of the behavior Disorders. *Annual Review of Psychology, 18,* 373–406. Palo Alto, CA: Annual Reviews.

Zubin, J. (1984). Inkblots do not a test make. *Contemporary Psychology, 29,* 153–154.

Zung, W. W. K. (1965). A self-rating depression scale. *Archives of General Psychiatry, 12,* 63–70.

3

The Role of Standard Cognitive and Personality Tests in Neuropsychological Assessment

GORDON J. CHELUNE, WILLIAM FERGUSON, and KURT MOEHLE

INTRODUCTION

Clinical neuropsychology has become one of the fastest growing specialty areas in psychology (Golden & Kuperman, 1980), and is now recognized as a distinct and legitimate area of specialized practice. Central to this growth, and well documented throughout this volume, has been the success of neuropsychological testing procedures in detecting the presence and localization of brain dysfunction. The purpose of this chapter is to examine the role of standard cognitive and personality tests in the neuropsychological assessment of adults. The specific assessment devices covered are those measures generally considered to be part of the standard battery employed by traditionally trained clini-

---

This chapter was written while the authors were affiliated with West Virginia University Medical Center.

GORDON J. CHELUNE • San Diego Veterans Administration Medical Center and Department of Psychiatry, University of California School of Medicine, San Diego, CA.   WILLIAM FERGUSON • Behavioral Medicine Service, St. Joseph's Hospital, Lancaster, PA 17604.   KURT MOEHLE • Department of Psychology, Purdue University School of Science, 1125 E. 38th St. P.O. Box 647, Indianapolis, IN 46223.

cians for routine assessment purposes (e.g., the Wechsler tests, Bender-Gestalt, Rorschach, Minnesota Multiphasic Personality Inventory, etc.). The utility of these testing procedures is first examined in a historical perspective that acknowledges the differences between the clinical psychological and neuropsychological approaches to the assessment of brain dysfunction. The strengths and weaknesses of these tests for neurodiagnostic purposes are next reviewed. Finally, the potential utility of standard cognitive and personality tests as adjuncts to other neuropsychological procedures is discussed in the context of the emerging emphasis on issues of everyday living.

NEUROPSYCHOLOGY AND THE ROLE OF ASSESSMENT

Psychological assessment has been a mainstay of clinical psychology since its formal origins around the turn of the century. By developing standardized methods for sampling theoretically relevant behavior, psychologists have sought to identify differences between individuals. However, individual differences are relevant for psychologists only insomuch as they reliably provide clinically useful information. Anastasi (1982, p. 3) observes that "one of the first problems that stimulated the development of psychological tests was the investigation of the mentally retarded." Beginning with the publication of the first practical test of intelligence in 1905 by Binet and Simon (Matarazzo, 1972), the assessment of intellectual functioning has remained an important application of psychological tests, and has contributed greatly to the early growth of clinical psychology as a profession (Shaffer & Lazarus, 1952). Likewise, interest in intelligence from a biological perspective has served as a seminal issue in the development of neuropsychology (Halstead, 1947).

Neuropsychological assessment, particularly in the United States, owes much to the clinical tradition of psychometric testing. As noted by Davison (1974, p. 3), clinical neuropsychology has its "roots in academic psychology, behavioral neurology, and, especially, the mental measurement or psychometric field of psychology." There are further similarities as well. The clinical neuropsychologist and the clinical psychodiagnostician both frequently serve as consultants to other professionals, and are asked to render opinions as to the nature of a patient's difficulties. Often using some of the same tests, they both gather standardized observations of behavior from which to make inferences regarding the patient. Although the clinical neuropsychologist and the clinical psychologist may employ some of the same testing procedures to study similar clinical problems in a parallel attempt to aid in patient care, they differ widely in their "bodies of assumptions and techniques concerning the assessment of the behavioral effects of brain damage" (Davison, 1974, p. 19).

Despite the many superficial similarities, including the shared use of standard cognitive and personality tests, the differences in conceptual and theoretical orientations of the traditionally trained psychologist and clinical neuropsychologist have been substantial (Cleeland, 1976; Davison, 1974; S. Goldstein, 1976). Although many academic and clinical training settings are currently integrating neuropsychology components into their programs (Lubin & Sokoloff, 1983; Noonberg & Page, 1982), differences in theoretical bias are still apparent in daily clinical practice. In order to appreciate the role standard cognitive and personality tests have had in neuropsychological assessment, it is first important to explore the historical antecedents of the differences in orientation to the construct of brain dysfunction.

Historical Antecedents: The Search for Organicity

As noted previously, the history of clinical psychology is closely tied to the development of diagnostic testing procedures. For many years, the principal contribution that clinical psychologists made to mental health teams was a careful and systematic diagnostic assessment of each patient in terms of relevant individual differences (Phares, 1979). Given the diversity of potential individual differences, many psychologists sought to develop a standard battery of tests that could be routinely used as a single diagnostic tool for the majority of assessment problems. The classic example and prototype for many clinicians was the battery employed by Rapaport and his colleagues at the Menninger Foundation in the 1940s (Rapaport, Gill, & Schafer, 1968). More recently, Davison (1974) observed that

> the most commonly used tests are: the Wechsler series of intelligence scales, the Rorschach, the Thematic Apperception Test, the Minnesota Multiphasic Personality Inventory, the Bender Gestalt figure drawings and the Draw-A-Person task. (p. 13)

Armed with what Cronbach (1969) has called "large-bandwidth" test batteries, clinical psychologists sought to provide their consumers with useful diagnostic information relevant to a wide variety of clinical problems. A particularly frequent referral question posed to clinicians concerned the etiology of apparent intellectual deterioration among patients. It was well accepted at that time that decrements in intellectual functioning could result from either psychiatric disturbance or from injuries to the brain. "Throughout the 1930's and 1940's and well into the 50's, most clinicians treated brain damage as if it were a unitary phenomenon" (Lezak, 1983, p. 16) that produced a relatively homogenous change in behavior known as "organicity." The origins of this view were rooted in the experimental work of Lashley and the clinical observations of Kurt Goldstein. Lashley's (1929) animal research led him to formulate the law of mass action and the principle of equipotentiality, which suggested that cognitive impairment covaried with the amount of cortical damage relatively independent of its localization. Similarly, K. Goldstein's (1939) clinical work with brain-injured patients led him to propose that brain dysfunction invariably resulted in a single and basic change in behavioral functioning characterized as a loss of "abstract attitude." Although loss of abstract ability was not uniformly accepted (Shaffer & Lazarus, 1952), it was believed that organicity could be defined along similar unitary behavioral dimensions.

In attempting to make differential diagnoses of organicity, two divergent approaches emerged among psychodiagnosticians. Some clinicians turned to their standard batteries of psychological tests to "look for an 'organic' syndrome or pattern of functioning considering the profile of performance over all the tests used" (Davison, 1974, p. 14). Others sought to supplement their basic test batteries by developing specific, singular measures of organicity that could successfully discriminate organics from psychiatric patients and normal control subjects. Due to these early efforts, the unitary conceptualization of brain dysfunction and its behavioral effects became firmly entrenched in traditional psychodiagnostics. However, because both approaches were based on what we now know are faulty assumptions, neither approach succeeded in effectively dealing with the complexities of the problem of organicity, although the legacies of both are still very much with us.

Efforts to find the single best test of organicity resulted in the development of many ingenious and creative assessment devices (for a review of many of these devices, see Lezak, 1983), which in turn generated a sizable research literature. However, when psychologists attempted to evaluate this literature, it became apparent that no one behavioral dimension was consistently associated with brain dysfunction. In their early review, Haynes and Sells (1963) summarize their impression of this assessment literature by concluding that "approaches to the diagnosis of organic brain damage reflect diverse concepts of the nature of brain damage and its behavioral effects" (p. 367). In many respects, the early failures to find the single best test of organicity helped to delineate the complexities of the potential effects brain lesions could have on behavior, which is now a basic tenet of modern neuropsychology. Furthermore, the work of Reitan during the 1950s and 60s challenged the belief that brain dysfunction was a unitary phenomena. His research (Reitan, 1966) firmly established that the magnitude and patterning of the behavioral effects resulting from cerebral lesions were dependent on such factors as lateralization, localization, lesion characteristics (e.g., type and velocity) and various patient attributes (e.g., age, education, premorbid health).

Although the fervor to develop the ultimate measure of organicity gradually waned, there were still those who sought to find

> evidence of reduced level of performance, qualitative indications of brain damage (such as concreteness of verbal expression), pathognomic signs of brain damage, and diagnostic patterns on intelligence tests and on personality tests. (Davison, 1974, p. 14)

Although brain dysfunction was still treated as essentially a unitary phenomenon, it was seen as a measurable, although nonspecific, condition. Lezak (1983) notes that this view of organic pathology "remains a vigorous concept, reflected in the many test and battery indices, ratios, and quotients that purport to represent some quality or relative degree of 'organicity'" (p. 17).

NEUROPSYCHOLOGY: TOOLS VERSUS PERSPECTIVE

Drawing from the experiences of their earlier clinical colleagues, neuropsychologists have abandoned the unidimensional view of brain dysfunction and its behavioral effects. Current thinking reflects an understanding that the behavioral manifestations of brain dysfunction are multifaceted and multidimensional in nature, and as such require appropriate multidimensional assessment approaches. However, because of the rapid growth and popularization of neuropsychology, there are many practitioners that equate neuropsychological assessment with the administration of neuropsychological tests or batteries. Despite the best efforts of presenters, it is not uncommon for many clinicians attending their first neuropsychological assessment workshop to develop an inflated view of the test procedures. The situation is similar to the aura of magic and mystique early psychodiagnosticians were endowed with by their psychiatric colleagues by virtue of the successes achieved with standard psychological tests (Phares, 1979).

Although there is considerable work being done to establish standards for training and credentialing in clinical neuropsychology (Meier, 1981), it is useful to remind ourselves that tests, be they psychological or neuropsychological, are merely tools for gathering standardized samples of behavior. As Anastasi (1982, p. 23) aptly states "the *diagnostic* or *predictive value* of a psychological test depends on the degree to which it serves as an indicator of a relatively broad and significant area of behavior." Test data only provide referents around which interpretations are made. Interpretative statements do not arise from test scores, but from the factual and theoretical knowledge of the test user. Any test measure can be viewed as a neuropsycholgical test provided it is interpreted within a neuropsychological framework. In the following section we will examine how many so-called standard tests have been used and interpreted from a neuropsychological perspective for the purpose of neurodiagnosis.

NEURODIAGNOSIS AND THE STANDARD BATTERY

Although the composition of standard batteries is likely to vary among clinicians depending on their personal biases, these batteries generally consist of two parts: one designed to assess cognitive abilities and another to evaluate personality attributes. Theoretically, these two aspects may be considered to be independent, yet in reality they are often overlapping and interactive. It is not uncommon for practitioners to utilize measures, such as the Wechsler Intelligence Scales or the Rorschach, as data bases for evaluating both cognitive and personality dimensions (e.g., Allison, 1978; Exner & Clark, 1978). However, for our discussion of the use of standard tests in adult neurodiagnosis, we have maintained the theoretical distinction between cognitive and personality tests. The measures we have selected for coverage are, for the most part, those mentioned by Davison (1974): the Wechsler Intelligence Scales, the Bender-Gestalt, Minnesota Multiphasic Personality Inventory, Rorschach, and Thematic Apperception Test. We have also included the Wechsler Memory Scale and the Wide Range Achievement Test because of their frequent use in traditional and neuropsychological assessment (Hartlage, Chelune, & Tucker, 1982).

Cognitive Measures

We have grouped the Wechsler Intelligence Scales, Wechsler Memory Scale, Wide Range Achievement Test, and Bender-Gestalt under the category of cognitive measures for three reasons. First, each of these tests requires a patient to actually produce a behavioral sample, that is, to demonstrate a skill or capacity to perform a specific task. Second, these tests all presuppose a comparison standard against which an individual's performance can be evaluated. This standard may be normative, as in the case of the Wechsler Intelligence Scales, or species-wide as with the Bender-Gestalt. Finally, a deficit-measurement paradigm is used in the interpretation of all of these measures for neurodiagnosis (Lezak, 1983).

The Wechsler Intelligence Scales

The assessment of intellectual functioning is an integral component of all standard batteries, and the most frequently used tests of intelligence are the scales devised by David Wechsler (Golden, 1979). The original Wechsler-Bellevue (WB) (Wechsler, 1939) was extensively revised in 1955, yielding the Wechsler Adult Intelligence Scale (WAIS) (Wechsler, 1955). The WAIS was subsequently revised again in 1981 in the form of the current Wechsler Adult Intelligence Scale-Revised (WAIS-R) (Wechsler, 1981). The WB, WAIS, and WAIS-R (collectively referred to here as the Wechsler Intelligence Scales) were designed to assess what Wechsler described as "the aggregate or global capacity of the individual to act purposefully, to think rationally, and to deal effectively with the environment" (Matarazzo, 1972, p. 79). Because of Wechsler's emphasis on the global or aggregate nature of intelligence, it is not surprising that his scales were quickly adopted by those who subscribed to the unitary notion of organicity. Even today, the Wechsler Intelligence Scales are the single most frequently used tests in neuropsychological assessment (Hartlage et al., 1982).

Psychometric Properties. Despite the variations in content and format among the different revisions, the basic structure and psychometric properties of the Wechsler Intelligence Scales have remained virtually unchanged. Wechsler (1939) had originally divided the 11 subtests of the WB into Verbal and Performance sections based on the intercorrelations among the subtests. Subsequent factor analytic studies of the WB and the WAIS (see Matarazzo, 1972, pp. 261–276 for a review) have generally supported this dichotomy. In addition to a general intellectual factor (i.e., g-factor), three additional second-order factors have been identified: a Verbal Comprehension factor composed of the Information, Vocabulary, Comprehension, and Similarities subtests; a Perceptual Organization factor consisting of the Picture Completion, Picture Arrangement, Block Design, and Object Assembly subtests; and a weaker Memory/Freedom from Distractibility factor made up of the Digit Span, Arithmetic and Digit Symbol subtests.

Since the publication of the WAIS-R in 1981 (Wechsler, 1981), a number of investigators have explored its factor structure (Blaha & Wallbrown, 1982; Glass, 1982; O'Grady, 1983; Parker, 1983). With the exception of O'Grady's (1983) study using a confirmatory maximum likelihood factor analysis, the WAIS-R has been found to have a similar factor composition as its predecessors (Glass, 1982; Parker, 1983).

Using a hierarchial factor analysis approach, Blaha and Wallbrown (1982) suggest that their "findings not only support the validity of the WAIS-R as a measure of general intelligence but also the validity of maintaining separate Verbal and Performance IQs" (p. 652). As we will see, general level of intellectual functioning and Verbal-Performance IQ discrepancies have been frequently used as indexes of neuropsychological impairment.

Intellectual Level and Brain Dysfunction. The earliest and most obvious use of intelligence tests in neurodiagnosis was in the evaluation of intellectual deterioration. Because intelligence was defined as a global or aggregate capacity, it was believed that generalized decrements in intellectual functioning due to brain dysfunction would be reflected in quantitative changes on the Wechsler Intelligence scales. Subsequent research has consistently shown that heterogenous groups of patients with documented cerebral pathology do obtain lower IQ scores than matched normal subjects (e.g., Fitzhugh, Fitzhugh, & Reitan, 1962; Reitan, 1959; Vega & Parsons, 1967; Vogt & Heaton, 1977). The Wechsler summary IQ scores have also been shown to be sensitive to the chronicity of brain lesions (Becker, 1975; Fitzhugh *et al.*, 1962; Mandleberg, 1976) and to the severity of injury during the initial phase of recovery (Mandleburg & Brooks, 1975). Whereas Reitan (1959) found that the Halstead Impairment Index was more sensitive to brain damage than any of the Wechsler-Bellevue IQ summary scores using rank-order differences, other researchers report that discriminant function procedures yield comparable discrimination rates for the Wechsler summary scores and two of the most frequently used neuropsychological test batteries, the Halstead-Reitan and Luria-Nebraska Batteries (Kane, Parsons, & Goldstein, 1985). Furthermore, Chelune (1982) has noted that the Wechsler IQ scores account for approximately 67% of the shared variance between these two batteries.

Although intellectual functioning as measured by the Wechsler scales appears to covary with the organic integrity of the brain, intellectual level alone is of limited value as a diagnostic indicator in clinical practice. Factors other than brain dysfunction (e.g., poor genetic endowment, low socioeconomic background, and psychiatric disturbances) can result in low Wechsler IQ scores (Matarazzo, 1972). Interpretation is also complicated by the psychometric nature of the Wechsler lntelligence Scales. By design, Wechsler constructed his intelligence scales so that their scores would be normally distributed. Thus, 50% of the general population can be expected to obtain IQ scores below 100, and 10% will earn scores below 80.

> To be considered a clinically useful measure of intellectual deterioration, an index must reflect a loss of functioning rather than a demonstration that the

subject's IQ differs significantly from that of a nonimpaired subject. (Leli & Filskov, 1979, p. 702)

In individual assessment, the clinical significance of a patient's observed Wechsler scores must be evaluated against an appropriate individual comparison standard (i.e., an estimate of premorbid functioning). Two approaches to this problem are described in the next sections.

Deterioration Ratios and Indexes. The most direct method of documenting intellectual deterioration would be to compare a patient's premorbid intellectual scores with current scores. However, premorbid scores are rarely available, and investigators have sought indirect methods of estimating premorbid intellectual functioning. Based on early research using samples of brain-injured patients with hetereogenous lesion characteristics, investigators observed that some Wechsler subtests appeared more sensitive to brain dysfunction (as a unitary phenomenon) than others, and attempted to design various indexes of intellectual deterioration (see Table 1). The earliest deterioration index and prototype for subsequent attempts was the *Mental Deterioration Index* (MDI) devised by Wechsler (1944). Based on his observations of the effects of normal aging on the patterns of performance among the

Table 1. Wechsler Adult Intelligence Scale Deterioration Indexes

Author	Deterioration index
Wechsler (1944) MDI	("hold"—"don't hold")/"hold" ≥ .20 where "hold" = (I + OA + PC + C) and "don't hold" = (DS + DSy + A + BD).
Wechsler (1958) DQ	("hold"—"don't hold")/"hold" ≥ .20 where "hold" = (I + PA + PC + V) and "don't hold" = (DS + DSy + S + BD).
Hunt (1948)	("hold"—"don't hold")/"hold" where "hold" = (I + C) and "don't hold" = (DSy + BD).
Allen (1948)	(I + C) ≥ (5 + DS + DSy)
Saunders (1961)	DSy ≤ ((I + C + DS + A + S + PA + PC + BD − OA)/9)) − 2
Hewson (1949)	Ratio I = (PC + PA)/(A + DSy) ≥ 1.3; Ratio II = (I + C)/A ≥ 3.0; Ratio III = (I + C)/(DS + DSy) ≥ 1.7; Ratio IV = (I + C)/(PA + DSy) ≥ 1.6; Ratio V = (I + C)/DSy ≥ 3.5; Ratio VI = (I + C)/(DS + PA + DSy) ≥ 1.1; Ratio VII = (DS + DSy)/(S + BD) ≥ 1.1; Ratio VIII = (C + PA)/DSy ≥ 3.0.

Note. All indexes are scored in the direction of organicity.
Abbreviations: Mental Deterioration Index = (MDI), Deterioration Quotient = (DQ), Information (I), Comprehension (C), Similarities (S), Vocabulary (V), Block Design (BD), Arithmetic (A), Digit Span (DS), Picture Completion (PC), Digit Symbol (DSy), Object Assembly (OA), Picture Arrangement (PA).

elderly on the Wechsler-Bellevue subtests, Wechsler hypothesized that decrements in intellectual performance that exceeded normal age expectations must be the result of an organic disturbance. His MDI formula reflects a ratio between those subtests that "hold" or change very little with aging (Information, Comprehension, Object Assembly, and Picture Completion) and those that "don't hold" or are the most susceptible to aging effects (Digit Span, Arithmetic, Block Design, and Digit Symbol). Implicit in Wechsler's (1944) formulation of the MDI is the notion that brain dysfunction results in a condition similar to that of early senility.

With the publication of the WAIS (Wechsler, 1955), the MDI was revised in the form of the *Deterioration Quotient* (DQ) (Wechsler, 1958). Unfortunately, neither the MDI or DQ ever correctly classified patients beyond 75% (Lezak, 1983). Other investigators (Allen, 1948; Hunt, 1949; Saunders, 1961) have attempted to produce more robust deterioration indexes, but these appear to be merely variations on Wechsler's original formula (see Table 1). One exception was the approach taken by Hewson (1949). Rather than a single index of deterioration, she used different combinations of Wechsler subtests to yield eight different ratios in an attempt to account for the heterogenous effects of diverse types of cerebral pathology. However, like the single index approaches, the Hewson ratios tend to "misclassify too many cases for clinical application" (Lezak, 1983, p. 251). In a comparative study of eight different deterioration approaches based on the Wechsler Intelligence Scales, Vogt and Heaton (1977) report that the various indexes generally discriminated between impaired and nonimpaired patients at a level beyond chance, but that the correct classification rates were quite modest, ranging from 51% to 69%. In fact, these investigators found that a simple cutoff of 100 on the Full Scale IQ yielded a correct classification rate of nearly 87%.

Demographic Approaches. Attempts to estimate Wechsler intelligence scores from characteristics other than current test scores have relied on demographic information. The assumption is that "since adult onset disease should have little effect on demographic status, the accuracy of such estimates should be limited only by the correlation between IQ and the demographics" (Wilson et al., 1978, p. 1554). Age, educational attainment, sex, occupational status, and race have all been found to be associated with performance on the Wechsler Intelligence Scales (Matarazzo, 1972). Using the 1955 WAIS standardization data, Wilson and his colleagues (1978) derived multiple regression equations for predicting Verbal, Performance, and Full Scales IQ scores using age, sex, race, occupation, and education as predictor variables. Wilson, Rosenbaum, and Brown (1979) subsequently employed these equations as estimates of premorbid intellectual functioning and were able to correctly classify nearly 75% of their subjects as impaired or nonimpaired. In a related approach, Leli and Filskov (1979) developed a discriminant function equation based on comparisons of whether a patient's observed Full Scale IQ was above, below, or compatible with expected IQ scores for his or her occupation and educational background, using the tables presented by Matarazzo (1972, pp. 166–167, 178). The authors were able to correctly classify over 80% of their patients in the derivation and cross-validation samples using this procedure. Although less than perfect, estimates of premorbid functioning based on extratest data appear to be a superior individual comparison standard for determining intellectual deterioration than test-based formulas.

Pattern Analysis Approaches. A particularly difficult problem in clinical assessment is the differential diagnosis of organic and schizophrenic conditions (Heaton, Baade, & Johnson, 1978). Early attempts to distinguish schizophrenics from brain-injured patients on the basis of overall level of test performance on the Wechsler Intelligence scales met with little success (Spreen & Benton, 1965). For this reason, a number of investigators attempted to distinguish these two diagnostic groups on the basis of their patterns of performance on the Wechsler subtests. DeWolfe (1971) used the relationship between the Digit Span (DS) and Comprehension (C) subtests as a means for discriminating between samples of chronic schizophrenics and nonlateralized brain-damaged patients at two age levels. Patients obtaining scores on C less than DS were classified as schizophrenic, whereas those with C greater than DS were classified as brain damaged. In cases where C was equal to DS, Vocabulary was substituted for C. Using these simple criteria, DeWolfe (1971) reported a 78% correct hit rate among his older subjects and a 72% correct classification rate among the younger patients. Watson (1972) attempted to cross-validate DeWolfe's criteria in two separate hospital settings where patients were also classified according to length of hospitalization. DeWolfe's (1971) findings were supported at both chronicity levels in one hospital setting, but not at the other. Watson (1972) concluded that the DS-C pattern differences for distinguishing between schizophrenic and brain-damaged patients should be used cautiously and only after local cross-validation.

Ipsative approaches to pattern analysis have also been employed in an attempt to differentiate schizophrenics from organic patients. Davis, Dizzonne, and DeWolfe (1971) noted that research aimed at identifying characteristic patterns of Wechsler subtest performance for brain-damaged and schizophrenic patients is often confounded by initial differences in mean level of performance. To control for differences in mean level of performance, they employed a procedure called "deficit pattern analysis," which relates each subject's test score to his or her own mean across subtests. Davis et al. (1971) used this procedure to examine the Wechsler patterns of performance of process and reactive schizophrenics at two levels of institutionalization. They found that the reactive group had higher scores on the Block Design and Digit Symbol subtests relative to their overall mean than did the process group. Using the same procedure, DeWolfe, Barrell, Becker, and Spenner (1971) found Wechsler pattern differences between groups of chronic schizophrenics and nonlateralized brain-damaged patients at two age levels. Compared to the brain-damaged patients, schizophrenics obtained lower deficit pattern scores on the Comprehension subtest, but higher scores on the Block Design subtest across both age levels. Among the older groups, the schizophrenics tended to do better on Block Design but worse on Picture Completion than did the brain-damaged patients. Davis, DeWolfe, and Gustafson (1972) extended their use of deficit pattern analysis to study the intellectual pattern differences among groups of brain-damaged and process and reactive schizophrenic patients using cross-validation procedures. Independent of mean differences, the combined schizophrenic groups consistently outperformed the brain-damaged subjects on the Similarities, Block Design, and Object Assembly subtests. The reactive group also obtained higher deficit pattern scores than either the brain-damaged or process groups.

The discriminative contributions of both mean level and pattern of WAIS performance among carefully screened schizophrenic and chronic diffusely brain-damaged patients were examined by Chelune, Heaton, Lehman, and Robinson (1979). The schizophrenic group obtained higher WAIS scores than the acute brain-damaged group, but were not different from the chronic brain-damaged patients. However, in contrast to the reports of Davis et al. (1972) and DeWolfe et al. (1971), no deficit pattern differences were found. Using discriminant function procedures, Chelune et al. (1979) were able to correctly classify only 68% of the patients as either brain-damaged or schizophrenic using both mean level and pattern scores from the WAIS. Furthermore, when Mahalanobis distances were computed for each step of their discriminant analysis, it was determined that the pattern differences did not contribute significantly to the diagnostic discrimination achieved by level alone. Based on their results, Chelune et al. (1979) concluded that "mean level of performance can be used to discriminate clearly defined schizophrenic and diffusely brain-damaged groups, but that pattern analysis offers little additional information."

The Wechsler Scales and Laterality of Dysfunction. The relationship between the verbal and performance sections of the Wechsler Scales has been the focus of considerable research in neuropsychology, especially with respect to lateralized cerebral dysfunction. Anderson (1951) was the first to observe discrepancies between the Verbal and Performance sections of the Wechsler-Bellevue among patients with focal unilateral lesions. Subsequent research by Reitan (1955a) further demonstrated that patients with right-hemisphere brain lesions tend to obtain lower Performance IQ (PIQ) scores than Verbal IQ (VIQ) scores, whereas patients with left-hemisphere lesions obtain lower VIQ than PIQ scores. With few exceptions, this relationship between VIQ-PIQ discrepancies and laterality of cerebral dysfunction has stood the test of time for both the Wechsler-Bellevue and the WAIS (for an excellent review of the literature see Bornstein and Matarazzo, 1982), and more recently for the WAIS-R (Bornstein, 1983). Intuitively, this relationship fits well with neuropsychological theory; for most individuals, verbal tasks are mediated by the left hemisphere whereas visual-spatial tasks are primarily mediated by the right hemisphere.

Although the relationship between VIQ-PIQ discrepancies and laterality of cerebral dysfunction appears quite robust, its clinical significance in the individual case must be interpreted cautiously. Examination of the magnitude of the VIQ-PIQ discrepancies among the 28 patient groups reviewed by Bornstein and Matarazzo (1982) reveals only modest discrepancies, with only four studies reporting mean discrepancies of 15 points or more (all among right-hemisphere lesion groups). Because 18% of the WAIS standardization sample (Matarazzo & Matarazzo, 1984) and nearly 20% of the WAIS-R standardization sample (Grossman, 1983) obtained VIQ-PIQ discrepancies of 15 points or more, even fairly large discrepancies need not be associated with unilateral brain damage. Filskov and Leli (1981) point out that factors such as overall intellectual level, educational background, and specialized occupational skills can result in substantial VIQ-PIQ differences independent of cerebral pathology. Conversely, lateralized brain dysfunction is not always associated with the expected VIQ-PIQ differences. Duration and type of brain lesion can be major moderating variables. The typical VIQ-PIQ patterns are much more evident among patients with acute lesions than chronic ones (Fitzhugh et al., 1962; Russell, 1972, 1979; Vega & Parsons, 1969). Among patients with intrinsic tumors and acute cerebrovascular lesions, Russell (1979) observed minimal VIQ-PIQ differences in his left-hemisphere group, supporting Wechsler's (1958) original claim that any organic damage produces a decrement in PIQ. Rate of recovery among head-injury patients also affects potential VIQ-PIQ differences, with VIQ recovering to levels near that of controls within 6 to 12 months post-injury, whereas

PIQ shows continued improvement for up to three years (Becker, 1975; Mandleburg, 1976; Mandleburg & Brooks, 1975).

A factor further complicating the clinical use of VIQ-PIQ differences as neurodiagnostic indicators emerges from research that suggests that there are sex differences in the effects of unilateral brain lesions. In the early 1960s, Lansdell (1962) noted that the patterning of cognitive deficits following unilateral temporal lobectomies varied as a function of both surgical site and patient sex. Using the Wechsler Intelligence scales, McGlone (1977, 1978) found a similar interaction between sex and lesion lateralization. Specifically, McGlone reported that only her male patients exhibited a significant lateralization effect with respect to VIQ-PIQ differences; female patients showed a general lowering of both VIQ and PIQ irrespective of the side in which the lesion occurred. Subsequent reviews of the literature examining the effects of unilateral brain lesions on VIQ-PIQ differences (Bornstein & Matarazzo, 1982; Inglis & Lawson, 1981) have provided further support for McGlone's findings that "the specific effects on Verbal versus Performance IQ appear more prominently in males" (Bornstein Matarazzo, 1982, p. 319).

Despite the general validity of VIQ-PIQ differences as indexes of unilateral brain damage, at least among men, Bornstein (1983, p. 779) cautions that

> VIQ-PIQ discrepancies in isolation are ineffective indexes of cerebral dysfunction and that patterns of performance must be viewed in the context of a complete neuropsychological examination and relevant medical and educational historical data.

Still, attempts continue to be made to devise useful Wechsler-derived indexes of cerebral impairment. For example, Lawson and his colleagues (Lawson & Inglis, 1983, Lawson, Inglis, & Stroud, 1983) have generated a Laterality Index based on the factor structure of the Wechsler subtests to assess cognitive deficits following unilateral brain injuries. Cluster analytic techniques have also been employed (Clark, Crockett, Klonoff, & MacDonald, 1983) that have identified groups of patients that differ in terms of both level of intellectual deficit and type of cognitive deficit (defined in terms of various combinations of factor-derived dimensions: verbal, visual-spatial, and attention/memory). The clinical utility of these empirically derived measures remains to be determined, but they appear to hold promise.

Summary. It is obvious from the foregoing discussion that the Wechsler Intelligence Scales have a rich, although not always productive, history in neuropsychological assessment. It is quite likely that the Wechsler Intelligence Scales will continue to be used in the clinical assessment of brain dysfunction until some radically new and superior measure of intelligence is designed. However, some caution is indicated. As Matarazzo and Matarazzo (1984) state

> some forty-five years of research on Wechsler Intelligence Scale performance in relation to brain damage has made it clear that the clinician should rarely, if ever, attempt to assess the likelihood of brain damage or the cognitive deficits associated with known brain damage exclusively on the basis of scores on a Wechsler scale, including the WAIS-R. (p. 94)

The Wechsler Memory Scale

The clinical assessment of memory in many ways parallels that of intellectual assessment in that the early emphasis was on differentiating organic from functional memory impairment (Williams, 1978). Noting that psychologists working in mental hospitals were frequently asked to assess a patient's "memory," particularly as it related to the rest of the patient's functioning, Wechsler (1945) sought to develop a "rapid, simple, and practical memory examination" (p. 87) that would "be useful in detecting special memory deficits in individuals with specific organic injuries" (p. 90). Wechsler's (1945) work over a 10-year period resulted in the publication of the Wechsler Memory Scale (WMS), which yields an overall Memory Quotient (MQ) that is directly comparable with the IQ scores from the Wechsler Intelligence Scales. Although memory is currently viewed as a highly complex phenomenon that is intricately associated with other aspects of behavior (Russell, 1981; Williams, 1978), the WMS remains the most frequently used standardized measure of memory functioning.

Psychometric Properties. The WMS consists of seven subtests: Personal and Current Information, Orientation, Mental Control, Logical Memory, Digit Span, Visual Reproduction, and Associate Learning. Provisional norms were based on "approximately" 200 normal men and women between the ages of 25 and 50. Wechsler (1945) provides normative data by subtest for 96 subjects at two age levels, but scaled scores for each subtest were never developed. The composite MQ was empirically derived from a sample of 100 subjects by summing the subject's raw subtest scores and adding an age correction constant such that the subject's MQ approximated their Full Scale IQ on the Wechsler-Bellevue.

Unlike Wechsler's Intelligence Scales, the WMS has never been restandardized, and Wechsler's provisional norms have remained essentially the only norms for an adult population (Prigatano, 1978). However, the restricted age range of Wechsler's normative data has prompted other investigators to establish norms for both younger (Ivinskis, Allen, & Shaw, 1971) and older (Cauthen, 1977; Haaland, Linn, Hunt, & Goodwin, 1983; Hulicka, 1966) populations. Stone, Girdner, and Albrecht (1946) developed an alternate form of the WMS (Form II), but most research has focused on Wechsler's (1945) original version (Form I; here referred to as the WMS).

The psychometric properties of the WMS have been criticized on several counts. Prigatano (1978) points to the lack of standard scores for the individual subtests, poor scoring criteria for Logical Memory, limited norms for the standardization sample, a lack of test–retest reliability for the MQ among normal subjects, and the need to restandardize the distribution of MQ scores using the WAIS and/or WAIS-R. Despite these limitations, factor analytic studies of the WMS have revealed a fairly consistent three-factor pattern (Bachrach & Mintz, 1974; Kear-Colwell, 1973, 1977; Skilbeck & Woods, 1980). The first factor appears to be an immediate learning and recall dimension with loadings on the Logical Memory, Visual Reproduction, and Associate Learning subtests. The second factor has been labeled attention/concentration, with loadings on the Mental Control and Digit Span substests. The third factor has loadings on the Personal and Current Information and Orientation subtests, and is seen as an orientation and long-term information recall factor. This 3-factor structure was found to be essentially the same for both normal and combined psychiatric/neurological samples, suggesting considerable stability in the factor structure of the WMS (Kear-Colwell & Heller, 1978).

The Wechsler Memory Scale and Brain Dysfunction. Early research on the adequacy of the WMS as a general indicator of brain dysfunction yielded mixed results (Prigatano, 1977, 1978). This is not surprising given that the organic groups were often heterogenous in nature and memory was treated as a global variable without an appropriate comparison standard. More encouraging results emerged when comparison standards (intratrest and extratest) were used in the context of homogenous patient groups. For example, because the MQ was designed to parallel IQ scores (an extratest standard), the discrepancy between IQ and MQ has been used as an index of memory impairment. Significant IQ-MQ discrepancies in the order of 10 or more points have been reported among alcoholics with Korsakoff's syndrome (Victor, Herman, and White, 1959), patients with temporal lobe (psychomotor) seizure disorders (Milner, 1975; Quadfasel and Pruyser, 1955), commissurotomy patients (Zaidel and Sperry, 1974), and individuals who sustained closed head injuries (Black, 1973). "These findings add considerable strength to the use of the MQ score relative to IQ as an indication of a memory deficit" (Prigatano, 1978, p. 824).

In addition to the IQ-MQ discrepancy, other researchers have attempted to capitalize on the apparent differential sensitivity of the WMS subtests to cerebral pathology to devise intratest indexes of memory impairment. In their study of psychiatric patients with mild to moderate "cerebral dysfunction," Bachrach and Mintz (1974) observed that only the Personal and Current Information, Logical Memory, Visual Reproduction, and Associate Learning subtests discriminated their organic group from controls matched on age, IQ, and range of "ego weakness." Visual Reproduction was found to be as sensitive as a multiple regression equation composed of all the WMS subtests, and correctly classified 89% of the patients using a cutoff score of 10. Using a "hold–don't hold" methodology, Kljajic (1975) devised two predictive equations: (1) Digit Span score $- 1 >$ Associate Learning score, and (2) Information + Orientation score $\geq$ Associate Learning score. Both equations correctly classified 70% of the subjects. Kear-Colwell and Heller (1980) found that whereas all of the WMS subtests were related to the presence or absence of dysfunction among head-injury patients, factor analytic techniques revealed that the most marked deficits occurred on the verbal learning and immediate recall factor (Logical Memory and Associate Learning).

Unilateral Brain Dysfunction. Because six of the seven WMS subtests are verbal in nature, the MQ and derived IQ-MQ differences are generally thought to be essentially measures of verbal memory deficit (Milner, 1975). Research summarized by Prigatano (1978) tends to support this observation. However, when patterns of deficit on the WMS subtests are examined, differences among patients with documented unilateral lesions have been noted. Bornstein (1982) found that although his right- and left-hemisphere groups did not differ on MQ, IQ, or IQ-MQ, the "left lesion patients performed poorly on the logical memory and associate learning subtests, while the right lesion patients performed poorly on visual reproduction" (p. 391). Bornstein's empirical results are consistent with the a priori thinking that had earlier led Russell (1975a) to propose a rationally based, modified version of the WMS that would theoretically be more equally sensitive to unilateral lesions and would incorporate a broader definition of memory than that implied by the traditional WMS. Because of its growing clinical use, Russell's (1975) revision will be presented separately in the next section.

Russell's Revised Wechsler Memory Scale. Guided by earlier factor analytic research and observed differences in the differential sensitivity of various WMS subtests to mild cerebral impairment. Russell (1975) chose the Logical Memory and Visual Reproduction subtests as his starting point. He further noted that since "one is a verbal test and the other a figural test, they presumably are lateralized functions" (p. 801). A delayed memory component for each subtest was added by readministering the subtest 1/2 hour after the initial administration. Six scores are obtained: a short-term memory score, a delayed-memory score, and a difference score (immediate − delayed) for both the semantic (Logical Memory) and figural (Visual Reproduction) subtests.

The validity of Russell's (1975a) Revised Wechsler Memory Scale (RWMS) scores as general indexes of organic deficit was demonstrated in two ways. First, each of the six RWMS measures are found to discriminate significantly between controls and a mixed neurological population consisting of both unilateral and diffuse/bilateral brain-injured patients. Second, each of the six memory scores were correlated with the Average Impairment Rating (Russell, Neuringer, & Goldstein, 1970), a summary impairment index derived from the Halstead-Reitan Battery. For the combined sample, the correlations ranged from −.49 for semantic percent retained to −.76 for delayed figural score; all correlations were significant. To examine the discriminative validity of the semantic and figural components of the RWMS to unilateral lesions, Russell (1975) first transformed the memory scores for his combined sample to a 6-point scale (0–5) using z-scores to parallel that of the Average Impairment Rating. The resulting scaled scores were then analyzed for patients with documented unilateral lesions in a group × test analysis of variance. For both the short-term and delayed memory scores, the results indicated that the right- and left-hemisphere groups were equally impaired across tests, and that both test components were equally sensitive across patient groups. More importantly, there was a significant interaction between the side of lesion and test scores, indicating a lateralization effect.

Several investigators have employed Russell's (1975a) RWMS in their study of elderly populations. RWMS has been found to discriminate between normal age-related memory changes and those associated with dementia (Brinkman, Largen, Gerganoff, & Pomara, 1983; Logue & Wyrick, 1979), with the demented patients being more impaired than controls on both the semantic and figural subtests. Normative data on the RWMS scores for a healthy elderly population (ages 65 to 80) have been presented by Haaland et al. (1983), and their results demonstrated a progressive decline in memory abilities (figural greater than semantic) with age.

Summary. Despite its poor psychometric properties and global approach to the assessment of memory, the WMS has proved to be a useful neuropsychological instrument in a general context. In its traditional form, it is probably more of a measure of verbal than nonverbal memory impairment, and hence may be differentially sensitive to left hemisphere lesions. Russell's (1975a) RWMS appears to be a more balanced assessment instrument with respect to unilateral lesions, and has the advantage of a provision for the assessment of both short-term and delayed memory. Because the patterning of memory difficulties is important in the differential diagnosis of functional versus organic memory deficits (Williams, 1978), the RWMS is likely to have greater clinical utility than the WMS.

The Wide Range Achievement Test

Tests of academic achievement have frequently been included as part of standard assessment batteries to serve as adjuncts to tests of intelligence and personality functioning (Golden, 1979). Typically, these measures of educational skill are highly correlated with intelligence among the general population so that discrepancies between IQ and academic achievement may be clinically meaningful. Neuropsychologically, achievement tests such as the Wide Range Achievement Test (WRAT) (Jastak & Jastak, 1978) have been extensively used in the evaluation of children (Hynd & Obrzut, 1981) where early brain lesions can significantly affect the normal acquisition of a broad range of skills (Chelune & Edwards, 1981).

In adult assessment, the WRAT and other measures of academic achievement serve a different purpose. The WRAT is often used, assuming a person's normal development, as a comparison standard because basic academic skills are generally well retained except in cases of severe impairment. As Golden (1979, p. 51) observes, "in most brain injuries there will be a gap between IQ and standard scores on the WRAT which can be used to estimate the degree of impairment." In this respect, the WRAT is useful because of its relative lack of sensitivity to many forms of cerebral pathology in adults.

Composition and Psychometric Properties of the WRAT. The WRAT was first published in 1936 by J. Jastak and has subsequently been revised several times, with the most recent revision being in 1978 (Jastak & Jastak, 1978). It consists of two age levels (ages 5 to 12 and 12 through adulthood), with each having a spelling, reading, and arithmetic subtest. In the adult level, spelling is tested by having the subject produce written responses from dictation; reading is essentially an oral recognition task; and arithmetic requires the subject to perform written arithmetic involving a variety of computational operations. It is important to note that the WRAT's mode of administration/response may differ from that of other achievement tests. For example, the Peabody Individual Achievement Test (PIAT) (Dunn & Markwardt, 1970) assesses the same academic skills, but employs a multiple choice/visual recognition format for spelling and math. Intertest comparisons may reveal significant differences that can be potentially useful for diagnostic work.

The WRAT subtests are highly correlated with WAIS Full Scale IQ scores (range .71 to .77), and are more strongly related to Verbal IQ (range .76 to .87) than Performance IQ (range .54 to .67). The factor structure of the WRAT yields a strong general factor and a verbal factor largely represented by the Reading and Spelling subtests. Arithmetic contributes little to the verbal factor, but appears to have a "motivational component" (Lezak, 1983).

Summary. In adult neuropsychological assessment, the WRAT is often used as a "hold-type" of test because it provides an estimate of premorbid functioning. This estimate can then be used as a comparison standard to evaluate the adequacy of other, more brain-sensitive measures. Intertest comparisons with other measures of achievement that vary in presentation/response formats may also provide potential insights into the nature of some ability deficits. Lezak (1983) further suggests that an examination of the type of errors a patient makes on the arithmetic subtest can be helpful in determining whether the person's difficulties are "due to a dyscalculia of the spatial type, a figure or number alexia, or an anarithmetria in which number concepts or basic operations have been lost" (p. 294).

The Bender-Gestalt

Bender's (1938) interest in the European theories of gestalt psychology led her to develop the Visual Motor Gestalt Test, a figure-drawing test consisting of nine designs that could be used for clinical diagnosis. Bender's (1938) basic assumption for this test, which is now simply refered to as the Bender-Gestalt (B-G), was that

> the Gestalt function may be defined as that function of the integrated organism whereby it responds to a given constellation of stimuli as a whole; the response itself being a constellation, a pattern or Gestalt. (p. 3)

Any pathological process that disrupted the functioning of the "integrated organism" would be reflected in a disturbed capacity to reproduce the gestalten of the figures. Although Bender (1938) was primarily interested in the maturation of visual-motor gestalt functioning in children, the B-G was quickly adopted by clinicians as a measure of organicity, psychiatric disorder, and mental retardation.

Today, the B-G is one of the most frequently used psychological tests (Brown & McGuire, 1976; Wade & Baker, 1977). As a neuropsychological procedure, Golden and Kuperman (1980) found that the B-G was being taught in approximately 68% of the clinical/counseling graduate programs offering coursework in clinical neuropsychology. Likewise, a survey of the charter members and first-year additions to the Division of Clinical Neuropsychology of the American Psychological Association revealed that the B-G was routinely used by approximately 50% of the respondents (Hartlage et al. 1982). The popularity of the B-G has given rise to a number of scoring systems (e.g., Hain, 1964; Hutt, 1977; Koppitz, 1964; Pascal & Suttell, 1951) and modifications (e.g., Canter, 1966, 1976; Schraa, Jones, & Dirks, 1983), and has been used as both a general test of brain dysfunction and a measure of specific function (i.e., visuo-constructional ability). As Golden (1979, p. 129) notes, "extensive research has been both positive and negative, offering support to nearly any position one wishes to adopt in the use of the Bender."

The Bender-Gestalt as a General Measure of Organicity. In contrast to the other cognitive measures we have examined, performance on the B-G is *not* normally distributed in the general adult population. Pascal and Suttell (1951) indicate that by the age of 11 all of the B-G designs can be correctly reproduced, and with the exception of "mental defectives," the raw B-G scores are not strongly correlated with general intelligence. Thus, it is assumed that difficulties in reproducing the B-G designs are pathognomonic of some kind of disturbance (Lezak, 1983).

Clinical research has generally suported the discriminative validity of the B-G as a neurodiagnostic instrument (e.g., Bensberg, 1952; Billingslea, 1963; Brilliant & Gynther, 1963; Goldberg, 1959; Hain, 1964; Hutt, 1977; Lacks, Harrow, Colbert, & Leorne, 1970; McGuire, 1960; Mermelstein, 1983; Pascal & Suttell, 1951). Interpretation of this literature is complicated by the diversity of the scoring systems employed. However, most systems concur that signs of perseveration, reversal, rotation, distortion, angulation, and substitution are more frequent among organic patients than other groups. Of these signs, rotation of one or more of the B-G designs is perhaps the strongest correlate of neurological dysfunction (Billingslea, 1963; Griffith, & Taylor, 1960; Hannah, 1958). Hannah (1958) suggested that organic patients might be particularly susceptible to the difference in orientation between the stimulus cards (horizontal) and the test paper (vertical). Because adequacy of brain function has been found to be related to field dependence (Neuringer, Goldstein, & Gallaher, 1975), Hannah's observation may have merit.

Although the research literature indicates that the B-G has anywhere from a 60% to 90% correct hit-rate as a neurodiagnostic instrument (Bigler & Ehrfurth, 1981; Golden, 1979), its adequacy as a singular, objective measure of organicity has been seriously questioned. In an early study, Goldberg (1959) found no differences in diagnostic accuracy among staff psychologists, psychology trainees, psychology secretaries, and the Pascal and Suttell (1951) Objective Index, although all were able to discriminate organic from nonorganic patients at a better than chance level. Goldberg's (1959) results raise concerns about the severity of the organic groups used in the early B-G validation studies. If only the most extreme cases of organic pathology are included such that untrained secretaries can do as well as experienced psychologists and an objective scoring system, the sensitivity of the B-G to milder forms of cerebral impairment must be questioned. In his review of the diagnostic accuracy of the B-G, Golden (1979, p. 136) indicates that "the [hit] rate increases in more severe brain injured groups and less impaired control groups irrespective of the system of scoring employed." Further concern is raised by Bigler and Ehrfurth (1981), who comment on the interpretation of B-G classification rates. Citing the 80% correct classification rate reported by Hain (1964), Bigler and Ehrfurth (1981) examined the false-positive and false-negative rates among the brain-damaged and control groups. They found that while approximately 90% of the control group was correctly classified, only 59% of the brain-damaged group was correctly identified. Bigler and Ehrfurth (1981, p. 567) conclude that

> given the importance of accurate diagnosis in these cases, to rely on a single measure of screening of overall neurological status, particularly when that measure has a demonstrated rate of false negatives in the neighborhood of 40% or worse, is without question poor practice.

In summary, the B-G does appear to be sensitive to cerebral pathology, but its clinical sensitivity as a general index may vary as a function of the severity of brain dysfunction. Even Pascal and Suttell (1951, p. 40) acknowledge that "the B-G test cannot, in the absence of other data, answer that question [organicity], except occasionally in extreme cases which are also clinically apparent." At best, the B-G should probably be considered as a pathognomonic test; that is, when the reproduction of the B-G designs is disturbed some kind of pathological condition (organic or functional) is likely to be present, but the absence of difficulties does not necessarily rule out the possibility of brain dysfunction. Furthermore, when disturbances are present in the B-G record, it may not be possible to adequately discriminate between organic and psychiatric etiologies, especially chronic schizophrenia (Golden, 1979; Heaton et al., 1978). Because functional disturbances can affect the capacity to accurately reproduce the B-G designs, several modifications have been developed in an attempt to increase the differential sensitivity of the B-G.

The Canter Background Interference Procedure. Frustrated with the problem of false positives among psychiatric patients, Canter (1966, 1976) developed the Background Interference Procedure (BIP) to enhance the usefulness of the B-G as a neuropsychological screening instrument. The procedure consists of two parts. The B-G is first administered in the standard manner. On completion, the patient is again administered the B-G, but this time must reproduce the designs on a sheet of paper with intersecting sinusoidal lines, which provides the background interference. Both protocols are scored using Pascal and Suttell's (1951) scoring system, and the score for the normal administration is subtracted from the BIP score, preserving the sign. The greater the number and size of positive differences, relative to the overall quality of the standard B-G score, the greater the likelihood of brain dysfunction.

In his summary of the validity literature, Canter (1976) indicates that the BIP substantially reduces the number of false-positives among psychiatric patients who produce poor quality B-G's, while at the same time also reducing the number of false-negatives among neurological patients who produce high quality B-G records. Golden (1979) estimates that the use of the BIP with the B-G adds about 15% accuracy to that of the Bender. In their review of 94 studies in which neuropsychological tests were used to discriminate between cerebral dysfunction and nonorganic psychiatric disorders, Heaton et al. (1978) found that the combination of B-G and BIP performed better with more samples than did other neuropsychological tests, and had a median correct classification rate of 84% for 11 studies. Only two studies employed the BIP with chronic/process schizophrenics, but both had respective hit rates (60% and 85%) that were better than the 54% median hit rate observed for the 34 studies reviewed involving schizophrenics.

Bender-Gestalt Recall Procedures. Another frequently used modification of the B-G involves having the patient recall the designs immediately after the standard copy condition. Schraa, Jones, and Dirks (1983) recently collated the data from 26 studies reporting normative data on B-G recall for various diagnostic groups. Although they acknowledged methodological differences in scoring procedures and diagnostic criteria among the studies, Schraa and his colleagues suggest that the data indicate that recall of 6 to 7 designs can be considered average, whereas recall scores of 4 or less are impaired. Among psychiatric patients, those groups with major psychiatric diagnoses (e.g., psychoses, schizophrenia) recalled fewer designs than groups with less severe diagnoses (e.g., neurosis, character disorders). Patients with an organic diagnosis obtained lower recall scores than other diagnostic categories with the means ranging from 2.5 to 4.9. Schraa et al. (1983, p. 150) indicate that in the seven independent sets of data reporting Bender recall which included organics, the groups of organic patients always ranked last in the number of designs recalled in comparison with groups of patients with functional disorders.

Holland and Wadsworth (1979) found that B-G recall and recall after Canter's (1976) BIP significantly discriminated between brain-damaged and schizophrenic groups independently of IQ, whereas the BIP did not. Recall procedures were also found to be superior to the standard B-G in differentiating among the offspring of Huntington's patients who later became symptomatic and those who remained asymptomatic (Lyle & Quast, 1976).

The Bender-Gestalt as a Measure of Specific Function. As a measure of specific function, the B-G is generally viewed as a visual-motor constructional task "requiring accurate visual perception and proper motor execution and planning in order to reproduce the figure[s]" (Filskov & Leli, 1981). Errors in the accurate reproduction of the B-G designs are considered indicative of constructional dyspraxia, and are most frequently associated with parietal lobe lesions, especially in the right hemisphere (e.g., Benson & Barton, 1970; Benton & Fogel, 1962; Garron & Cheifetz, 1965). However, more recent research (e.g., Arena & Gainotti, 1978; DeRenzi, 1978) indicates that constructional dyspraxia can occur with lesions in either hemisphere, but that there are qualitative differences in the types of difficulties manifested. Rotation, fragmentation, and distortion are more common in right-hemisphere lesions, suggesting disturbed visual-spatial processes, whereas angulation, simplication, and size reductions occur more frequently with left-hemisphere lesions, implicating perceptual-motor and motor planning difficulties (Filskov & Leli, 1981).

When the BIP is used with the B-G, a visual interference component is added that requires figure-ground discrimination. Severe BIP effects may reflect optic dysgnosia (Luria, 1973), and tend to be more frequent in right-hemisphere lesions (Nemec, 1978). B-G recall procedures can be viewed as a measure of visual or figural memory, which is primarily mediated by the right-temporal and hippocampal areas (Russell, 1975a; Schraa et al., 1983). Rogers and Swenson (1975) have demonstrated that B-G recall scores correlate reasonably well with overall Wechsler Memory Scale scores (.74), and specifically with the Visual Reproduction subtest (.76).

Summary. As a general screening instrument, the B-G shows a statistically significant sensitivity to cerebral pathology. However, its high false-negative hit rates among neurological patients and low false-positive rates among normals suggest that it is best used as a pathognomonic measure rather than a general test of brain dysfunction. Furthermore, disturbances evident on the standard B-G tend to be nonspecific with respect to organic versus functional etiology. Modifications such as the BIP appear to enhance the differential sensitivity of the B-G, and should probably be included when the B-G is being used as a screening device for suspected brain dysfunction. As a measure of specific function, the B-G and its modifications can yield useful qualitative information regarding visual-perceptual, visual-motor, visual-memory, and motor planning functions.

PERSONALITY MEASURES

The personality component of the standard assessment battery typically includes a combination of both objective and projective tests. By far the most frequently used objective personality test is the Minnesota Multiphasic Personality Inventory (MMPI), whereas the Rorschach and Thematic Apperception Test (TAT) are the most popular projective techniques (Wade & Baker, 1977). Because questions of differential diagnosis (i.e., functional versus organic) are frequently posed to psychologists, it was natural for clinicians to turn to the personality tests in their standard batteries for assistance in their diagnostic evaluations. The expectation was that if brain dysfunction resulted in personality changes, "tests of personality functions could be used to predict the presence of brain pathology. The obvious assumption in this argument is that some 'organic' personality pattern will emerge" (Filskov & Leli, 1981, pp. 558–559). In the following sections we selectively review the research on the MMPI, Rorschach, and TAT as measures of neuropsychological functioning.

The Minnesota Multiphasic Personality Inventory

The Minnesota Multiphasic Personality Inventory (MMPI) (Hathaway & McKinley, 1951) is a 566-item self-report inventory that was designed to provide a rapid and effective method of objective psychodiagnostic assessment. The MMPI items were selected on the basis of statistical criteria to optimize discrimination between normal subjects and patient groups with known psychiatric diagnoses. Its construction lends itself well to principles of actuarial prediction, and a variety of codebooks and computerized scoring and interpretation systems are currently available. Researchers carried over these actuarial principles of personality assessment in their attempts to use the MMPI as a neuropsychological screening devise. Their work led to the development of a number of special scales, profile code-types, and keys for the general discrimination of organic and psychiatric patients. Other investigators have sought to use the MMPI to identify personality characteristics associated with localized brain lesions. These approaches are reviewed in the following sections.

MMPI Special Scales and Indexes of Brain Dysfunction. In an item analysis of MMPI protocols from patients on a neurology service, Hovey (1964) identified five items dealing with physical complaints that discriminated between normal patients and those with brain dysfunction. Using a cutoff score of 4 items or more, Hovey was able to correctly identify 50% of the organic group and 82% of the normal group. When used only with patients who had a raw K-scale score of 8 or above, the false-positives were reduced to 7% among the controls. Subsequent attempts to cross-validate the Hovey scale have been unsuccessful in discriminating organic patients from schizophrenics (Watson, 1971), mixed psychiatric patients (Maier & Abidin, 1967), and normals (Weingold, Dawson, & Kael, 1965). The only exception to these negative results appears to be among patients with multiple sclerosis (MS); Jortner (1965) was able to correctly classify 65% of his MS patients, and Hovey (1967) correctly identified 69% of his MS patients.

Shaw and Matthews (1965) developed a 17-item *Pseudo-Neurologic* (PsN) scale after comparing the MMPI results of patients with documented cerebral lesions with those who, after extensive neurological evaluation, were found to have no neurological disease. Using a cutting score of 7, the PsN scale was able to correctly identify 81% of the pseudoneurological group and 75% of brain damaged group. On cross-validation the PsN was again able to significantly discriminate similar groups, but only correctly identified 67% of the pseudoneurologic patients and 78% of the patients with demonstrable cerebral pathology. The error rates associated with the PsN scale severely limit its potential clinical utility in individual assessment.

Because schizophrenics and brain-damaged patients frequently obtain similar scores on cognitive tests, Watson (1971) reasoned that "personality-oriented measures may be more useful in separating members of the two groups" (p. 121). Watson performed an item analysis of MMPI records from 61 male organics and 65 male schizophrenics at a VA hospital, and produced a pool of 80 items that discriminated the two groups at the .05 level or better. From these items, three *Schizophrenia-Organicity* (Sc-O) scales were developed: an unweighted long-form, a weighted long-form that gave additional weights to items that discriminated at higher probability levels, and a weighted short-form that consisted of 30 items discriminating at a .01 level or better. The three Sc-O scales resulted in discrimination rates between 78% and 83% for the schizophrenics and correctly identified 89% of the organics. Cross-validation was carried out in another VA hospital and at a state mental hospital, and revealed somewhat lower overall classification rates (58% to 71%) for the VA population. Hit rates for the males from the state hospital ranged from 68% to 74%, but the Sc-O scale failed to replicate for women in the state institution (range of 41% to 59%). Neuringer, Dombrowski, and Goldstein (1975) cross-validated the three Sc-O scales in another VA setting, and also obtained significant discrimination rates (65% to 76%), although these were somewhat lower than those reported by Watson (1971). Other validity studies (Ayers, Templer & Ruff, 1975; Holland, Lowenfeld, & Wadsworth, 1975) have likewise produced significant, albeit marginal, results with the Sc-O scales. Based on these studies, the Sc-O scales should be used cautiously in individual assessment, and restricted to male populations.

An alternative approach to discriminating schizophrenic from organic patients was developed by Russell (1975b) that involves a key consisting of a series of decision-making rules for identifying schizophrenic patients. Using this MMPI key, Russell was able to correctly identify 75% of his patients as either brain-damaged or schizophrenic. However, in a later study Russell (1977) found that a cutoff T-score of 80 on the Sc (Schizophrenia) scale yielded hit rates that nearly equaled those achieved by the key.

Several organic code-types have been suggested in MMPI interpretative guides. Among those codes most frequently cited as being associated with organicity are the "139," "19," "29," and "98" codes (Gilberstadt & Duker, 1965; Lacher, 1974; Marks, Seeman, & Haller, 1974). Wooten (1983) found that the "89/98" and "29/92" code combinations did occur significantly more frequently in his sample of 303 neuropsychological referrals than among a large sample of general outpatients. However, the incidence of "139" and "19" profile codes did not differ between the two groups. Furthermore, the "89/98" code type occurred in only 10.6% of the neuropsychological groups, and the "29/92" combination appeared in only 3.6% of the sample. The low incidence of the organic "29" profile among brain-damaged patients was also noted by Russell (1977). However, Marks et al. (1974) claim that when the "29/92" high-point code does occur, "the possibility of a brain syndrome should at least be considered, since about a third of adult patients with 9-2 or 2-9 profiles are ultimately so diagnosed" (p. 31).

MMPI and Pathoanatomical Characteristics. There have been a number of attempts to identify MMPI differences among neurological patients on the basis of lesion localization and lateralization. Anderson and Hanvik (1950) examined MMPI differences among patients with frontal and parietal lesions, and suggested that patients with frontal-lobe lesions displayed more of a hysteroid reaction whereas those with parietal lobe lesions had MMPI profiles characteristic of an anxiety neurosis. Williams (1952) constructed a 36-item Caudality (Ca) scale to differentiate patients with temporo-parietal lesions from those with frontal lobe lesions. Using a cutting score of 11, the Ca scale correctly identified 78% of the frontal-lobe patients, a figure considerably higher than the 50% base rate in his sample. Meier and French (1964) compared the Ca scale scores of patients before and after temporal lobectomy, and found that the Ca scores decreased as would be expected after surgery. However, subsequent research has been equivocal in its support of the MMPI's sensitivity to differential personality characteristics associated with caudality dimension (Black & Black, 1982; Dikmen & Reitan, 1974b, 1977; Reitan, 1976).

Research examining the MMPI correlates of lateralized brain lesions has produced mixed results. When a lateralization effect has been observed, left-hemisphere patients have generally been found to have elevated Depression (2) and Schizophrenia (8) scales (Black, 1975; Gasparrini, Satz, Heilman, & Coolidge, 1978). These elevations were interpreted as being consistent with the observations of Gainotti (1972) who has suggested that left-hemisphere lesions often result in catastrophic emotional reactions whereas right-hemisphere lesions lead to emotional indifference. In contrast, Dikmen and Reitan (1974a, 1977) and others (Osmon & Golden, 1978) argue that the MMPI is more closely associated with the adequacy of adaptive functions than it is with either lesion localization or lateralization. In a recent study examining the relationship between subjective complaints of disability and scores on the MMPI and Halstead-Reitan Battery, Chelune, Heaton, and Lehman (1985) similarly found that patients' MMPI elevations on scales 2 and 8 were associated with perceived disability, relatively independent of level of neuropsychological impairment.

Finally, numerous attempts have been made to identify MMPI correlates of epilepsy. The results of such efforts have been for the most part disappointing, and no consistent MMPI differences have been reported that reliably differentiate among seizure patients or between seizure patients and other groups (e.g., Dikmen, Hermann, Wilensky, & Rainwater, 1983; Francy, 1950; Jordan, 1963; Lachar, Lewis, & Kupke, 1979; Stevens, Milstein, & Goldstein, 1972). These negative results cannot be simply attributed to the insensitivity of the MMPI, as Dikmen et al. (1983) demonstrated that the MMPI was differentially sensitive to varying degrees of psychopathology among seizure patients.

Summary. Research indicates that the MMPI is not an effective instrument for identifying brain dysfunction or its pathoanatomical characteristics. Emotional responses concomitant with cerebral pathology represent a complex interaction between the severity, type and location of brain pathology, premorbid personality, and the subsequent level of environmental stress that the individual must cope with. (Filskov & Leli, 1981, p. 560)

However, the MMPI can provide valuable information regarding level of adjustment and how patients are coping with their cognitive limitations, thereby contributing to a more complete understanding of patients' overall clinical picture. For this reason, Russell (1977) recommends the sequential use of the MMPI with other neuropsychological tests.

The Rorschach

Efforts to use projective techniques in the differential diagnosis of brain dysfunction were underway long before comparable work began with the MMPI. As early as 1931, Oberholzer, a student of Rorschach's, developed a number of Rorschach test (Rorschach, 1942) response characteristics in an attempt to identify patients with organic pathology (cited in Piotrowski, 1937). Despite its early origins, interest in the use of the Rorschach as a neuropsychological procedure has waned among American investigators, and the few recent reports that have appeared (e.g., Gordon, Greenberg, & Gerton, 1983) have reported negative results.

As a neuropsychological procedure, the Rorschach was originally presumed to be "an aid in neuropsychiatric diagnosis on the assumption that irreversible somatic disturbances of the central nervous system affect the personality markedly and specifically" (Piotrowski, 1937, p. 525). Current thinking (Lezak, 1983) recasts the Rorschach as a measure of cognitive and perceptual ability because the patient must perceptually organize an ambiguous stimulus and verbally encode the percept in a meaningful manner. Much of the work that has been done with the Rorschach has employed a combination of quantitative and qualitative signs, and these are reviewed in the following sections.

Rorschach Signs and Organicity. Perhaps the most widely recognized Rorschach sign approach to identifying brain dysfunction was developed by Piotrowski (1937). His 10 signs are presented in Table 2, and are based on both formal test characteristics (e.g., F+%, P%) and the patient's extratest behavior (e.g, perplexity, impotence). Piotrowski (1937) stressed that "no single sign alone points to abnormality in the psychiatric sense, to say nothing of organic involvement of the brain" (p. 529). Thus, he established that five or more signs must be present before making a diagnosis of organicity. In his original sample of 18 patients with "involvement of the brain" and 15 patients with either

Table 2. Rorschach Signs of Organicity

	Piotrowski's signs		Aita et al.'s signs
(R)	Total responses <15	(Inflx)	Inflexibility—unable to produce alternative interpretations to card
(T)	Average time/response <1 minute		
(M)	Total "movement" responses ≤1	(Act Obj)	Reacts to card as actual objects
(Cn)	Color naming	(CR)	Concrete responses
(F+%)	Percentage of "good" form responses <.70	(Def)	Unclear definitions
		(Catas)	Overt catastrophic reactions
(P%)	Percentage of "popular" responses <.25	(Edj)	Edging
		(IC)	Irrelevant comments
(Rpt)	Repetition of responses	(CvrCds)	Covers portion of card
(Imp)	Impotence—patient recognizes poor quality of response but cannot improve it.	(W & R)	Withdrawal and re-attack
(PLx)	Perplexity—patient expresses distrust of own abilities and requests reassurance.		
(AP)	Automatic phrases—frequent use of "pet phrases."		

Note: Data in the left column are from "The Rorschach Inkblot Method in Organic Disturbances of the Central Nervous System" by Z. Piotrowski, 1937, in *Journal of Nervous and Mental Disease, 86*. Copyright 1937 by Williams & Wilkins. Adapted by permission. Data in the right column are from "Rorschach's Test as a Diagnostic Aid in Brain Injury" by J. A. Aita, R. M. Reitan, and J. M. Ruth, 1947, in *American Journal of Psychiatry, 103*. Copyright 1947 by American Psychiatric Association. Adapted by permission.

"non-involvement" or a diagnosis of conversion hysteria, Piotrowski found that 94% of the organic patients had five or more signs, whereas none (100%) of his control patients had more than three signs.

Aita, Reitan, and Ruth (1947) examined Piotrowski's signs as well as nine additional signs (see Table 2) in a comparative study of 60 posttraumatic head injury patients and 100 controls. The brain-injury group consisted of men who sustained missile injuries during combat, and were divided into three levels of severity on the basis of length of unconsciousness, amnesia, and clinical and neuroradiological impressions. Although traditional quantitative indexes (e.g., determinents, content, etc.) were not found to be helpful in distinguishing between groups, all 19 signs were found to occur with greater frequency in the brain-damaged group than among the controls. There was also a clear, but not statistically tested, trend for the incidence of any given sign to increase with the severity of brain injury. Among Piotrowski's signs, those found to be most discriminating were Impotence (Imp), Perplexity (Plx), Automatic Phrases (AP), Repetition (Rpt), and Color Naming (Cn), whereas among Aita et al.'s (1947) signs, Unclear Definition (Def) and Actual Object (Act Obj) were found to be most helpful followed by Inflexibility (Inflx), Concrete Responses (CR), Catastrophic Reactions (Catas), and Edging (Edg).

Hughes (1950) provided independent support for the validity of Piotrowski's (1937) and Aita et al.'s (1947) signs in his factor analytic study of organic, schizophrenic, and neurotic patients. Hughes identified 14 signs as being associated with brain damage, and assigned weights to the signs on the basis of their factor loadings to create a brain-damage scale ranging from -7 to $+17$. Using a cutting score of $+7$, he was able to correctly identify 81.2% of the organics versus schizophrenics, and 93.8% of the organics versus neurotics. He also obtained a point biserial correlation of .79 between his brain-damage scale and the presence or absence of brain dysfunction for the combined sample.

Not all research on Rorschach signs has produced positive results. In what is now a classic study, Birch and Diller (1959) made a distinction between organicity and brain damage, defining *organicity* as a behavioral (as opposed to a pathophysiological) consequence of brain damage. They reasoned that organicity almost always occurred in the presence of brain damage, but that not all brain damage results in organicity. To demonstrate their proposition, they divided their sample into a 2×2 matrix on the basis of overt medical signs of perceptual difficulties and the presence or absence of documented brain damage. On the basis of their results, they concluded that the Rorschach did not appear to be sensitive to brain damage *per se*, but rather to organicity, one of the behavioral consequences of brain damage.

Echardt (1961) failed to find support for Piotrowski's signs in differentiating brain-damaged patients from schizophrenics. Diers and Brown (1951) report being unsuccessful using Hughes' (1950) signs among patients with multiple sclerosis, and suggested that the signs appeared to be related to intelligence rather than organic symptomatology. In a carefully controlled cross-validation of Aita et al. (1947), Reitan (1955b) found that five of Piotrowski's signs and five of Aita et al.'s signs discriminated brain-damaged patients from controls at the .05 level. However, Reitan noted that Rorschach indicators of brain damage were dependent on the type of lesion, and found that the Halstead Impairment Index and 8 of 10 Halstead tests "differentiated the groups at much more extreme confidence levels than were obtained with the Rorschach" (p. 450). Interestingly, Reitan (1955b) also observed that those Rorschach signs that were most discriminating were not necessarily indigenous to the Rorschach per se (e.g., Impotence, Perplexity, Catastrophic Reaction).

The Rorschach and Pathoanatomical Correlates. Relatively few attempts have been made to examine the Rorschach in the context of localized or lateralized lesions. Reitan (1954) compared the Rorschach responses of organic patients with and without language disturbances (aphasia) with those of nonorganic controls. No differences were found on the quantitative Rorschach scores, suggesting that verbal expression does not strongly influence test results.

Dorken and Kral (1952) examined seven quantitative Rorschach signs among normal controls, psychiatric controls, and organic groups with diffuse pathology and localized lesions in the cortex and/or adjacent white matter, diencephalon, striate body, and pallidum-nigra system. These authors report being able to correctly identify 92.9% of the organic patients and 83.3% of the nonorganic controls. Furthermore, Dorken and Kral observed trends in Rorschach performance that appeared related to lesion localization. Specifically, patients with diencephalic and striate lesions showed distinct disturbances in emotionality (i.e., color responses) compared to those individuals with either coritical or diffuse pathology.

Using discriminant function techniques, Hall, Hall, and Lavoie (1968) compared the Rorschach responses of 50 patients with well-documented unilateral and midline/bilateral lesions. Although these investigators reported no significant intrahemispheric differences, they did find a highly significant lateralization effect. Not only did patients with unilateral lesions differ from those with bilateral/midline involvement, but the right- and left-hemisphere groups showed distinctive and divergent differences. Left-hemisphere lesions appeared to result in a limited and constricted ideational style, whereas right-hemisphere lesions seemed to produce an expansive and uncritically innovative mode of ideation. Hall et al. (1968) interpreted these differences as reflecting "an exaggeration, in pathological form, of characteristics which operate harmoniously in the normal brain" (p. 531).

Summary. Early research using various sign approaches has demonstrated that the Rorschach may be a useful neuropsychological assessment tool in some cases. However, Exner (personal communication, December, 1983) reports that in several recent unpublished studies involving at least 125 brain-damaged patients, efforts to validate previously established signs have been disappointing. He suggests that traditional Rorschach signs may appear among severely impaired patients, but that they are not reliable indicators of brain dysfunction in milder cases. Furthermore, many of the most sensitive Rorschach signs (e.g., perplexity, impotence, catastrophic reactions) are extratest in nature, and can also be observed on other test measures.

The Thematic Apperception Test

Murray (1938) developed the Thematic Apperception Test (TAT) on the premise that individuals tend to interpret ambiguous human interactions in accordance with their own experiences and drives. Although the TAT is a frequent component of standard assessment batteries (Wade & Baker, 1977), there is a dearth of literature on its usefulness as a measure of brain dysfunction. Reviews concerning the clinical utility of the TAT (e.g., Bellak, 1975; Stein, 1978) typically do not even mention it as a potential neuropsychological procedure. Nonetheless, Lezak (1983) suggests that brain-damaged patients produce TAT responses that parallel the qualitative characteristics seen in the Rorschach protocols of these patients. Specifically, she notes that patients with cerebral pathology are apt to use few words and ideas in telling stories, respond slowly, be satisfied with simple description of discrete features of the stimulus pictures, manifest tendencies toward confusion, perseverate, and express feelings of self-doubt. However, these characteristics are extratest behaviors that are not idiosyncratic to the TAT *per se*, and like the Rorschach signs, they are likely to be more evident in severely impaired groups than among those with more subtle brain dysfunction.

ASSESSMENT VERSUS SCREENING

It is clear from our review of the literature concerning the role of standard cognitive and personality tests in neurodiagnosis that these tests have been used extensively, and that some have demonstrated a moderate degree of validity. However, although these tests may offer potentially useful information, they are not considered to be as effective as those tests and batteries that have been explicitly designed for neurodiagnostic purposes (Bigler & Ehrfurth, 1981; Reitan, 1955a, 1959; Matarazzo & Matarazzo, 1984). These findings have implications for the way we view the role of such tests in neuropsychological assessment.

Tests such as the Wechsler Intelligence Scales, Bender-Gestalt, and Rorschach were not designed to provide a comprehensive assessment of brain–behavior relationships. Their potential utility, especially when used in combination, lies in their widespread availability and use. Whereas use of specialized, comprehensive batteries of ability tests (e.g., Halstead-Reitan and Luria-Nebraska Batteries) for the evaluation and diagnosis of brain dysfunction is clearly desirable, such procedures are not cost-effective for general screening purposes, in many situations where the base rate of neurological disorders is low. In these situations it may be efficacious to use traditional cognitive and personality tests to determine whether more expensive and comprehensive neuropsychological tests are warranted. For screening purposes, traditional tests can provide useful, inexpensive, and readily obtainable data to assist the neuropsychologically minded clinician in making this decision.

> Screening tests should not be used to diagnose brain damage, nor should they be used to attempt to describe the nature and extent of any possible neuropathological process. It should also be obvious that a clinician who is led, on the basis of poor performance on a screening test, to suspect possible neurological involvement is not necessarily declaring the patient to be permanently and immutably brain damaged. (Boyd, 1982, p. 294)

Conversely, the astute clinician must take into account the ratio of false-positives to false-negatives associated with traditional tests. In the context of a screening evaluation, one might have good confidence that a patient obtaining a "bad" score is indeed "impaired," but for many cognitive and personality tests, a "good" score does not necessarily mean that the patient is "unimpaired." Given these cautions, we feel that standard cognitive and personality tests have a place in neuropsychological assessment as screening procedures when they are interpreted within a neuropsychological framework.

Whereas clinical neuropsychology is likely to continue to make valuable contributions to the neurodiagnostic process, changing trends in theory development and consumer needs are leading neuropsychology into new frontiers that require "an entirely new way of structuring brain–behavior relationships" (Rourke, 1982, p. 3). There is currently a growing interest among clinical neuropsychologists in the assessment of adaptive functioning within the context of everyday living (Chelune & Moehle, 1985; Heaton & Pendelton, 1981). In this respect, the central criterion has shifted from the organic integrity of the brain to the capacity to perform everyday tasks and activities in the environment. Patients' residual strengths as well as their deficits are important targets for assessment (Chelune, 1983), and traditional cognitive and personality tests may contribute to a more complete understanding of the patient within this future trend.

FUTURE DIRECTIONS

Neuropsychological assessment has developed within a context of evolutionary change in theory. In a historical overview of model building in clinical neuropsychology, Rourke (1982) has outlined three phases in which the focus of neuropsychological investigation has progressed from an emphasis on the brain to one of how individuals approach everyday tasks. This shift presents an important challenge to neuropsychologists. Specialized neuropsychological tests have been developed to detect the presence and localization of cerebral lesions within a deficit measurement paradigm (Lezak, 1983). However, prediction of a patient's potential success in everyday tasks and activities requires a comprehensive evaluation of the patient's *strengths and deficits,* as well as environmental demands and resources (Chelune & Moehle, 1985). These factors must be further considered in conjunction with such moderating variables as motivation, psychiatric status, frustration level, and attitude toward disability, which could interact with a patient's ability deficits to produce poorer social and vocational adjustments than would be predicted on the basis of ability test scores alone.

Because the standard battery of cognitive and personality tests was designed for "large-bandwidth" (Cronbach, 1969) assessment purposes, it provides a general picture of a patient's overall clinical status. This information can be integrated with specific neuropsychological test findings to yield "a complete understanding of the impact of impaired functioning on the individual" (Filskov & Leli, 1981, p. 545). Without this knowledge, the prediction of everyday functioning is very difficult, as there is substantial evidence that patients' ultimate capacity to function in the environment is multiply determined, and not solely dependent on their residual neuropsychological abilities (Chelune & Moehle, 1985). Traditional cognitive and personality tests can serve as useful adjuncts to neuropsychological procedures when the task of assessment is to predict everyday living skills.

Recent reviews of literature concerning the relationship between neuropsychological assessment procedures and everyday functioning clearly document the potential utility of traditional psychological tests. Of the 40 studies reviewed by Heaton and Pendleton (1981), two-thirds used IQ scores as the primary predictor of various aspects of everyday functioning. Similarly, intelligence scores and personality measures were found to be frequently used in combination with specific neuropsychological tests in the additional 56 studies reviewed by Chelune and Moehle (1985). Clearly, the inclusion of so-called standard psychological tests as adjuncts to neuropsychological tests is gaining empirical support.

Although research has supported the empirical utility of standard cognitive and personality tests with respect to everyday functioning, some caution must be exercised when interpreting these tests for the purpose of individual clinical assessment. Unlike specialized neuropsychological measures that were designed explicitly for use among neurological populations, traditional psychological tests were developed for use among more general populations, and the clinical descriptors associated with their specific test scores may need to be interpreted differently for brain-injured individuals. For example, MMPI elevations on Depression (2) and Schizophrenia (8) are frequently observed among both general outpatients and neuropsychological referrals (Chelune et al., 1985; Wooten, 1983). Psychiatric interpretation of this profile pattern would suggest a possible psychotic thought disturbance (Marks et al., 1974), yet among neurological patients this profile configuration could simply reflect reduced cognitive efficiency. Thus, just as information derived from standard batteries can be helpful in modifying the inferences drawn from neuropsychological test results, the neuropsychological status of the patient must also be considered when interpreting the individual's performances on traditional psychological tests.

SUMMARY AND CONCLUSIONS

In this chapter we have examined the utility of a number of traditional psychological tests for the purpose of neurodiagnosis. Although many of the studies reviewed report positive results, the adequacy of these test procedures for the comprehensive assessment and diagnosis of neurological disorders falls short when compared to the diagnostic accuracy achieved by specialized neuropsychological batteries. Considering the theoretical assumptions (e.g., organicity) that guided their modification for neurodiagnostic purposes, this is not surprising. Still, there is substantial evidence that these standard clinical tests are at least moderately sensitive to the organic integrity of the brain. Given their widespread use and potential cost-effectiveness in clinical settings where there is a low incidence of neurological disorders, the use of these tests as screening devices may be warranted if appropriate caution is exercised.

As neuropsychology enters what Rourke (1982) has called its *dynamic* phase, increasing emphasis will be placed on issues of everyday functioning. Within this phase, traditional cognitive and personality tests are apt to serve as adjuncts to neuropsychological assessment procedures in providing a comprehensive picture of a patient's clinical status. There is already a growing body of evidence that suggests that the comprehensive assessment of patients' strengths and deficits can be used to infer adequacy in many daily tasks and activities. In this respect, standard cognitive and personality tests compliment the information derived from specialized neuropsychological tests, providing a more complete understanding of the impaired individual.

REFERENCES

Aita, J. A., Reitan, R. M., & Ruth, J. M. (1947). Rorschach's test as a diagnostic aid in brain injury. *American Journal of Psychiatry, 103,* 770–779.

Allen, R. M. (1970). A note on the use of the Wechsler-Bellevue Scale Mental Deterioration Index with brain-injured patients. *Journal of Clinical Psychology, 26,* 71–73.

Allison, J. (1978). Clinical contribution of the Wechsler Adult Intelligence Scale. In B. B. Wolman (Ed.), *Clinical diagnosis of mental disorders.* New York: Plenum Press.

Anastasi, A. (1982). *Psychological testing,* [5th ed.]. New York: Macmillan.

Anderson, A. L. (1951). The effect of laterality localization of focal brain lesions in the Wechsler-Bellevue subtests. *Journal of Clinical Psychology, 7,* 149–153.

Anderson, A. L., & Hanvik, L. J. (1950). The psychometric localization of brain lesions: The differential effect of frontal and parietal lesions on MMPI profiles. *Journal of Clinical Psychology, 6,* 177–180.

Arena, R., & Gainotti, G. (1978). Constructional apraxia and hemispheric locus of lesion. *Cortex, 14,* 463–473.

Ayers, J., Templer, D. I., & Ruff, C. F. (1975). The MMPI in the differential diagnosis of organicity versus brain damage: Empirical findings and a somewhat different perspective. *Journal of Clinical Psychology, 32,* 685–686.

Bachrach, H., & Mintz, J. (1974). The Wechsler Memory Scale as a tool for the detection of mild cerebral dysfunction. *Journal of Clinical Psychology, 30,* 58–60.

Becker, B. (1975). Intellectual changes after closed head injury. *Journal of Clinical Psychology, 31,* 307–309.

Bellak, L. (1975). *The Thematic Apperception Test, the Children's Apperception Test and the Senior Apperception technique in clinical use* (3rd ed.). New York: Grune & Stratton.

Bender, L. A. (1938). A visual motor gestalt test and its clinical use. *American Orthopsychiatric Association Research Monographs,* No. 3.

Bensberg, G. J. (1952). Performance of brain injured and familial mental defectives on the Bender-Gestalt test. *Journal of Consulting Psychology, 16,* 61–64.

Benson, F. D., & Barton, M. I. (1970). Disturbances in constructional ability. *Cortex, 6,* 19–46.

Benton, A. L., & Fogel, M. L. (1962). Three-dimensional constructional praxia. *Archives of Neurology, 7,* 347–354.

Bigler, E. D., & Ehrfurth, J. W. (1981). The continued inappropriate singular use of the Bender Visual Motor Gestalt Test. *Professional Psychology, 12,* 562–569.

Billingslea, F. Y. (1963). The Bender-Gestalt: A review and a perspective. *Psychological Bulletin, 60,* 233–251.

Birch, H., & Diller, L. (1959). Rorschach signs of "organicity:" A physiological basis for perceptual disturbances. *Journal of Projective Techniques, 23,* 184–197.

Black, F. W. (1973). Cognitive and memory performance in subjects with brain damage secondary to penetrating missile wounds and closed head injuries. *Journal of Clinical Psychology, 29,* 441–442.

Black, F. W. (1975). Unilateral brain lesions and MMPI performance: A preliminary study. *Perceptual Motor Skills, 40,* 87–93.

Black, F. W., & Black, I. L. (1982). Anterior-posterior locus of lesion and personality: Support for the caudality hypothesis. *Journal of Clinical Psychology, 38,* 601–605.

Blaha, J., & Wallbrown, F. H. (1982). Hierarchical factor structure of the Wechsler Adult Intelligence Scale-Revised. *Journal of Consulting and Clinical Psychology, 50,* 652–660.

Bornstein, R. A. (1982). Effects of unilateral lesions of the Wechsler Memory Scale. *Journal of Clinical Psychology, 38,* 389–392.
Bornstein, R. A. (1983). Verbal IQ-Performance IQ discrepancies on the Wechsler Adult Intelligence Scale-Revised in patients with unilateral or bilateral cerebral dysfunction. *Journal of Consulting and Clinical Psychology, 51,* 779–780.
Bornstein, R. A., & Matarazzo, J. D. (1982). Wechsler VIQ versus PIQ differences in cerebral dysfunction: A literature review with emphasis on sex differences. *Journal of Clinical Neuropsychology, 4,* 319–334.
Boyd, J. L. (1982). Reply to Rathbun and Smith: Who made the Hooper blooper? *Journal of Consulting and Clinical Psychology, 50,* 284–285.
Brilliant, P. J., & Gynther, M. D. (1963). Relationships between performances on three tests for organicity and selected patient variables. *Journal of Consulting Psychology, 27,* 474–479.
Brinkman, S. D., Largen, J. W., Gerganoff, S., & Pomara, N. (1983). Russell's Revised Wechsler Memory Scale in the evaluation of dementia. *Journal of Clinical Psychology, 39,* 989–993.
Brown, W. R., & McGuire, J. M. (1976). Current psychological assessment practices. *Professional Psychology, 7,* 475–484.
Canter, A. (1966). A background interference procedure to increase sensitivity of the Bender-Gestalt test to organic brain disorder. *Journal of Consulting Psychology, 30,* 91–97.
Canter, A. (1976). *The Canter Background Interference Procedure for the Bender Gestalt Test: Manual for administration, scoring, and interpretation.* Los Angeles: Western Psychological Services.
Cauthen, N. R. (1977). Extension of the Wechsler Memory Scale norms to older age groups. *Journal of Clinical Psychology, 33,* 208–211.
Chelune, G. J. (1982). A re-examination of the relationships between the Luria-Nebraska and the Halstead-Reitan Batteries: Overlap with the WAIS. *Journal of Consulting and Clinical Psychology, 50,* 578–580.
Chelune, G. J., & Edwards, P. (1981). Early brain lesions: Ontogenetic-environmental considerations. *Journal of Consulting and Clinical Psychology, 49,* 777–790.
Chelune, G. J., & Moehle, K. A. (1985). Neuropsychological assessment and everyday functioning. In D. Wedding, A. M. Horton, and J. Webster (Eds.), *Handbook of clinical and behavioral neuropsychology.* New York: Springer.
Chelune, G. J., Heaton, R. K., Lehman, R. A., & Robinson, A. (1979). Level versus pattern of neuropsychological performance among schizophrenic and diffusely brain-damaged patients. *Journal of Consulting and Clinical Psychology, 47,* 155–163.
Chelune, G. J., Heaton, R. K., & Lehman, R. A. (1985). Relation of neuropsychological and personality test results to patients' complaints of disability. In G. Goldstein (Ed.), *Advances in clinical neuropsychology* (Vol. 3). New York: Plenum Press.
Clark, C., Crockett, D., Klonoff, H., & MacDonald, J. (1983). Cluster analysis of the WAIS on brain-damaged patients. *Journal of Clinical Neuropsychology, 5,* 149–158.
Cleeland, C. S. (1976). Interferences in clinical psychology and clinical neuropsychology: Similarities and differences. *Clinical Psychologist, 29,* 8–10.
Cronbach, L. J. (1969). *Essentials of psychological testing.* New York: Harper & Row.
Davis, W. E., DeWolfe, A. S., & Gustafson, R. C. (1972). Intellectual deficit in process and reactive schizophrenia and brain injury. *Journal of Consulting and Clinical Psychology, 38,* 146.
Davis, W. E., Dizzonne, M. F., & DeWolfe, A. S. (1971). Relationships among WAIS subtest scores, patient's premorbid history, and institutionalization. *Journal of Consulting and Clinical Psychology, 36,* 400–403.

Davison, L. A. (1974). Introduction. In R. M. Reitan & L. A. Davison (Eds.) *Clinical neuropsychology: Current status and application.* New York: Wiley.

DeRenzi, E. (1978). Hemispheric asymmetry as evidenced by spatial disorders. In M. Kinsbourne (Ed.), *Asymmetrical function of the brain.* Cambridge, England: Cambridge University Press.

DeWolfe, A. S. (1971). Differentiation of schizophrenia and brain damage with the WAIS. *Journal of Clinical Psychology, 27,* 209–211.

DeWolfe, A. S., Barrell, R. P., Becker, B. C., & Spanner, F. E. (1971). Intellectual deficit in chronic schizophrenia and brain damage. *Journal of Consulting and Clinical Psychology, 36,* 197–204.

Diers, W., & Brown, C. (1951). Rorschach "organic signs" and intelligence level. *Journal of Consulting Psychology, 15,* 343–345.

Dikmen, S., Hermann, B. P., Wilensky, A. J., & Rainwater, G. (1983). Validity of the Minnesota Multiphasic Personality Inventory (MMPI) to psychopathology in patients with epilepsy. *Journal of Nervous and Mental Disease, 171,* 114–122.

Dikmen, S., & Reitan, R. M. (1974a). Minnesota Multiphasic Personality Inventory correlates of language disturbance. *Journal of Abnormal Psychology, 83,* 675–679.

Dikmen, S., & Reitan, R. M. (1974b). MMPI correlates of localized cerebral lesions. *Perceptual and Motor Skills, 39,* 831–840.

Dikmen, S., & Reitan, R. M. (1977). MMPI correlates of adaptive ability deficits in patients with brain lesions. *Journal of Nervous and Mental Disease, 165,* 247–254.

Dorken, H., & Kral, V. A. (1952). The psychological differentiation of organic brain lesions and their localization by means of the Rorschach test. *American Journal of Psychiatry, 108,* 764–770.

Dunn, L. M., & Markwardt, F. C. (1970). *Peabody Individual Achievement Test manual.* Circle Pines, Minnesota: American Guidance Service.

Echardt, W. (1961). Piotrowski's signs: Organic or functional: *Journal of Clinical Psychology, 17,* 36–38.

Exner, J. E., & Clark, B. (1978). The Rorschach. In B. B. Wolman (Ed.), *Clinical diagnosis of mental disorders.* New York: Plenum Press.

Filskov, S. B., & Leli, D. A. (1981). Assessment of the individual in neuropsychological practice. In S. B. Filskov and T. J. Boll (Eds.), *Handbook of clinical neuropsychology.* New York: Wiley.

Fitzhugh, L. C., Fitzhugh, K. B., & Reitan, R. M. (1962). Wechsler-Bellevue comparisons in groups of "chronic" and "current" lateralized and diffuse brain lesions. *Journal of Consulting Psychology, 26,* 306–310.

Francy, R. (1950). A study on the epileptic personality. *Canadian Journal of Psychology, 4,* 81–87.

Gainotti, G. (1972). Emotional behavior and hemispheric side of lesions. *Cortex, 8,* 41–55.

Garron, D. C., & Cheifetz, D. I. (1965). Comment on "Bender Gestalt discernment of organic pathology." *Psychological Bulletin, 63,* 197–200.

Gasparrini, W. G., Satz, P., Heilman, K. M., & Coolidge, F. L. (1978). Hemispheric asymmetries of effective processing as determined by the Minnesota Multiphasic Personality Inventory. *Journal of Neurology, Neurosurgery, and Psychiatry, 41,* 470–473.

Gilberstadt, H., & Duker, J. (1965). *A handbook for clinical and actuarial MMPI interpretation.* Philadelphia: Saunders.

Glass, A. (1982). *Factor structure of the WAIS-R.* Paper presented at the 90th annual meeting of the American Psychological Association, Washington, D.C.

Goldberg, L. R. (1959). The effectiveness of clinicians' judgments: The diagnosis of organic brain disease from the Bender-Gestalt Test. *Journal of Consulting Psychology, 23,* 25–33.

Golden, C. J. (1979). *Clinical interpretation of objective psychological tests.* New York: Grune & Stratton.

Golden, C. J., & Kuperman, S. K. (1980). Graduate training in clinical neuropsychology. *Professional Psychology, 11,* 55–63.

Goldstein, K. H. (1939). *The organism.* New York: American Book Company.

Goldstein, S. G. (1976). The neuropsychological model: Interface and overlap with clinical psychology. *Clinical Psychologist, 29,* 7–8.

Gordon, M., Greenberg, R. P., & Gerton, M. (1983). Wechsler discrepancies and the Rorschach experience balance. *Journal of Clinical Psychology, 39,* 775–779.

Griffith, R. M., & Taylor, V. H. (1960). Incidence of Bender-Gestalt figure rotation. *Journal of Consulting Psychology, 24,* 189–190.

Grossman, F. M. (1983). Percentage of WAIS-R standardization sample obtaining Verbal-Performance discrepancies. *Journal of Consulting and Clinical Psychology, 51,* 641–642.

Haaland, K. Y., Linn, R. T., Hunt, W. C., & Goodwin, J. S. (1983). A normative study of Russell's variant of the Wechsler Memory Scale in a healthy elderly population. *Journal of Consulting and Clinical Psychology, 51,* 878–881.

Hain, J. D. (1964). The Bender Gestalt Test: A scoring method for identifying brain damage. *Journal of Consulting Psychology, 28,* 34–40.

Hall, M. M., Hall, G. C., & Lavoie, P. (1968). Ideation in patients with unilateral or bilateral midline brain lesions. *Journal of Abnormal Psychology, 73,* 526–531.

Halstead, W. C. (1947). *Brain and intelligence.* Chicago: University of Chicago Press.

Hannah, V. (1958). Causitive factors in the production of rotation on the Bender Gestalt designs. *Journal of Consulting Psychology, 22,* 398–399.

Hartlage, L., Chelune, G. J., & Tucker, D. (1982). Survey of professional issues in the practice of clinical neuropsychology. *Behavioral Neuropsychology Newsletter, 4,* 3–5.

Hathaway, S. R., & McKinley, J. C. (1951). *The Minnesota Multiphasic Personality Inventory manual.* New York: Psychological Corporation.

Haynes, J. R., & Sells, S. B. (1963). Assessment of organic brain damage by psychological tests. *Psychological Bulletin, 60,* 316–325.

Heaton, R. K., Baade, L. E., & Johnson, K. L. (1978). Neuropsychological test results associated with psychiatric disorders in adults. *Psychological Bulletin, 85,* 141–162.

Heaton, R. K., & Pendleton, M. G. (1981). Use of neuropsychological tests to predict adult patients' everyday functioning. *Journal of Consulting and Clinical Psychology, 49,* 807–821.

Hewson, L. R. (1949). The Wechsler-Bellevue Scale and the substitution test as aids in neuropsychiatric diagnosis. *Journal of Nervous and Mental Disease, 109,* 158–183.

Holland, T. R., Lowenfeld, J., & Wadsworth, H. M. (1975). MMPI indices in the discrimination of brain-damaged and schizophrenic groups. *Journal of Clinical Psychology, 43,* 426.

Holland, T. R., & Wadsworth, H. M. (1979). Comparison and combination of recall and Background Interference Procedures for the Bender Gestalt Test with brain-damaged and schizophrenic patients. *Journal of Personality Assessment, 43,* 123–127.

Hovey, H. B. (1964). Brain lesions and 5 MMPI items. *Journal of Consulting Psychology, 28,* 78–79.

Hovey, H. B. (1967). MMPI testing for multiple sclerosis. *Psychological Reports, 21,* 599–600.

Hughes, R. M. (1950). A factor analysis of Rorschach diagnostic signs. *Journal of General Psychology, 43,* 85–103.
Hulicka, I. M. (1966). Age differences in Wechsler Memory Scale scores. *Journal of Psychology, 109,* 135–145.
Hunt, W. L. (1949). The relative rates of decline of the Wechsler-Bellevue "Hold" and "Don't Hold" tests. *Journal of Consulting Psychology, 13,* 440–443.
Hutt, M. L. (1977). *The Hutt adaptation of the Bender-Gestalt Test.* New York: Grune & Stratton.
Hynd, G. W., & Obrzut, J. E. (1981). *Neuropsychological assessment and the school-age child.* New York: Grune & Stratton.
Inglis, J., & Lawson, J. S. (1981). Sex differences in the effects of unilateral brain damage on intelligence. *Science, 42,* 693–695.
Ivinskis, A., Allen, S., & Shaw, E. (1971). An extension of Wechsler Memory Scale norms to lower age groups. *Journal of Clinical Psychology, 27,* 354–357.
Jastak, J. F. (1936). *Wide Range Achievement Test manual.* Wilmington, Delaware: C. L. Story.
Jastak, J. F., & Jastak, S. R. (1978). *The Wide Range Achievement Test manual.* Wilmington, Delaware: Guidance Associates.
Jordan, E. J. (1963). MMPI profiles of epileptics: A further evaluation. *Journal of Consulting Psychology, 27,* 267–269.
Jortner, S. A. (1965). A test of Hovey's MMPI scale for CNS disorders. *Journal of Clinical Psychology, 21,* 285.
Kane, R. L., Parsons, O. A., and Goldstein, G. (1984). Statistical relationships and discriminative accuracy of the Halstead-Reitan, Luria-Nebraska and Wechsler IQ scores in the identification of brain damage. *Journal of Clinical and Experimental Neuropsychology, 7,* 211–223.
Kear-Colwell, J. J. (1973). The structure of the Wechsler Memory Scale and its relationship to "brain damage." *British Journal of Social and Clinical Psychology, 12,* 384–392.
Kear-Colwell, J. J. (1977). The structure of the Wechsler Memory Scale: A replication. *Journal of Clinical Psychology, 33,* 483–485.
Kear-Colwell, J. J., & Heller, M. (1978). A normative study of the Wechsler Memory Scale. *Journal of Clinical Psychology, 34,* 354–357.
Kear-Colwell, J. J., & Heller, M. (1980). The Wechsler Memory Scale and closed head injury. *Journal of Clinical Psychology, 36,* 782–787.
Kljajic, I. (1975). Wechsler Memory Scale indices of brain pathology. *Journal of Clinical Psychology, 31,* 698–701.
Koppitz, E. M. (1964). *The Bender Gestalt Test for younger children.* New York: Grune & Stratton.
Lachar, D. (1974). *The MMPI: Clinical assessment and automated interpretation.* Los Angeles: Western Psychological Services.
Lacher, D., Lewis, R., & Kupke, T. (1979). MMPI in differentiation of temporal lobe and non-temporal lobe epilepsy: Investigation of three levels of test performance. *Journal of Consulting and Clinical Psychology, 47,* 186–188.
Lacks, P. B., Harrow, M., Colbert, J., & Leorne, J. (1970). Further evidence concerning the diagnostic accuracy of the Halstead organic test battery. *Journal of Clinical Psychology, 26,* 480–481.
Lansdell, H. (1962). A sex difference in the effect of temporal-lobe neurosurgery on design preference. *Nature, 194,* 852–854.
Lashley, K. S. (1929). *Brain mechanisms and intelligence: A quantitative study of injuries to the brain.* Chicago: University of Chicago Press.
Lawson, J. S., & Inglis, J. (1983). A laterality index of cognitive impairment after hemi-

spheric damage: A measure derived from a principal-components analysis of the Wechsler Adult Intelligence Scale. *Journal of Consulting and Clinical Psychology, 51*, 832–840.

Lawson, J. S., Inglis, J., & Stroud, T. W. (1983). A laterality index of cognitive impairment derived from a principal-components analysis of the WAIS-R. *Journal of Consulting and Clinical Psychology, 51*, 841–847.

Leli, D. A., & Filskov, S. B. (1979). Relationship of intelligence to education and occupation as signs of intellectual deterioration. *Journal of Consulting and Clinical Psychology, 47*, 702–707.

Lezak, M. D. (1983). *Neuropsychological assessment* (2nd ed.). New York: Oxford University Press.

Logue, P., & Wyrick, L. (1979). Initial validation of Russell's Revised Wechsler Memory Scale: A comparison of normal aging versus dementia. *Journal of Consulting and Clinical Psychology, 47*, 176–178.

Lubin, B., & Sokoloff, R. M. (1983). An update of the survey of training and internship programs in clinical neuropsychology. *Journal of Clinical Psychology, 39*, 149–152.

Luria, A. R. (1973). *The working brain.* New York: Basic Books.

Lyle, O., & Quast, W. (1976). The Bender Gestalt: Use of clinical judgment versus recall scores in prediction of Huntington's disease. *Journal of Consulting and Clinical Psychology, 44*, 229–232.

Maier, L. R. & Abidin, R. R. (1967). Validation attempt of Hovey's five-item MMPI index for central nervous system disorders. *Journal of Consulting Psychology, 31*, 542.

Mandleburg, I. A. (1976). Cognitive recovery after severe head injury. *Journal of Neurology, Neurosurgery and Psychiatry, 39*, 1001–1007.

Mandleburg, I. A. & Brooks, D. N. (1975). Cognitive recovery after severe head injury: Serial testing on the Wechsler Adult Intelligence Scale. *Journal of Neurology, Neurosurgery and Psychiatry, 38*, 1121–1126.

Marks, P. A., Seeman, W., & Haller, D. L. (1974). *The actuarial use of the MMPI with adolescents and adults.* Baltimore, MD: Williams & Wilkins.

Matarazzo, J. D. (1972). *Wechsler's measurement and appraisal of adult intelligence* (5th ed.). Baltimore, MD: Williams & Wilkins.

Matarazzo, R. G., & Matarazzo, J. D. (1984). Assessment of adult intelligence in clinical practice. In P. McReynolds and G. Chelune (Eds.), *Advances in Psychological Assessment* (Vol. 6, pp. 77–108) San Francisco: Jossey-Bass.

McGlone, J. (1977). Sex differences in the cerebral organization of verbal functions in patients with unilateral brain lesions. *Brain, 100*, 775–793.

McGlone, J. (1978). Sex differences in functional brain asymmetry. *Cortex, 14*, 122–128.

McGuire, F. (1960). A comparison of the Bender-Gestalt and flicker fusion as indicators of central nervous system involvement. *Journal of Clinical Psychology, 16*, 276–278.

Meier, M. J. (1981). Education for competency assurance in human neuropsychology: Antecedents, models, and directions. In S. B. Filskov & T. J. Boll (Eds.), *Handbook of clinical neuropsychology.* New York: Wiley.

Meier, M. J. & French, L. A. (1964). Caudality scale changes following unilateral temporal lobectomy. *Journal of Clinical Psychology, 20*, 464–467.

Mermelstein, J. J. (1983). A process approach to the Bender-Gestalt Test and its use in differentiating schizophrenic, brain-damaged, and medical patients. *Journal of Clinical Psychology, 39*, 173–182.

Milner, B. (1975). Psychological aspects of focal epilepsy and its neurosurgical management. *Advances in Neurology, 8*, 299–321.

Murray, H. A. (1938). *Explorations in personality.* New York: Oxford Press.

Nemec, R. E. (1978). Effects of controlled background interference on test performance by right and left hemiplegics. *Journal of Consulting and Clinical Psychology, 46,* 294–297.

Neuringer, C., Dombrowski, P. S., & Goldstein, G. (1975). Cross-validation of an MMPI scale for differential diagnosis of brain damage from schizophrenia. *Journal of Clinical Psychology, 31,* 208–271.

Neuringer, C., Goldstein, G., & Gallagher, R. B. (1975). Minimal field dependency and minimal brain dysfunction. *Journal of Consulting and Clinical Psychology, 43,* 20–21.

Noonberg, A. R., & Page, A. H. (1982). Graduate neuropsychology training: A later look. *Professional Psychology, 13,* 252–257.

O'Grady, K. E. (1983). A confirmatory maximum likelihood factor analysis of the WAIS-R. *Journal of Consulting and Clinical Psychology, 51,* 826–831.

Osmon, D. C. & Golden, C. J. (1978). Minnesota Multiphasic Personality Inventory correlates of neuropsychological deficits. *International Journal of Neuroscience, 8,* 112–122.

Parker, K. (1983). Factor analysis of the WAIS-R at nine age levels between 16 and 74 years. *Journal of Consulting and Clinical Psychology, 51,* 302–308.

Pascal, G. R., & Suttell, B. J. (1951). *The Bender-Gestalt Test: Quantification and validity for adults.* New York: Grune & Stratton.

Phares, E. J. (1979). *Clinical psychology: Concepts, methods and profession.* Homewood, IL: Dorsey Press.

Piotrowski, Z. (1937). The Rorschach inkblot method in organic disturbances of the central nervous system. *Journal of Nervous and Mental Disease, 86,* 525–537.

Prigatano, G. P. (1977). The Wechsler Memory Scale is a poor screening test for brain dysfunction. *Journal of Clinical Psychology, 33,* 772–777.

Prigatano, G. P. (1978). Wechsler Memory Scale: A selective review of the literature. *Journal of Clinical Psychology, 34,* 816–832.

Quadfasel, A. F. & Pruyser, P. W. (1955). Cognitive deficit in patients with psychomotor epilepsy. *Epilepsia, 4,* 80–90.

Rapaport, D., Gill, M. M., & Schafer, R. (1968). *Diagnostic psychological testing,* (Rev. ed. by R. R. Holt). New York: International Universities Press.

Reitan, R. M. (1954). The performance of aphasic, non-aphasic and control subjects on the Rorschach test. *Journal of General Psychology, 51,* 199–212.

Reitan, R. M. (1955a). Certain differential effects of left and right cerebral lesions in human adults. *Journal of Comparative and Physiological Psychology, 48,* 474–477.

Reitan, R. M. (1955b). Validity of Rorschach test as measure of psychological effects of brain damage. *A.M.A. Archives of Neurology and Psychiatry, 73,* 445–451.

Reitan, R. M. (1959). The comparative effects of brain damage on the Halstead Impairment Index and the Wechsler-Bellevue scale. *Journal of Clinical Psychology, 15,* 281–285.

Reitan, R. M. (1966). A research program on the psychological effects of brain lesions in human beings. In N. R. Ellis (Ed.), *International review of research in mental retardation* (Vol. 1). New York: Academic Press.

Reitan, R. M. (1976). Neurological and physiological bases of psychopathology. *Annual Review of Psychology, 27,* 189–216.

Rogers, D. L. & Swenson, W. M. (1975). Bender-Gestalt recall as a measure of memory versus distractibility. *Perceptual and Motor Skills, 40,* 919–922.

Rorschach, H. (1942). *Psychodiagnostics: A diagnostic test based on perception.* (P. Lemkau & B. Kronenberg, Trans.). New York: Grune & Stratton.

Rourke, B. P. (1982). Central processing deficiencies in children: Toward a developmental neuropsychological model. *Journal of Clinical Neuropsychology, 4,* 1–18.

Russell, E. W. (1972). WAIS factor analysis with brain-damaged subjects using criterion measures. *Journal of Consulting and Clinical Psychology, 39*, 133–139.

Russell, E. W. (1975a). A multiple scoring method for the assessment of complex memory functions. *Journal of Consulting and Clinical Psychology, 43*, 800–809.

Russell, E. W. (1975b). Validation of a brain-damage versus schizophrenic MMPI key. *Journal of Clinical Psychology, 31*, 659–651.

Russell, E. W. (1977). MMPI profiles of brain-damaged and schizophrenic subjects. *Journal of Clinical Psychology, 33*, 190–193.

Russell, E. W. (1979). Three patterns of brain damage on the WAIS. *Journal of Clinical Psychology, 35*, 611–620.

Russell, E. W. (1981). The pathology and clinical examination of memory. In S. B. Filskov & T. J. Boll (Eds.), *Handbook of clinical neuropsychology*. New York: Wiley.

Russell, E. W., Neuringer, C., & Goldstein, G. (1970). *Assessment of brain damage: A neuropsychological key approach*. New York: Wiley.

Saunders, D. R. (1961). Digit Span and alpha frequency: A cross-validation. *Journal of Clinical Psychology, 17*, 165–167.

Schraa, J. C., Jones, N. F., & Dirks, J. E. (1983). Bender-Gestalt recall: A review of the normative data and related issues. In J. N. Butcher & C. D. Spielberger (Eds.), *Advances in personality assessment* (Vol. 2). Hillsdale, NJ: Lawrence Erlbaum.

Shaffer, G. W. & Lazarus, R. S. (1952). *Fundamental concepts in clinical psychology*. New York: McGraw-Hill.

Shaw, D. J., & Matthews, C. G. (1965). Differential MMPI performance of brain-damaged versus pseudo-neurologic groups. *Journal of Clinical Psychology, 21*, 405–408.

Skilbeck, C. E., & Woods, R. T. (1980). The factorial structure of the Wechsler Memory Scale: Samples of neurological and psychogeriatric patients. *Journal of Clinical Neuropsychology, 2*, 293–300.

Spreen, O., & Benton, A. L. (1965). Comparative studies of some psychological tests for cerebral damage. *Journal of Nervous and Mental Disease, 141*, 323–333.

Stein, M. I. (1978). Thematic Apperception Test and related methods. In B. B. Wolman (Ed.), *Clinical diagnosis of mental disorders*. New York: Plenum Press.

Stevens, J. R., Milstein, J., & Goldstein, S. (1972). Psychometric test performance in relation to the psychopathology of epilepsy. *Archives of General Psychiatry, 26*, 532–538.

Stone, C. P., Girdner, J., & Albrecht, R. (1946). An alternate form of the Wechsler Memory Scale. *Journal of Psychology, 22*, 199–206.

Vega, A., & Parsons, O. A. (1967). Cross-validation of the Halstead-Reitan tests for brain damage. *Journal of Consulting Psychology, 31*, 619–625.

Vega, A., & Parsons, O. A. (1969). Relationship between sensorimotor deficits and WAIS Verbal and Performance scores in unilateral brain damage. *Cortex, 5*, 229–241.

Victor, M., Herman, K., & White, E. E. (1959). A psychological study of Wernike-Korsakoff syndrome. *Quarterly Journal of Studies on Alcohol, 20*, 467–479.

Vogt, A. T., & Heaton, R. K. (1977). Comparison of WAIS indices of cerebral dysfunction. *Perceptual and Motor Skills, 45*, 607–615.

Wade, T. C., & Baker, T. B. (1977). Opinions and use of psychological tests: A survey of clinical psychologists. *American Psychologist, 32*, 874–882.

Watson, C. G. (1971). An MMPI scale to separate brain-damaged from schizophrenic men. *Journal of Consulting and Clinical Psychology, 36*, 121–125.

Watson, C. G. (1972). Cross-validation of a WAIS sign developed to separate brain-damaged from schizophrenic patients. *Journal of Clinical Psychology, 28*, 66–67.

Wechsler, D. (1939). *The measurement of adult intelligence*. Baltimore, MD: Williams & Wilkins.

Wechsler, D. (1944). *The measurement of adult intelligence* (3rd Ed.). Baltimore, MD: Williams & Wilkins.
Wechsler, D. (1945). A standardized memory scale for clinical use. *Journal of Psychology, 19*, 87–95.
Wechsler, D. (1955). *Manual for the Wechsler Adult Intelligence Scale.* New York: Psychological Corporation.
Wechsler, D. (1958). *The measurement and appraisal of adult intelligence* (4th ed.). Baltimore, MD: Williams & Wilkins.
Wechsler, D. (1981). *Manual for the Wechsler Adult Intelligence Scale—Revised.* New York: Psychological Corporation.
Weingold, H. P., Dawson, J. G., & Kael, H. C. (1965). Further examination of Hovey's "index" for identification of brain lesions: Validation study. *Psychological Reports, 16*, 1098.
Williams, H. L. (1952). The development of a caudality scale for the MMPI. *Journal of Clinical Psychology, 8*, 293–297.
Williams, M. (1978). Clinical assessment of memory. In P. McReynolds (Ed.), *Advances in Psychological Assessment* (Vol. 4). San Francisco: Jossey-Bass.
Wilson, R. S., Rosenbaum, G., & Brown, G. (1979). The problem of premorbid intelligence in neuropsychological assessment. *Journal of Clinical Neuropsychology, 1*, 49–54.
Wilson, R. S., Rosenbaum, G., Brown, G., Rourke, D., Whitman, D., & Grisell, J. (1978). An index of premorbid intelligence. *Journal of Consulting and Clinical Psychology, 46*, 1554–1555.
Wooten, A. J. (1983). MMPI profiles among neuropsychology patients. *Journal of Clinical Psychology, 39*, 392–406.
Zaidel, D., & Sperry, R. W. (1974). Memory impairment after commissurotomy in man. *Brain, 97*, 263–272.

4

The Flexible Battery in Neuropsychological Assessment

HAROLD GOODGLASS

INTRODUCTION

This chapter describes an approach referred to as "The Flexible Battery," in which the examiner chooses test instruments on the basis of the patient's history and presenting symptoms, and on the basis of the outcome of any prior testing that may have been done. This can also be called "patient-centered testing." This approach to testing, which this author has held over the years, has only slightly been modified by the appearance of such packages as the Halstead-Reitan and the Luria-Nebraska Batteries. The rationale for this approach arises from the fact that observations of brain-injured patients over the last century have revealed deficits, often of a very specific nature, associated with focal lesions, as well as other deficits that are really quite nonspecific in type. For example, among the highly specific disorders are aphasic language disorders, disorders of facial recognition, disorders of purposeful movement, and hemispatial neglect. Among the nonfocal disorders are deficiencies in abstract thinking, slowness and stickiness of mentation. Memory disorders may, in some cases, be quite nonfocal

---

HAROLD GOODGLASS • Boston Veterans Administration Medical Center and Department of Neurology, Boston University School of Medicine, Boston, MA 02130. The preparation of this chapter was supported in part by the Medical Research Service of the Veterans Administration and in part by USPHS Grant NS 06209.

and, in other cases, rather localizing, depending on the characteristics of the memory disorder.

When we use these verbal tags, we have to be aware that they only approximate the true nature of the underlying process that may be impaired. These verbal tags are terms that are available in our vocabulary, that fit into our conceptual system, and that permit us to talk about our observations. However, one of the goals of neuropsychology is to reach successively better approximations of the true nature of the processes that we are observing and of their deficits, and gradually, to improve the repertory of terms that we have available to describe impaired functions. During this process, as we arrive at a redefinition of an observed deficit, we operationalize it in the form of a new or improved examination procedure, which comes closer to tapping the process as we now understand it. The last 20-odd years have seen the development of a large number of these specifically targeted tests, among them Benton's test of facial recognition (Benton & Van Allen, 1968) and Albert and Butters' (1979) test for a temporal gradient in retrograde amnesia. Among other procedures are tests of attention, such as Mirsky's Continuous Performance Test (Mirsky, Primac, Ajmone, Harsan, Rosvold, & Stevens, 1960; Mirsky & Van Buren, 1965), tests of spatial neglect, and so on. In the area of visual spatial function, there has been an effort to develop procedures that would distinguish between right- and left-hemisphere contributions to failure in this general area of performance. Much of this series of developments came out of the insights over the last 10 years from research on the performance of split-brain patients.

Among the innovations in specifically targeted test procedures in recent years, many have their origin in Luria's writings and are due to the vast inventiveness that he displayed in exploring the deficits of brain-injured patients. Many of these tests absolutely require attention to the qualitative as well as to the pass–fail aspect of the patient's performance.

In addition to tests that are specifically targeted to identify very well defined neuropsychological deficits, there are a number of widely used complex and multifactorial procedures, such as the Wechsler Intelligence Scale (1955) and the Wechsler Memory Scale (1945), whose scores do not usually denote a highly selective deficit but that have become so familiar that they are easily incorporated into the evaluation of the patient's functioning. These familiar multifactorial tests, such as the WAIS, can become anchor points of, and screening devices for, the choice of a follow-up battery of more selectively choosen and specifically targeted test procedures. In addition to these objective tests, there are a variety of additional procedures that cannot be precisely quantified and these, of course, involve the interview techniques and the observation that the examiner carries out. These may be aimed at probing for behavioral indexes of particular syndromes, as in the case of the patient who may have, or who clearly has, right hemisphere disease, where the examiner should be attuned to observe the emotional flattening that is commonly seen in these cases. In the case of temporal-lobe epilepsy, the examiner may probe for those personality changes that must be elicited by interview, such as a tendency towards hypergraphia, towards hypermorality, or changes in religious attitudes, which are associated with this disorder. So, given this picture of the ever developing status of neuropsychology, it would seem self-destructive for the field to freeze its procedures in accordance with a battery based on the thinking of the 1950s or even on the thinking of the 1970s.

It was therefore gratifying to note Golden's (1982) reference to the Luria-Nebraska Battery as something that is still evolving. It should be stressed, however, that the evolving nature of tests should not simply consist of juggling the method of statistical analysis or of making minor changes in the rules for administering the subtests. They should evolve also through radical changes or elimination of existing procedures or introduction of new ones on the basis of the findings from ongoing basic research.

At another level, there is the consideration of what it is about this field that makes neuropsychology an intellectually attractive and challenging profession. To the extent that the examiner works with a ready-made battery, which has as its goal the detection of organicity and which may be met by consulting norms, the examiner's function tends to be converted to that of a technician. To the extent that the examiner develops a particular battery on the basis of the medical history and combines qualitative observations, timed scores, and correct/fail information, the examiner is functioning more at a professional level. But, in order to function at such a level, the neuropsychologist must bring certain prior knowledge to the job. Although this requirement has been mentioned previously, it should be highlighted at this point. As one moves from being a general clinical psychologist, who has some exposure to problems of brain injury, to the stage of becoming competent in neuropsychology, it is essential to acquire a basic knowledge of brain anatomy and brain function, as well as of the gamut of clinical manifestations of various types of brain injury and recognized syndromes as they affect higher mental processes. Not only is it necessary to develop an internalized schema of the dimensions of cognitive changes related to brain injury, but by the time the neuropsychologist is ready to go into unsupervised practice, it is necessary to have a broad acquaintance with various typical forms taken by closed-head injury, by various types of cerebral accident, brain tumor, the dementing diseases, such as Alzheimer's, to mention only a part of the range. One has to know something about their underlying physiology, their typical course of development, and how their behavioral manifestations relate to the brain structures that are involved. With this kind of knowledge, one is not rooted to experience with a particular set of tests. As one of our colleagues jokingly remarked, a good neuropsychologist is one who can find himself on a desert island and make up an adequate neuropsychological examination with whatever sticks and objects he finds lying around. The point is that training in neuropsychology must begin, not with exposure to a workshop and a particular set of tests, but with firsthand exposure to the clinical manifestations of various organic syndromes as they affect cognition and clinical behavior. Obviously, this is not going to be done in a one or two week workshop. Much of the publicity promoting workshops for the various fixed batteries includes either an explicit or implicit statement that the participants will be acquiring competence in neuropsychology. This phenomenon is a response to the intense craving in the psychological community for a quick solution to the problem of gaining such competence. The unrealistic aspiration that appears to be rampant among many in our profession that they can with a very brief continuing-education exposure become competent in neuropsychology seems to have drawn a positive response from some administrators of psychology programs. They espouse the policy that the clinical psychologist is expected to be a totally rounded individual with equal competence in all specialties. Of course this is not a realistic goal.

The remainder of this chapter will be devoted to a discussion of the dimensions of cognitive impairment and the range of techniques at our disposal to explore these various dimensions. Reference to the dimensions of cognitive impairment is not identical with taking a syndrome approach. The intent is to convey that there are factors in cognitive performance distilled out of observations of the syndromes, as well as from observations of organically damaged patients who may not fit any of the well known or named syndromes. This approach contrasts with that of the analysis of "syndromes," which represent combinations of deficits in which one or more of the dimensions of cognitive ability may be impaired in extremely dramatic fashion, or in which one or more of these dimensions may be remarkably well preserved in the context of other impairments.

NONFOCAL COGNITIVE DEFICITS AND THEIR EVALUATION

To begin with, we will discuss those areas of deficit that are usually not selectively associated with a focal lesion in any region of the brain, but tend to be present to some degree no matter where the locus of brain injury is. Among these, however, are a number of features that have often been attributed to frontal-lobe damage. It is our experience that there is virtually no one of the so-called frontal-lobe signs that may not also occur with extensive lesions any place in the brain.

The first of these variables is the speed of mental operation during the maintenance of a simple response set. This might also be considered a measure of attention. Among the measures of this capacity, Mirsky's Continuous Performance Test is a remarkably clean measure because there is a minimal demand on skills relating to language, memory, or visual spatial ability. The only form in which memory is involved is that the patient is required to maintain his response orientation and to remember what the target response is. In the basic form of this examination, a series of letters appears in random order on a screen and the patient's task is simply to press a button each time a particular target letter, such as an X, appears. In a modified and more difficult version of the test, the subject responds on a contingency basis only when the target letter has been preceeded by another specified letter. But the beauty of the Continuous Performance Test is that it can be adapted to all kinds of modalities. Such adaptations include a tape recorded auditory version of this test and one in graphic form where the letters are displayed on a sheet of paper and the patient is required to cross out all of the instances of a specified target. In one written form of the examination, the patient may draw a continuous line from one target to the next one that the patient spots, leaving behind a trace of the pattern of scanning the page, which provides additional valuable information. This later adapation has been introduced by Dr. Edith Kaplan. Still another variant of this procedure that was introduced in our center was called "Auditory Trail Making." In this procedure a tape-recorded series of letters is presented in which the letters of the alphabet occur with various randomly interspersed letters, the patient's task being to respond by raising his finger each time he recognizes the next letter of the alphabet in the series of perceived letters.

If one moves up the scale to examinations that require a more complex response set there is again a wide choice of procedures. These procedures are useful because they open up the possibility of observing error patterns. Particularily in the case of frontal-lobe-damaged patients, one has the opportunity of observing what we call "stimulus bound" behavior.

The most familiar of the procedures that involve a complex stimulus set in continuous performance is the Digit Symbol substitution test of the Wechsler examination as well as Smith's version of this procedure, the Symbol Digit Modalities Test (1973). Another test in this category is the multiple-loop test, in which the patient is asked to produce a series of multi-looped figures resembling a "3," but with three instead of two loops, and to continue drawing these figures across the page. The task in this instance involves resisting the pull to continue adding loops to each figure as a result of a tendency to motor perseveration. Motor perseveration that leads to the addition of one or more surplus loops in this procedure is characteristic of frontal-lobe patients. It is commonly seen in the writing of right frontal-lobe patients who add extra loops to any ms or ns that they encounter in their writing as well as reduplicating recurring portions of other letters of the alphabet. This, of course, is a type of writing difficulty that is not necessarily associated with aphasia at all and should not be confused with an aphasic agraphia.

There is another class of procedures for examining the nonspecific effects of brain injury in which the examiner does not define the response set that the patient must maintain. The easiest to give, and one that certainly should be a part of any examination procedure is what this author and colleagues have called "wordlist generation" and that in other settings has been called "word fluency." It is not clear whether this procedure predated the Binet examination but it certainly entered the test repertory early in the history of mental testing, as it appears at the 10-year level in the Binet examination (Terman & Merrill, 1937), where the child is asked to name all the words he can think of. An alternate form of this procedure was to have them list all the animals they could think of. This procedure was subsequently picked up by Spreen and Benton (1969) as the "Word Fluency Test," in which patients were asked to list all the words they could that begin with the letters F, A, and S, for one minute each. Word-list generation is extremely sensitive to frontal lobe disorders, we believe, because it requires patients to carry out a self-initiated search of their memory and, having run out of one list of associations, to shift to another line of associations. Self-initiation of successively changing mental sets is a type of cognitive operation that appears to be particularly vulnerable to frontal lobe damage, although it is not exempt from impairment by other lesions as well.

Another procedure related to the foregoing one is the well known Wisconsin Card Sorting Test. For those readers who are not familiar with the procedure, the examiner has a deck of cards on which there appear geometric forms that vary in number from one to four, that vary in shape among triangles, crosses, circles, and stars, and that vary in color among green, yellow, red, and blue. The patient is required to sort each successively presented card according to the criterion of color, number, or form, using only the examiner's indication of right or wrong as a basis for arriving at a sorting criterion. Patients must also be ready to hunt for another criterion when after a series of correct assignments one is suddenly told that a response is no longer correct. One of the reasons why this and the preceding procedures appear particularly sensitive to frontal-lobe injury is that these patients often seem relatively intact otherwise, because injuries confined to prefrontal areas may produce no conspicuous defects in language or memory or visual-spatial skills at an elementary level.

We now proceed to discuss the problem of memory disorder. Memory impairment of some type is almost universal after brain injury and, in this sense, can be regarded as a nonspecific indicator of brain damage. However, it should be remembered that there are a number of syndromes involving primarily subcortical damage, such as the injury of the mamillary bodies and the dorsomedial thalamic nuclei associated with Korsakoff's disease, or bilateral damage to the hippocampi, which is an occasional by-product of closed head injury. Both of these injuries produce selective damage to recent memory, often referred to as "amnesic syndromes." The amnesic syndromes are different in character from the nonspecific memory disorders, both because of their severity and because of their relative restriction to a time span extending from a few minutes before the period of testing to several years. Because of the dramatic nature of these disorders, which may be very severe and unaccompanied by other cognitive damage, they have features that identify them as among the focal rather than among the nonspecific and nonlocalizing deficits. One of the features that is valuable in differentiating between the nonspecific and the highly focal memory disorders is the shape of the retrograde amnesia curve. Albert and Butters (1979) have introduced a test that demonstrates that Korsakoff patients have a retrograde amnesia gradient that becomes milder and milder with the increasing number of years prior to the time of the onset of the illness. However, beyond this factor it must be emphasized that there are critical aspects of memory that are impossible to determine with any kind of a preset battery. These factors depend on the examiner's ability to improvise at the moment of the interview. There are no objective test items in any memory test that serve to determine the extent of a retrograde amnesia preceding trauma or preceding a memory disorder of acute onset. Here the questions must be tailored by the examiner to the patient's own history of ongoing events just prior to the injury or to using public information that was current at the time of the patient's injury.

Now there is no lack of memory tests available to the neuropsychologist who wishes to make a choice among them. The Wechsler Memory Scale is certainly useful, even though it is not as sensitive as one would like to some features such as recent and remote memory.

The Rey Auditory Verbal Learning Test (1964) is a type of super-span test with repeated administrations that is very sensitive to disorders of new verbal learning. Of course one cannot rely on tests of verbal learning alone because there are patients who have been well described who have impairment in verbal learning but not in visual memory and vice versa. Consequently, any examination for memory that pretends to be at all thorough must have such tests of visual memory as the designs of the Wechsler Memory Scale or of the Benton Visual Retention Test (1963). An additional 20-minute delay condition for the reproduction of the Rey figure has been incorporated as a result of Edith Kaplan's initiative.

A discussion of the nonspecific aftereffects of brain injury would be incomplete without reference to impairment of abstract thinking. This is of course represented in the Halstead-Reitan battery by the Category test and is undoubtedly the factor that makes this test so sensitive to the effects of brain damage. There are alternative procedures available to psychologists, however. The Wechsler Adult Intelligence Scale provides both verbal similarities and proverbs. Proverbs are a more rigorous test of abstracting ability but, unfortunately, the Wechsler Scale has only a small sampling for proverb interpretation. Those examiners who have not developed their own set of proverbs have available the Gorham multiple choice proverb interpretation test (1956), which is of some value.

FOCAL COGNITIVE DEFICITS AND THEIR EVALUATION

We come now to the problem of focal deficits in the evaluation of the effects of brain injury and, in particular, the examination of these deficits from the point of view of the patient-centered approach. The first area to be discussed is that of constructional disorders. Quite a few years back, the term *constructional apraxia* was invented to refer to disordered execution of visual-spatial tasks that involved a motor component. This, for example, would be manifested in carrying out geometrical block constructions, stick designs, and paper-and-pencil drawing of geometric shapes or pictures. The word "apraxia" was borrowed by analogy with the disorders of purposeful movement of a nonverbal nature that are manifested in the inability to carry out pretended actions on command. In fact, however, there is no connection between constructional apraxia and an apraxia of the limbs or the face. It should be noted that the most dramatic instances of constructional apraxia occur in the case of patients with right-brain injury whereas the occurrence of limb apraxia or facial apraxia is virtually entirely restricted to patients with left-brain injury. In fact, it is rarely observed in patients who are not also aphasic. A consideration to bear in mind in the evaluation of constructional disorders is that quantitative scoring has very little localizing value except in cases of very extreme disorganization. It is in cases of extreme disorganization of visual-spatial performance that the right-parietal lobe is almost always implicated. However, this is not the case in milder degrees of constructional difficulty. Outside of those extremely low-scoring individuals whose performance is indicative of right-parietal damage, quantitative scores that indicate deficits of a milder degree are of little more value than as indicators of the presence of probable organic brain damage. The most fruitful basis for scoring in cases of constructional difficulty is based on the character of the errors. The scoring of qualitative characteristics of errors is not necessarily subjective because the features to be identified may be defined quite clearly and tallied numerically.

The first major feature, of course, is unilateral neglect of the side of the visual field opposite the lesion as expressed by partial or total omission of details on the left side of the drawing. Unilateral neglect is common in patients who have suffered a right-hemisphere injury. In a typical performance by a patient with a right-hemisphere lesion, we find that in the drawing of a flower to copy there is a remarkable omission of the petals on the left hand side of the flower. On the other hand, when the patient draws a flower to command, without reference to the model on the page, the drawing may be quite complete, showing no signs of neglect. This dissociation between drawing to copy and drawing to command is often seen, the right-brain-damaged patient demonstrating more severe neglect when drawing to copy than in drawing to command. However, some patients will neglect the left side of space even in their spontaneous drawing. Obviously, it would be a horrendous loss of information if the drawing were simply scored quantitatively for its degree of accuracy without a notation of the presence of neglect on the left side. We would have lost an extremely powerful indicator of right-hemisphere damage and, usually, right-parietal-lobe damage. Unilateral spatial neglect, although most often seen with right-brain damage, where it is manifested by neglect or omissions in the left-visual field, may also occur with left-brain damage. In this case, neglect is observed in the right-visual field. However, instances of neglect on the right are less frequent and the manifestations are less dramatic than is the case for the other side.

Whereas unilateral neglect, when it does occur, is an extremely powerful indicator, there are other indexes of laterality of lesions that produce constructional deficiencies that are useful though less obvious. These are procedures that also take more effort on the part of the examiner to categorize, and require counting instances of errors of various types. One of these indexes is the side on which errors occur. This index has clearly been demonstrated to apply in the case of the Wechsler Block Design, where patients with unilateral brain damage typically produce more misplacements of blocks, whether these misplacements are self-corrected or not, on the side of the design opposite the lesion. Similar effects can be observed in the drawing of designs and the building of stick constructions. In the case of block designs, the application of this method requires a careful recording of the location of misplacements of blocks during the performance of each of the designs.

A second feature that helps in categorizing errors that have lateralizing implications is the direction of procedure from left to right or right to left followed by the patient. The patient with left-brain injury more typically begins a construction on the left and proceeds in a left-to-right direction in the construction of block designs. Although reversal of direction is not universal in right-brain-injured patients, a reversal of direction is a significant feature that should be noted. Another feature in the construction of block designs is the frequency with which the patient's block placement ignores or violates the square configuration of the design. The loss of the square configuration is overwhelmingly more frequent in patients with right-brain injury than with left and this may be observed even on designs that are not extremely difficult. The breaking of the configuration is somewhat less significant when it occurs on very difficult designs, where subjects who are not brain injured may momentarily lose the sense of the two-by-two or three-by-three pattern of the design.

Whereas the performances reviewed up to this moment are in the realm of constructional difficulties, there is another class of visual-spatial problems that are essentially perceptual in nature. Here we discuss a variable in visual-spatial perception that has arisen out of basic research on cerebral dominance and it relates to the variable of the verbal codability or absence of verbal codability in the percept. We will see that there are visual-spatial configurations that, though not obviously namable, are easily coded with a tag to which a verbal label may be assigned. In one such study, patients were required to identify two configurations as target items out of a set of four. Among the group from which the patient had to identify the targets were horizontal orientation, vertical orientation, diagonal to the right and diagonal to the left. When subjects were presented with one of the two targets, for instance, a right diagonal or horizontal line, they proved to be more proficient in responding to them correctly when they were presented in the right visual field than in the left. On the other hand, when the set of stimuli were composed of diagonals that were placed at intermediate angles, for which a verbal label was not readily available, the targets were much more frequently correctly identified when presented in the left visual field than in the right. The variable in this case is whether the visual spatial orientation is one that can easily be tagged as upright, horizontal, left or right, or whether it is one that cannot be assigned a conceptual tag that is so easily categorized. In the latter case, the processing is carried out more efficiently by the left-visual field, which represents the activity of the right hemisphere. This line of research has taught us that just any old test of perception of line orientation is not necessarily going to be effective as a lateralizing indicator, because those particular items that are going to be discriminating for presence or absence of right-brain disease must be items that are not readily codable in terms of orientation that is easy to label.

This is an excellent example of the interface between basic research and clinical test construction. The Benton Line Orientation Test (Benton, Hannay, & Varney, 1975) is a product of this type of research. In this test the patient is presented with a card on which one sees two lines radiating at different angles from a center point and one must attempt to remember the angles of those two lines in order to select them from a multiple choice card in which a set of 11 lines are seen radiating at different angles from the center point. The arbitrary angles that cannot be named and which must be remembered in a purely visual-spatial sense make this task particularly sensitive to right-brain injury, rather than left.

Another type of complex visual stimulus that provides laterizing information on the basis of purely perceptual factors is that of face recognition. Benton and Van Allen (1969) as well as Milner have devised facial memory tests in which target faces are to be matched in a multiple-choice array of other unfamiliar faces. In these tests it is important to recognize that the familiarity of the face is a major feature in its effectiveness as a lateralizing indicator. Here again, basic research in cerebral dominance has shown that highly familiar faces are recognized better in the right-visual field, through the action of the left hemisphere. On the other hand, the matching of unfamiliar faces is very strongly a right-hemisphere, left-visual-field function.

The problem of face recognition is an illustration of a very fundamental problem in clinical neuropsychology. This problem arises in the relation between prosopagnosia and disorders in matching of unfamiliar faces. Prosopagnosia, as many readers know, is a highly selective disorder in which patients are unable to recognize familiar faces at all and in severe case may not even recognize themselves in a mirror, may not be able to recognize the members of their family, or occasionally not even be able to tell a male from a female face. This may exist as an isolated deficit. However, prosopagnosia is not simply a more severe form of a facial recognition disorder of this same type that is picked up within patients who have right-hemisphere disease. In the first place, prosopagnosia is a disorder involving the recognition of familiar faces. Second, patients with prosopagnosia may perform normally in the matching of unfamiliar faces, as on the Benton experiment. Further, prosopagnosia is almost invariably a bilateral disease of the occipital visual association areas. Although a number of cases of unilateral right occipital-lobe lesions have been identified by CT scan, all of those that have eventually come to autopsy so far have proved to have an additional lesion in the left-occipital lobe as well. Damage to the lateral cortex of the right-hemisphere that results in impaired performance in the matching of unfamiliar faces does not produce prosopagnosia at all. The lesson we learn from this is to avoid the assumption that a severe form of a symtom is merely a stepped up version of more subtle difficulties, and that such subtle problems can be measured by more stringent testing of the task in question. This is usually an incorrect assumption. We may be dealing, as we are in the case of facial recognition, with two entirely different phenomena.

Language disorders in the form of aphasia are obviously a major diagnostic problem in clinical neuropsychology, but the full treatment of aphasia is too specialized a topic to be handled in much detail here. The reader is referred to Chapter 11: "The Examination of Aphasia." It should be emphasized that there is a great deal to be learned about the site of lesions within the language zone of the left hemisphere when one attends to the appropriate dimensions of spoken or written language. In listening to the speech of a patient who is known to have or who is suspected of having aphasia, the examiner must learn to attend to such features as the effort and accuracy of articulation, the fluency in production of sequences of spoken words, the quality of the prosodic pattern of speech, the presence of selective difficulties in word finding, in auditory comprehension, in repetition, in the usage of grammatical automatisms of ordinary speech, and in a number of specific features or reading and writing. The adequate interpretation of aphasic symptoms for purposes of lesion localization and recommendation of treatment requires more specific direct exposure to clinical experience with aphasic patients than can be conveyed in a short article. The screening tests for aphasia that are included in both the Luria-Nebraska examination and the Halstead-Reitan examinations are fairly adequate for determining the presence or absence of a significant degree of impairment.

Finally, a word on the diagnosis of apraxia. Apraxia may be considered to be a form of motor disorder but it is a motor disorder of a higher intergrated level, quite different in nature from disorders of strength and coordination. Disorders of strength and coordination are produced by lesions of the hemisphere opposite the affected side, whereas disorders of praxis, which involve the ability to carry out representational movements on request, are bilateral. These disorders are produced by a lesion of the left hemisphere in almost all cases. In fact, in the case of patients who have hemiplegia, the presence of apraxia must be detected by the examination of the nonparalyzed limb, because the patient can rarely be expected to carry out pretended movements with the paralyzed arm. The examination for disorders of praxis is one that is not well known to most neuropsychologists and is not very adequately represented in any of the standard batteries. It is necessary to examine voluntary purposeful movements of the face and respiratory apparatus, such as coughing, blowing, licking the lips, and pretending to sip through a straw, as well as purposeful movements of the limbs, including conventionalized gestures like saluting, waving good bye, and pretended manipulation of objects as in hammering, brushing teeth, or combing one's hair. This disorder represents one that is not only strongly lateralizing in terms of providing evidence of a lesion site, but is also of the highest importance in evaluating a patient's handicap and rehabilitation potential. It is not one that is easy to quantify and hence plays little role in the standard neuropsychological fixed-test batteries.

In summary, the arguments in favor of adopting the flexible approach to the assessment of neuropsychological problems in brain-injured patients and the major dimensions of cognitive impairment and the tools that we have available at the present time to probe these deficits have been reviewed. Although this presentation has been framed in terms of a controversial position between the fixed-battery approach and the flexible battery, it is recognized that there are conditions under which the choice of a fixed battery is the most economical and appropriate direction to go. One such work situation is where there are large numbers of patients on whom there is little prior information available and whom it is necessary to screen for the presence or absence of signs of organic brain damage. Further, in the situation of the clinical psychologist who has little special background in neuropsychology, it is valuable to have an instrument at hand that will help to determine the presence and the extent of organic involvement. Clinical psychologists trained in the 40s and 50s will remember the day when we counted Piotrowski organic signs on the Rorschach. Certainly, this is something one has to do from time to time when more detailed training and test materials are not available.

However, it should be emphasized that when one moves in the direction of partial or full competence in clinical neuropsychology, the place to begin is with hands-on training in the variety of manifestations of organic psychopathology and in the acquaintance with tests that are designed to probe for specific aspects of cognitive deficits following brain injury. This is a field that is still evolving and it is incumbent on the practicing neuropsychologist to keep one's procedures current with ongoing research and to be open to new procedures and modifications of existing ones as we learn more about the field.

REFERENCES

Albert, M. S., Butters, N. M., & Levin, J. (1979). Temporal gradients in the retrograde amnesia of patients with alcoholic Korsakoff's disease. *Archives of Neurology, 36,* 211–216.

Benton, A. L. (1963). *The Revised Visual Retention Test.* New York: Psychological Corporation.

Benton, A. L., Hannay, H. J., & Varney, N. R. (1975). Visual perception of line direction in patients with unilateral brain disease. *Neurology, 25,* 907–910.

Benton, A. L., & Van Allen, M. W. (1968). Impairment in facial recognition in patients with cerebral disease. *Cortex, 4,* 344–358.

Golden, C. J. (1982). *The Luria-Nebraska Neuropsychological Battery.* Paper presented at the Clinical Application of the Halstead-Reitan and Luria Batteries Conference, Northport, New York.

Gorham, D. R. (1956). A Proverbs Test for clinical and experimental use. *Psychological Reports Monograph Supplement, 1*(2), 1–12.

Mirsky, A. F., & Van Buren, J. M. (1965). On the nature of "absence" in centrencephalic epilepsy. A study of some behavioral, electroencephalographic and autonomic factors. *Electroencephalography and Clinical Neurophysiology, 18,* 334–348.

Mirsky, A. F., Primac, D. W., Ajmone Marsan, C., Rosvold, H. E., & Stevens, J. R. (1960). A comparison of the psychological test performance of patients with focal and nonfocal epilepsy. *Experimental Neurology, 2,* 75–89.

Rey, A. (1964). *L'examen Clinique en Psychologie.* Paris: Presses Universitaires de France.

Smith, A. (1973). *The Symbol Digit Modalities Test.* Los Angeles: Western Psychological Services.

Spreen, O., & Benton, A. L. (1969). *The Neurosensory Center Comprehensive Examination for Aphasia.* Victoria, BC: University of Victoria, Neuropsychology Laboratory.

Terman, L. M., & Merrill, M. A. (1937). *Measuring intelligence.* Boston: Houghton Mifflin.

Wechsler, D. (1945). A standardized memory scale for clinical use. *Journal of Psychology, 19,* 87–95.

Wechsler, D. (1955). *The Wechsler Adult Intelligence Scale.* New York: Psychological Corporation.

5

Neuropsychological Batteries

RALPH E. TARTER and KATHLEEN L. EDWARDS

INTRODUCTION

The development and use of neuropsychological test batteries has progressed in tandem with our understanding of the functional organization of the brain. Initially, psychometric tests were employed to diagnose the presence or absence of brain pathology and, as such, were adjunctive to the clinical neurological examination and laboratory indexes of disturbance. Usually, the assessment of organicity entailed administering one or several tests of often questionable validity or lacking in quantification. For example, the Bender Visual Motor Gestalt Test has, for many years, been the most frequently employed test of organicity; however, depending on the idiosyncratic preferences of the particular clinician, other tests were also commonly utilized.

The single test approach, regardless of the specific instrument employed, has a number of limiting characteristics. First, depending on the specific test, this approach taps only one or at best a few psychological processes. Therefore, it is possible that the particular psychological process being measured is insensitive to the effects of the brain pathology for any given individual. Second, the use of a single test assumes that brain pathology will be reflected uniformly on measures of perception, memory, motor capacity, and language, regardless of the severity and location of the lesion. Third, obtaining only a single (or a

---

RALPH E. TARTER and KATHLEEN L. EDWARDS • Department of Psychiatry, University of Pittsburgh School of Medicine, Pittsburgh, PA 15213.

few) indexes of functioning prevents the opportunity to profile cognitive strengths and weaknesses, information that is critical for implementing comprehensive rehabilitation.

The conceptual and methodological weaknesses of the single-test assessment approach, combined with a rapid increase in our understanding of brain–behavior relationships, led to a recognition of the need for more extensive or inclusive cognitive measures. Although systematic clinical evaluation had been ongoing in the Soviet Union for several decades (Luria, 1966), the adoption of multiple measures in neuropsychological assessment came into being in only the 1950s in the United States. Without doubt, the major innovative advance came about through the efforts of Ralph Reitan, who was primarily responsible for expanding and validating Halstead's tests of biological intelligence to form the Halstead-Reitan Neuropsychological Battery (1955). Since then, clinical neuropsychological assessment procedures have rapidly evolved so as to encompass the measurement of brain-related behavioral changes associated with normal development and aging, and pathological conditions occurring during virtually any phase of the life cycle. These advances in our understanding of brain–behavior relationships were, however, catalyzed by an awareness of the need to adopt a multivariate assessment perspective that fully acknowledges the complexity of brain organization, that at the operational level, involved developing aggregates of tests, or a "battery," in order to elucidate the association between brain and behavioral processes.

CHARACTERISTICS OF TEST BATTERIES

A behavioral event is the endpoint in a chain of numerous complex biological events. Hence, a given behavioral event (e.g., motor speed) can be deleteriously affected by a variety of disturbed biological processes. Also, from this it follows that variations in performance on any given task (e.g., figure copying) can occur as the result of dysfunction to disparate neural systems. Thus, any battery of validated tests, regardless of the type and number of measures included, reflects the nonspecific effects of cerebral pathology; that is, shared variance among the tests, and depending on the particular battery utilized, the information obtained from the various tests will have particular sensitivity for detecting a localized lesion. By functionally analyzing the quantitative and qualitative aspects of performance, and from understanding the association between the neural substrate and the behavior it subserves directly and indirectly, it is possible to make valid inferences about the location of cerebral lesions using a battery of neuropsychological tests. Through careful inspection of the profile of scores in a test battery, it is possible to accurately infer lesion lateralization and localization. To do so, however, requires an understanding of the neural mechanisms by which the behaviors being measured are subserved, the method of behavioral assessment, and an undertanding of the functional organization of psychological processes in the brain.

As the number and diversity of behaviors being measured by a battery are increased, so too is the capacity to validly detect a localized lesion. Whereas single measures may be sensitive to detecting the presence of pathology or organicity, they are not capable of determining if one brain region is pathological with respect to normal functioning in other brain areas. An issue of paramount practical importance, therefore, is the optimal number of tests that are required to comprise a comprehensive battery; that is, a battery that enables the differential assessment of the functional status of the different brain systems and regions. Unfortunately, there is no consensus or universally accepted solution to this problem. Obviously, constraints in the breadth and, hence, duration of the test battery depend on the characteristics of the client population, available assessment resources and support staff, caseload volume, and objectives of the evaluation. The point to be made is that a battery approach to testing is a requisite for lesion localization and for profiling cognitive processes, but because of the substantial shared variance among measures, it is only through an appreciation of the relationship among the multiple scores that a pathological cerebral region or system can be implicated.

TYPES OF TEST BATTERIES

There are two basic types of test batteries. The first type is intermodal or multifunction. These batteries assess a variety of different modalities or psychological processes, including visual, haptic and auditory perception, memory, language, spatial, and psychomotor capacities. They are, in effect, pandemic approaches to neuropsychological assessment. The most widely known and utilized batteries of this type are the Halstead-Reitan and Luria-Nebraska neuropsychological batteries.

The second type of battery has a more restricted measurement focus. These batteries are intramodal, and attempt to comprehensively assess either one modality or one particular category of psychological functioning. For example, the Bruninks-Oseretsky Test of Motor Proficiency (Bruninks, 1978) measures diverse aspects of motor development, whereas the Boston Diagnostic Aphasia Examination (Goodglass

& Kaplan, 1972) evaluates language capacities. Numerous intramodal test batteries have been developed for measuring perception, memory, and language processes, although, with the possible exception of assessing aphasia, they have not been widely adopted for general clinical use by neuropsychologists. Also, most such tests are less than satisfactory. For example, in relation to what is known in the experimental psychology literature about the mechanisms and parameters involved in information storage and retrieval, there is unfortunately presently not available a comprehensive test of learning and memory that is suitable for broad clinical application. Given the need to identify aspects of neuropsychological functioning that are specific enough to target systematic rehabilitation efforts, it is argued that the comprehensive intramodal type of assessment is of primary importance. Whereas the intermodal or general examination may reveal areas of impairment, further intensive intramodal or categorical assessment of particular psychological processes is necessary for implementing targeted interventions. A procedure whereby both broad spectrum or intermodal assessment and intramodal or categorical assessment can be integrated into a comprehensive assessment is discussed in some detail in a subsequent section of this chapter.

THEORETICAL BIAS IN BATTERY DEVELOPMENT

In assembling a neuropsychological test battery, several factors, stemming primarily from the theoretical position of the test developer, will determine the ultimate composition of the battery and the type of information obtained. A test battery can be devised based on the conceptualization or theoretical model of brain functioning advocated by the test developers. For example, the Bender-Gestalt Test (Bender, 1938) and the Goldstein-Sheerer Object Sorting Test (Goldstein & Sheerer, 1941) measure behaviors such as visual-perceptual integration and abstract reasoning. These psychological processes are of cardinal importance to adherants of the gestalt or equipotential model of functional brain organization. The proponents of this model, most notably Kurt Goldstein (1939), and Karl Lashley (1931) have argued that psychological processes are not localizable and that all psychological processes equally depend on the whole cortex.

Even though the gestalt or equipotential brain model has been seriously challenged, and is today not commonly accepted, the point to be made is that the nature of the behavioral measurements comprising a battery reflect the developer's theoretical orientation. So it is also with

Luria's (1966) examination procedure, which is best described as a regional equipotential model, being a compromise between Pavlov's connectionist (S-R) model and the equipotential model (gestalt) of brain functional organization. Luria's examination is based on the assumption that there are specific brain regions, histologically differentiated into primary, secondary, and tertiary areas. Thus, according to Luria, there is a high degree of specificity between the locus of a brain lesion and type of behavioral deficit. In devising, selecting, and utilizing behavioral measures, Luria was guided by this neo-Pavlovian conceptualization of brain organization in conducting clinical examinations of patients who mostly suffered from traumatic brain injuries. An attempt to systematize this evaluation was first made by Christensen (1979), whose test format was subsequently quantified and standardized by Golden and colleagues (Golden, Hammike, & Purisch, 1980). The culmination of this effort, the Luria-Nebraska Battery, reflects an attempt to satisfy the requirements of a psychometrically sound instrument that attempts to be theoretically consistent with Luria's model of brain functional organization.

Perhaps the least theoretical battery in common use is the Halstead-Reitan Neuropsychological Battery. This composite of tests consists of a variety of measures, some of which were originally developed for purposes other than lesion identification and localization, but through systematic validation studies have been found to have diagnostic value. For example, the Tactual Performance test is essentially the Seguin formboard that was initially developed to measure intelligence. The Seashore Rhythms test was originally developed to assess musical aptitude. The Weschler scales, which are psychometric measures of intelligence, were similarly found to contain information that has value in localizing and lateralizing cerebral lesions. The point to be made is that Reitan's approach to battery construction was guided foremost by empirical considerations, which, other than accepting a localization perspective of brain functional organization, made no additional assumptions about how psychological processes are cortically organized.

Thus, the process of development of neuropsychological test batteries, though meeting the psychometric criteria of reliability and validity, additionally is influenced by the developer's conceptual model of brain functional organization. This latter factor determines both *what* is to be measured and *how* the measurement is conducted. As will be seen shortly these two aspects of the assessment process are at the center of controversy surrounding the method and purpose of neuropsychological evaluation.

What Should Neuropsychological Batteries Assess?

Neuropsychological tests, as measures of cerebral integrity, are sensitive to pathology that is either diffuse or focalized. Depending on the information required, a battery can be composited to emphasize one or the other, or attempt a balance between focal and diffuse pathology measurement. The major obstacle is the number of tests needed and time available to the clinician, because the diagnosis of focal pathology can be arduous and requires an extensive battery, such as those discussed in other chapters of this volume.

A variety of brief tests are available that are sensitive to detecting the presence of brain pathology without regard to lesion location. Among the most commonly used tests are the Trailmaking Test (Reitan, 1958), Symbol Digit Modalities Test (Smith, 1973), Benton Visual Retention Test (Benton, 1974), and Purdue Pegboard Test (Purdue Research Foundation, 1948). By themselves, these latter tests cannot implicate focal pathology, but do yield information about overall integrity of cerebral functioning. Such information is, however, of particular value for clarifying the effects of systemic diseases (e.g., hepatic encephalopathy, uremia, etc.), or generalized conditions (e.g., hypoxia, dementia, etc.).

Tests sensitive to detecting lateralized pathology are of particular value in situations where there is acute neurological disorder (e.g., tumor, stroke). Motor and sensory tests, in addition to their brevity, are highly sensitive indicators of lateralized pathology. Laboratories equipped with a tachistoscope can also make exceptionally accurate diagnosis employing controlled visual stimulus input from the different visual fields. A battery that is balanced between verbal and nonverbal types of tests is, of course, a requisite for profiling cognitive status related to known or suspected lateralized cerebral pathology.

Including tests that are sensitive to focal pathology is a high-risk, high-gain strategy, if used in an initial screening battery. Such tests are, however, an essential ingredient in the comprehensive battery. With respect to screening batteries, the time required to administer such tests may be offset by the low frequency of hits. A systematic and economical evaluation process, therefore, would best proceed from the detection of cerebral disturbance irrespective of location, to the identification of a lateralized lesion and finally to the diagnosis, if possible, of a focal lesion.

In specifying the objectives of a neuropsychological battery, it is also important to consider specialized assessment objectives and the characteristics of the clientele. For example, a battery designed to measure the effects of frontal lobotomy has a different aim than a battery for evaluating brain dysfunction in a learning-disabled population. Assessment information should be linked to some intervention goal, diagnosis, or otherwise enhance our understanding of brain functional organization in the context of research inquiry.

Uniform and Flexible Batteries

An important theoretical issue concerns whether a fixed battery (the same tests are administered to all persons) or flexible battery (different tests to different individuals), approach is the preferred assessment strategy. Each of these two strategies has its merits and its limitations.

A flexible battery maximizes information accrual for clinical diagnosis, because whatever tests need to be administered are administered. On a case by case basis, no boundaries on the assessment process are imposed. Although there are obviously advantages in clinical work, this approach does not afford the opportunity to develop a systematic or unified data management system or a data base. This limitation inherently prevents clinicians from assessing their own clinical activities in the context of program evaluation, and also prevents the opportunity to acquire systematic data on persons with rarely occurring disorders (e.g., agenesis of the corpus callosum, Capgras Syndrome, etc.). Also, it is not possible to retrospectively analyze data (e.g., drug studies, treatment effects, etc.), that could serve as pilot data for undertaking prospective research, and for monitoring the impacts of the assessment process itself. In the flexible or idiosyncratic strategy, there is substantial intersubject variability in the assessment protocol, hence, systematic examination of group data is at best difficult and likely impossible. Perhaps more than any other subspecialty of applied psychology, the clinical neuropsychologist is trained within a well grounded empirical tradition in neuroscience and behavioral science, and for this reason, it can be argued that there is a professional mandate to conduct empirically sound clinical studies, as well as to evaluate the impact of neuropsychological assessment in the context of evaluation research, and the overall clinical process. Such is not possible if each client is administered a completely individualized battery of tests.

Since the advent of microcomputers, it is possible in virtually all clinical settings to establish a database management system. However, although affording the opportunity to conduct research, a rigid battery may not always be applicable for a particular patient. Also, once a clinician is familiar with and committed to a certain battery, there tends to be a reluctance to adopt new test procedures, because to do so effectively compromises the objectives and value of a uniform battery. Hence, a rigid battery may unwittingly contribute to and sustain obsolescence.

A DECISION APPROACH TO SYSTEMATIC ASSESSMENT

There is no simple resolution to the problem. The selection of either a flexible or uniform battery ultimately depends on the relative emphasis given to clinical and research activities. One compromise strategy is to adopt a uniform or standard battery that addresses core elements of a neuropsychological examination (attention, memory, perception, language, visual-spatial, and psychomotor processes) employing brief tests that, depending on the results obtained, would suggest additional tests for more individualized assessment. The second step, though not involving all clients, would nonetheless entail the administration of a standard subbattery to certain persons. For example, if the uniform initial assessment that is given to all individuals points to a language disorder, than the second step of the evaluation would require administration of a standard subbattery, such as the Boston Diagnostic Aphasia Examination. At the second stage of the assessment, therefore, there is attrition of the client pool because not all persons may need further examination, and where it is required, results in some intersubject variability with respect to the type of tests administered. Hence, some clients would receive a standard memory subbattery, whereas others would be tested more fully on other processes, such as language, psychomotor, or visual-spatial processes. However, the point to be recognized is that even at this second stage of the assessment, there still remains the opportunity to preserve the development and maintenance of a systematic database. If additional information is still required, than further testing can be conducted within a highly individualized format employing specialized tests.

This three-stage approach to assessment follows a logic-tree strategy in test administration and clinical decision making. This strategy is also particularly advantageous because it maximizes efficiency of time expenditure by the clinician. At each stage in the evaluation, the clinician decides if sufficient information has been obtained to meet the objectives of the assessment. Figure 1 depicts the general outline of the three stage assessment, which will be discussed in more detail shortly.

Screening Batteries

The approach described previously constitutes a sequential evaluation protocol in which a standard screening battery is followed by a

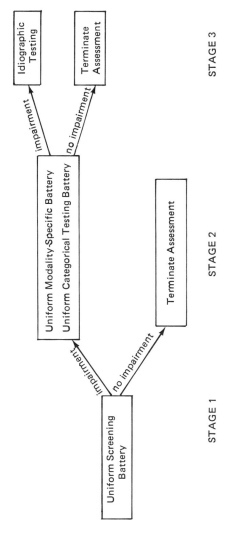

Figure 1. Three stages of neuropsychological assessment.

more detaled, but still structured, assessment. The composition of screening batteries is highly idiosyncratic, both in terms of the psychological processes assessed and specific instruments utilized. Benton (1975), Lezak (1976), Smith (1975), and others have proposed their version of a battery. Recently, Goldstein, Tarter, Shelly, and Hegedus (1983) developed a screening battery, the Pittsburgh Initial Neuropsychological Test System (PINTS), for use in a psychiatric population. What makes PINTS distinct from other batteries is that age-corrected scores are obtained in subjects between 13 and 70 years of age that are profiled on a standard T-scale across the various cognitive processes. The raw scores are transformed automatically by computer and printed out in the form of an age-corrected cognitive profile. On average the battery takes about an hour and a half to administer and yields close to 40 indexes of performance. The tests selected were based on the consensus opinion of a panel of neuropsychologists, with the one proviso being that the tests included in the battery be brief and have already demonstrated validity and norms.

As a screening battery, PINTS is rather comprehensive. It is highly efficient as indicated by the number of indexes obtained per unit of time. In addition to containing measures that are very sensitive to assessing the presence or absence of cerebral pathology, this battery also contains an array of tests that could implicate a lateralized or localized lesion. This screening assessment, followed by the selective administration of other tests, could ascertain the specific location of neurological involvement and pattern of the individual's cognitive strengths and weaknesses. Table 1 presents a typical computer generated PINTS printout. As can be seen, there not only is a fairly broad coverage of psychological processes, but also the performance scores are age referenced.

INTRAMODAL TESTS

From information obtained in the screening or initial evaluation, a decision is made to either continue the evaluation on a more in-depth basis, or to terminate the assessment at this juncture. A number of specialized tests and batteries are available that enable more extensive testing of specific processes. However, little research has been conducted on developing specialized test batteries that evaluate intensively specific modalities (e.g., audition, vision, etc.) or categorical processes (e.g., memory, perception, etc.). The dearth of such batteries notwithstanding, two examples of such tests are the Wechsler Memory Scale for evaluating memory (Wechsler, 1945); and, the Boston Diagnostic Aphasia Examination for evaluating language processes (Goodglass & Kaplan, 1972). These latter tests are already-packaged multiple

Table 1. Pittsburgh Initial Neuropsychological Test System

	Raw max	Raw score	T score	−	= MEAN =	+
Level of performance						
Peabody Picture Vocabulary-L	175	155	IQ 96		X	
Smith Symbol-Digit	110	31	31	X		
Trails A time	301*	52	46		X	
Trails B time	301*	139	45		X	
Trails B errors	*	0	57		X	
Memory functions						
WMS—mental control	9	7	48		X	
WMS—logical memory	23	9	50		X	
WMS—logical memory delay	46	14	39	X		
WMS—visual memory	14	8	52		X	
WMS—visual memory delay	14	13	63		X	X
WMS—associate learning easy	9	9				
WMS—associate learning hard	12	8				
WMS—associate learning total	21	17	57		X	
WMS—associate learning saving		5	83%			
Digit span forward	9	7	78%			
Digit span backward	8	4	50%			
Digit span forward + backward	17	11	50		X	
Digit span to + 1	5*	3				
Language functions						
BDAE animals		21	48		X	
BDAE phrase repetition—high	8	8	53		X	
BDAE phrase repetition—low	8	8	55		X	
BDAE confrontation naming	105	87	83%			
BDAE responsive naming	30	29	32	X		
Token test # correct	13	13	100%			
Reitan score	20*	3	85%			
Paragraph reading errors	38*	0				
Paragraph reading time	*	14				
Sentence reading errors	1*	0				
Paragraph interpretation	2	1				
Sentence writing	*	0				
Motor functions						
Star drawing DOM time	301*	29	45		X	
Star drawing DOM errors	*	0	56		X	
Star drawing N DOM time	301*	46	28	X		
Star drawing N DOM errors	*	4	40		X	
Peg board DOM	25	12	20	X		
Peg board N DOM	25	12	23	X		
Peg board both	25	10	22	X		

(continued)

Table 1 (Continued)

	Raw max	Raw score	T score	− = MEAN = +	
	Motor functions				
Finger tapping DOM hand		26	28	X	
Finger tapping N DOM hand		29	34		X
	Spatial-constructional functions				
Crosses	10[a]	2	57		X
Block design	24	16	53		X
WMS—visual copying	14	14	100%		

[a]Higher scores represent poorer performance.

measures of particular psychological processes. Given the number and diversity of neuropsychological tests that have been developed (see Lezak, 1983) other specialized batteries for assessing either a particular modality or psychological category can be readily developed by the clinician interested in examining in greater depth the nature of a cognitive impairment that cannot be otherwise determined by intermodal (pandemic) batteries or by screening batteries.

IDIOGRAPHIC TESTING

Numerous specialized tests, if required, can be administered at the third stage of the assessment process. This phase of the evaluation is restricted to the individualized assessment in which the cognitive nuances and particular neuropathological circumstances need to be assessed, but are not possible using a fixed-battery approach. At this stage of the assessment, clinician judgment and experience are crucial for selecting the most appropriate measures and for obtaining maximal information (objective psychometric, as well as subjective observational) from the client. The information gathered at this stage of the assessment confirms, as well as elaborates on, the findings generated from the first two stages of assessment. The qualitative features of the client's performance (e.g., trial and error vs. cognitive problem solving) are one important source of information obtained by the neuropsychologist that can contribute to a comprehensive assessment.

THE MEANING OF NEUROPSYCHOLOGICAL TEST RESULTS

All neuropsychological procedures are descriptive, correlative, and inferential as opposed to being explanatory, causal, and direct.

With respect to their descriptive properties, neuropsychological tests can be employed equally well to elucidate brain functional organization in normals and pathological cases and to profile cognitive and behavioral processes in suspected or known cases of neurological disorder. A variety of factors appear to contribute to how psychological processes are organized in the brain. For instance, the emerging evidence indicating that males and females perform differently on certain neuropsychological tests suggests that brain functional organization differs between the sexes. Similarly, left and right handers are sufficiently different on neuropsychological tests to suggest that the cerebral organization of psychological capacities is related to handedness. Findings such as these indicate that neuropsychological tests may be capable of describing brain and behavior relationships and discriminating subgroups in the population according to brain functional organizations. They do not, however, reveal how or why such groups are different, nor suggest the underlying mechanisms for these differences.

The previous findings illustrate that behavioral assessment can enhance our understanding of brain functional organization. Traditionally, neuropsychological tests have been employed in clinical settings to determine the presence and location of suspected neuropathology. With the development of very sophisticated neuroradiological techniques for assessing brain morphology (e.g., computerized tomography) and brain physiology (positron emission tomography and nuclear magnetic resonance), the use of neuropsychological tests solely for lesion localization is diminishing. Hence, a neuropsychological assessment, even though affording accurate lesion localization arguably is, by itself, not a sufficient reason to justify the cost involved. Additionally, a neuropsychological battery must be capable of describing cognitive strengths and weaknesses in a fashion that is clinically meaningful; that is, has treatment or rehabilitative applications. In this respect, a more useful application of neuropsychological testing is to delineate comprehensively the cognitive and behavioral sequelae of a known lesion so that treatment strategies of a targeted nature can be specified from the assessment and subsequently implemented.

The clinical assessment of brain–behavior relationships is exclusively a correlative process. It is correlative inasmuch as the attempt is made only to specify an association between neurological and behavioral phenomena. The point needs to be underscored that it is not necessarily the lesion itself that is causing the overt impairments, because neural disruption distal to the lesion site could also be responsible for the manifest changes. Therefore, there is no way in which a battery of tests can provide causal information such that an injury to a given brain area can be concluded to cause specific behavioral changes.

Neuropsychological tests are also inferential. In the absence of direct examination (e.g., CT scan), it is not possible to draw unequivocal conclusions. Therefore, unless neuropsychological test findings are buttressed with direct observations, the conclusions drawn must be necessarily tentative, and of an inferential nature.

Thus, all neuropsychological tests and batteries have the common shortcoming of being descriptive correlatives of neurological functioning. All conclusions drawn are necessarily inferential, because no actual observations of brain activity or structure are obtained.

DIMENSIONS OF NEUROPSYCHOLOGICAL ASSESSMENT

Up to this juncture, the discussion has focussed entirely on cognitive assessment. This is not particularly surprising because historically neuropsychological assessment has been almost entirely concerned with the effects of cortical pathology. Despite Kurt Goldstein's extensive discussions of the emotional concomitants and sequelae of neurological pathology, this aspect of psychological functioning has been either ignored or relegated to a position of minor importance in routine neuropsychological assessment.

In view of the emerging literature pointing to lateralization of emotion processes (Campbell, 1982), and neuropsychological impairment associated with certain psychiatric disorders, such as hysteria (Flor-Henry, Fromm-Auch, Tupper & Schopflucher, 1981), depression (Schweitzer, 1979), obsessive-compulsion (Flor-Henry, Yeudall, Koles, & Hourvath, 1979), and schizophrenia (Asarnow, 1983; Holzman & Tarter, 1982), it would appear to be essential to evaluate cognitive performance in tandem with noncognitive psychological processes. Virtually no research has been conducted on developing standardized tests of emotional disturbance for neurologically disturbed individuals. Therefore, in clinical practice, the techniques available are those that were developed and validated on nonneurologically impaired patients.

These limitations notwithstanding, it is possible to obtain meaningful information about emotional status employing psychometric tests (e.g., MMPI, SCL-90), and for purposes of making criterion diagnosis, structured interviews such as the Diagnostic Interview Schedule (Robins, Helzer, Croughan, & Ratcliff, 1981) and Schedule for Affective Disorders and Schizophrenia (Spitzer & Endicott, 1978) can be administered. These latter instruments have the advantage of not only elucidating current symptomatology, but additionally can provide information as to lifetime diagnoses, and whether the psychiatric problems were presently and in the past related to alcohol or drugs, as well as medical illness or injury.

Another component of the standardized neuropsychological assessment pertains to social adjustment. Inasmuch as there is some evidence linking neuropsychological capacity with social adjustment and employability (Heaton, Chelune, & Lehman, 1978), it is also desirable to assess this aspect of overall functioning, particularly with respect to utilizing neuropsychological information for rehabilitation planning.

A variety of social adjustment measures have been developed, although none have yet been adopted into routine clinical practice. One self-administered scale, the Sickness Impact Profile (Gibson, Bergner, Bobbitt & Carter, 1979), evaluates physical and social dysfunction in terms of percent of impairment. How daily routines (e.g., home management), activity (e.g., ambulation), and interpersonal relationships are affected, and to what extent, can be readily determined by this 15–30 minute questionnaire. For a more extensive assessment, or where valid information cannot be obtained from the client, the informant-administered Social Behavior Adjustment Schedule (SBAS) (Platt, Hirsch & Knights, 1981) is an alternative measure. This structured interview assesses social adjustment in the context of emotional well-being, and capacity to perform required social roles (e.g., working, parenting, etc.). Moreover, the SBAS delineates in quantified form the success of social adjustment in relation to the person's available environmental resources and limitations.

Thus, the neuropsychological battery, regardless of the specific composite of cognitive measures, could substantially contribute to the comprehensiveness of client evaluation by incorporating tests of psychiatric and social functioning. More often than not, cognitive, psychiatric, and social functioning are all affected in a person with neurological pathology; and, indeed, for certain conditions the psychiatric manifestations may even appear earlier and present a more florid display than the cognitive sequelae. For these reasons, the argument is advanced that the neuropsychological assessment include measures of psychiatric and social functioning as part of a comprehensive battery.

ASSESSMENT OF CHILDREN AND ADULTS: SPECIAL CONSIDERATIONS

Special issues face the clinician testing children. The period of cognitive development is rather lengthy and is not completed until mid-adolescence. It is critical, therefore, to determine if there is a reduced *rate* of cognitive development, or alternatively if there is a *constant magnitude of deficit* during the developmental period. In order to obtain this information at least two assessments must be conducted that are separated by a time period of sufficient duration to detect any changes in rate of cognitive development. A battery of tests is thus required that can be administered repetitively such that the results are not confounded by practice effects and where the applicable age range extends from childhood through adolescence. Alternatively, a battery consisting of two or more equivalent forms needs to be utilized.

At present, very little concern has been given to the need for serial clinical examinations. Yet, for children at known high risk for suffering cognitive impairment (e.g., those with perinatal injury) and for children with developmental disorders, the monitoring of cognitive development is of paramount importance. Rourke and Adams (1984) recently have addressed in detail the techniques and problems involved in the multivariate neuropsychological assessment of children, and presented a variety of tests that could be composited into a battery.

The neuropsychological evaluation of children also raises the question of the extent to which deficits, if any, influence the child's major task at hand; namely, education. A neuropsychological impairment is not necessarily associated with an intellectual deficit or educational difficulties. Thus, in order to determine if and how neuropsychological disturbance impacts on other spheres of functioning, such as educational achievement, it is important to include appropriate measures in the test battery.

The same general issues need to be considered in conducting clinical evaluations of patients at the other extreme of the life cycle. Thus, it is essential to evaluate whether a decline in cognitive capacity reflects normal aging, or is the result of a pathological process. Serial assessments, by evaluating the magnitude of change over time, afford the opportunity to ascertain if the person is experiencing the normal rate of decline or is suffering from a more insidious disorder (e.g., Alzheimer's disease). Also, with respect to the disorders of old age, one must be especially cognizant of organic-appearing presentations that actually are the consequence of a psychopathological condition. The most prominent of these disorders is pseudodementia, in which a clinical picture of dementia is observed but, in reality, is due to a severe depressive disorder that is reversible after appropriate intervention.

In summary, where a pathological process occurs in tandem with either cognitive growth or cognitive decline, it is indispensable to perform serial evaluations. In so doing, it is equally important in the course of a comprehensive assessment to clarify the reciprocal relationship between cognitive and other salient variables (e.g., education, psychiatric status, etc.) that are intimately involved in the overall adjustment process.

NORMATIVE AND CRITERION ASSESSMENT

Traditionally, neuropsychological assessment batteries have described performance capacity in normative terms; that is, as scores along a continuum ranging from superiority to normalcy to inferred neurological impairment. The score on any particular measure, therefore, reflects the performance of the individual relative to the normal population in inferring probabilistically the presence of a neurological disorder.

An alternative objective of the evaluation process, as yet not adopted in routine clinical practice, is to ascertain a level of performance required for optimal adjustment. This is a criterion oriented approach to assessment that attempts to specify the minimum level of competency needed for success in a particular situation or task. It is this assessment format that has the potential for establishing the ecological validity of neuropsychological test data; that is, the ability to extrapolate performance scores to predict performance in natural settings, such as school or job. For example, can a certain minimal level of performance on the Halstead-Reitan Categories test predict different types of job success? Is there a minimum performance level on the Purdue Pegboard that can predict assembly-line performance? These latter questions pertain to the utility of neuropsychological tests for devising vocational rehabilitation strategies that can be implemented from information accrued from a fixed or standardized battery. However, criterion-oriented assessment approaches have yet to be adopted into routine neuropsychological practice, although as can be readily appreciated, this is a most important and untapped area of applied research.

SUMMARY

Neuropsychological tests, measuring cognitive capacity and style, have evolved in number and sophistication during the past several decades. Research employing clinical neuropsychological measures has enhanced our understanding of brain functional organization and has increased our capacity for detecting the locus of cerebral lesions. This discussion argued that through a logic-tree format employing a uniform battery of tests, cognitive strengths and weaknesses can be assessed comprehensively. This type of neuropsychological assessment strategy has a number of advantages, the most important of which perhaps is the capacity to accrue systematic data such that the neuropsychologist can fulfill the dual responsibilities of clinical assessment and research contribution. In turn, this will facilitate elucidating brain–behavior relationships in a variety of situational contexts, and across a broad range of clinical conditions.

REFERENCES

Asarnow, R. (1983). Schizophrenia. In R. Tarter (Ed.), The child at psychiatric risk. New York: Oxford University Press.
Bender, L. (1938). A visual motor gestalt test and its clinical use. American Orthopsychiatric Association Research Monographs. No 3.
ton, A. (1955). Psychological tests for brain damage. In H. Freedman, H. Kaplan & B. Sadock (Eds.), Comprehensive textbook of psychiatry. Baltimore: Williams & Wilkins.
Benton, A. (1974). The Revised Visual Retention Test (4th ed.). New York: Psychological Corporation.
Bruninks, R. (1978). Bruninks-Oseretsky Test of Motor Proficiency. Examiner's manual. Circle Pines, MN: American Guidance Service.
Campbell, R. (1982) The lateralization of emotion: A critical review. International Journal of Psychology, 17, 211–229.
Christensen, A. L. (1979). Luria's Neuropsychological Investigation (2nd ed.). Copenhagen: Munksgaard.
Flor-Henry, P., Fromm-Auch, D., Tupper, M., & Schopflucher, D. (1981) A neuropsychological study of the stable syndrome of hysteria. Biological Psychiatry, 16, 601–626.
Flor-Henry, P., Yeudall, L., Koles, Z., & Hourvath, B. (1979). Neuropsychological and power spectral EEG investigations of the obsessive-compulsive syndrome. Biological Psychiatry, 14, 119–130.
Gibson, B., Bergner, M., Bobbitt, R., & Carter, W. (1979). The Sickness Impact Profile: Final development and testing. Seattle: Center for Health Service Research, Department of Health Services, University of Washington.
Golden, C., Hammeke, T., & Purisch, A. (1980). Manual for the Luria-Nebraska Neuropsychological Battery. Los Angeles: Western Psychological Services.
Goldstein, G., Tarter, R., Shelly, C., & Hegedus, A. (1983). The Pittsburgh Initial Neuropsychological Testing System (PINTS): A neuropsychological screening battery for psychiatric patients. Journal of Behavioral Assessment, 5, 227–238.
Goldstein, K. (1939). The organism. New York: American Book.
Goldstein, K., & Scheerer, M. (1941) Abstract and concrete behavior, an experimental study with special tests. Psychological Monographs, 53, (239).
Goodglass, H., & Kaplan, E. (1972). The assessment of aphasia and related disorders. Philadelphia: Lea & Febriger.
Heaton, R., Chelune, G., & Lehman, R. (1978). Using neuropsychological and personality tests to assess the likelihood of patient employment. Journal of Nervous and Mental Disease, 166, 408–415.
Holzman, A., & Tarter, R. (1982). Neuropsychological similarities of monozygotic twins discordant for schizophrenia symptomatology. Biological Psychiatry, 17, 1425–1433.
Lashley, K. (1931). Mass action in cerebral function. Science, 73, 245–254.

Lezak, M. (1983). *Neuropsychological assessment* (second edition). New York: Oxford University Press.
Luria, A. (1966). *Higher cortical functions in man.* New York: Basic Books.
Platt, S., Hirsch, S., & Knights, A. (1981). Effects of brief hospitalization on psychiatric patients' behavior and social functioning. *Acta Psychiatrica Scandinavica, 63,* 117–128.
Purdue Research Foundation (1948). *Examiner's manual for the Purdue Pegboard.* Chicago: Science Research Associates.
Reitan, R. (1955). Investigation of the validity of Halstead's measures of biological intelligence. *Archives of Neurology and Psychiatry, 73,* 28–35.
Reitan, R. (1958) Validity of the trailmaking test as an indicator of organic brain damage. *Perceptual and Motor Skills, 8,* 271–276.
Robins, L., Helzer, J., Croughan, J., & Ratcliff, K. (1981). National Institute of Mental Health Diagnostic Interview Schedule. *Archives of General Psychiatry, 39,* 381–389.
Rourke, B., & Adams, K. (1984). Quantitative approaches to the neuropsychological assessment of children. In R. Tarter & G. Goldstein (Eds.), *Advances in clinical neuropsychology* (vol. 2). New York: Plenum Press.
Schweitzer, L. (1979). Differences of cerebral lateralization among schizophrenic and depressed patients. *Biological Psychiatry, 14,* 721–733.
Smith, A. (1973). *A Symbol Digit Modalities Test.* Los Angeles: Western Psychological Services.
Smith, A. (1975). Neuropsychological testing in neurological disorders. In W. Friedlander (Ed.), *Advances in neurology* (vol. 2). New York: Raven Press.
Spitzer, R., & Endicott, J. (1978). *Schedule for affective disorders and schizophrenia* (3rd ed.). New York: Biometrics Research, New York State Psychiatric Institute.
Wechsler, D. (1945). A standardized memory scale for clinical use. *Journal of Psychology, 19,* 87–95.

6

Overview of the Halsted-Reitan Battery

OSCAR A. PARSONS

INTRODUCTION

In 1987 the Halstead Tests will be 40 years old. What has happened over the nearly four decades of their use? Certainly there is little doubt that the Halstead Neuropsychological Test Battery, now in the form of the Halstead-Reitan Battery (HRB), is the most widely used neuropsychological test battery in the United States and Canada. It has earned the respect of not only most psychologists but many representatives of the allied disciplines, such as neurologists, neurosurgeons, psychiatrists, pediatricians, rehabilitation workers, and the like. To a great extent the rise of clinical neuropsychology as an identifiable speciality has been concomitant with the increased establishment of neuropsychological laboratories in which the HRB has played an integral role (Parsons, 1984).

In this overview I consider the past, present, and future contributions of the HRB. In the first section I review my personal experiences with the battery and its developers; from this review three different epochs of the battery are described. In the second section I present the current state of the art clinical application of the battery by describing how we use it in our laboratory and by giving several case illustrations.

OSCAR A. PARSONS • Department of Psychiatry and Behavioral Sciences, University of Oklahoma Health Sciences Center, P.O. Box 26901, South Pavilion, 5th Floor, Oklahoma City, OK 73190.

Finally, in the third section, consideration is given to the direction for future clinical and research applications with the battery.

DEVELOPMENT OF THE HALSTEAD-REITAN BATTERY

Personal Experience with the Battery

My first acquaintance with the Halstead Tests came in 1951 when Dr. Halstead gave a talk at Worcester State Hospital, Worcester, MA, where I was working in research. Dr. Halstead traced his research starting in 1935 over the 12-year period culminating in the publication of his book *Brain and Intelligence: A Quantitative Study of the Frontal Lobes* in 1947. The results of his research enabled him to predict lesions in the brain with great accuracy, he averred, and he gave us examples of his techniques. Dr. Halstead was a rather impressive looking man who was also a very interesting speaker. I recall the net reaction of our psychology staff and students (interns and postdoctorals in clinical psychology) at the time was that this was an enthusiastic overselling of his research findings. Afterwards most of us shook our heads and said that it was too good to be true, that brain functions were not as localizable as he was portraying them and that there were too many other influences that could give rise to varying interpretations of tests. You must realize that most clinical psychologists at that time were committed to Kurt Goldstein's holistic approach to brain damage (Goldstein, 1959). Our tests and conceptualizations were based on his approach, that is, brain damage results in an impairment of the abstract attitude rather than deficits in specific cognitive and perceptual-motor functions.

Reitan was one of Halstead's doctoral students in the biological psychology program at the University of Chicago. After completing his doctoral program in 1950, he went to the University of Indiana College of Medicine and established a neuropsychology laboratory. In 1955 he published an article in which he corroborated Halstead's original findings; in a study of 50 brain-damaged individuals and matched controls, he was able to quantitatively separate almost all of the brain-damaged from their matched controls on Halstead's tests (Reitan, 1955). I was at the Duke University School of Medicine at the time providing consultative psychological services to medicine, neurology, and psychiatry when several of my colleagues who were interested in neonatal anoxyia and its effects on subsequent development asked me to identify a quantitative and valid test of neuropsychological functioning, I immediately brought the Reitan (1955) study to their attention.

After making appropriate arrangements, three of us (Drs. Lewis Cohen, David Arenberg, and I) took a train to Indianapolis in 1957 and spent several days with Ralph Reitan looking over his neuropsychology laboratory and discussing with him his methods and findings. Dr. Reitan was very generous with his time and I can recall at that time being extremely impressed by his achievements and by the potential contribution of the Halstead Battery tests. At the same time, I felt somewhat uncomfortable because of his base rate. The population that he used to make inferences about lesion location and the like came largely from neurology and neurosurgery. There were relatively few patients from psychiatry. My questions to him at the time had to do with differentiating schizophrenia, depression, states of high anxiety, character disorders, and the like, from the brain-damaged group. These were the patients that I was confronted with day in and day out on our psychological consultative service. At the time, Dr. Reitan did not have too much to say in response to my questions regarding differential diagnosis, largely because he was busily concerned with identification of the lesion, lateralization, localization, and ultimately, the prediction of type of lesion.

We did get the HRB at Duke and when I moved to the University of Oklahoma College of Medicine in 1959, I kept up my contacts and correspondence with Dr. Reitan. In the early 1960s, Dr. Arthur Vega and I had established a neuropsychology laboratory at the University of Oklahoma Health Sciences Center where we were administering the HRB routinely to patients in whom there was some question of brain dysfunction. (In the center of the dust bowl there was no question about what battery we would give!) In 1967 Dr. Vega and I published one of the first independent laboratory cross-validations (Vega & Parsons, 1967) of the HRB, basically replicating the findings reported by Reitan in 1955 article and by Halstead in 1947.

In 1968 Dr. Vega and I took a workshop with Dr. Reitan (one of the first he offered), and it became apparent that he had moved far beyond the original types of cases that he had discussed with us in 1957. I can recall vividly the fact that he was doing a study of patients with lung cancer and predicting whether or not brain metastases had occurred. I have never seen a publication of these data but some of the individual cases which he showed me impressed me greatly in terms of the sensitivity, at least in his hands, of the HRB in delineating the specific location of brain metastases. It was, of course, quite exciting to see Ralph Reitan at work doing cases "blindly" and making predictions not only about brain damage but lateralization, localization, and finally as to the nature of the disorder itself, that is, intrinsic tumor, cerebrovascular problem, degenerative disease, head trauma, and the like.

We returned to Oklahoma stimulated and challenged to apply our clinical neuropsychological skills to an even greater extent than previously.

My next major perspective on the HRB came with the publication of Russell, Neuringer, and Goldstein's (1970) book, the *Assessment of Brain Damage*, in which they describe their Neuropsychological Key approach and their modifications of the HRB. Incidentally, this group was from the University of Kansas, another dust bowl state. I later visited Dr. Goldstein's laboratory at Topeka VA in Kansas and saw first hand that the key was used to predict a number of aspects of brain dysfunction.

In addition to my clinical experience with the HRB over the last several decades, I have participated in a number of research projects using the HRB from studies of patients with clear-cut neurological diagnoses to alcoholics and to patients with chronic pulmonary disorders. Most recently, I have read some 50 work samples from neuropsychologists around the country, many of whom use the Halstead-Reitan Battery.

So much for the personal odyssey. However, out of the odyssey I have conceptualized the growth and development of the HRB into three major blocks of time: from 1935–1955 Halstead developed and presented the Halstead Tests to the scientific community; from 1955–1975 his student Reitan gradually became the dominant figure and the battery became the Halstead-Reitan Battery. From 1975 on the leadership has been divided but I believe most neuropsychological laboratories employ some version of the Russell, Neuringer, and Goldstein or other variation of the original HRB. Let us examine these periods more closely.

1935–1955: THE HALSTEAD YEARS

Ward Halstead started out to remedy a perceived lack in the psychometric devices of his day, especially with regard to the effects of frontal lobe damage, that is, that the latter frequently did not result in changes in the patient's standard intelligence test performance. Halstead sought measures of "biological intelligence" that he believed reflected the state of functioning of the central nervous system as opposed to psychometric measures of intelligence that reflected educational opportunities. Halstead (1947) gave a battery of 27 tests to 50 "healthy" adult males drawn from the head-injury service. All these cases were regarded as "medically recovered from a recent concussion type of head injury." (Given our knowledge today we would hardly consider this an adequate normative group.) From this battery 13 tests were selected for factor analysis. The factor analyses, performed by several leading factor analysts of the day, gave rise to four factors. These factors were C-central integrative; A-abstracting; P-power factor; and D-a "Modality Directionality" factor, and these factors constituted the major elements in biological intelligence. The application of these tests to 50 neurological cases (M age=37), the majority of whom were frontal lobe cases, and 30 controls (M age=32), was the basis for establishing the cut-off points, which have endured to this day (educational levels were not recorded in the data tables). Halstead showed that an Impairment Index, based on the number of tests on which the patient scored above an empirical cutting score, differentiated between the frontal-lobe patients (most impaired), posterior-damaged patients, and controls. The frontal lobes, declared Halstead (1947), had been shown to be of critical importance. "They are the organs of civilization—the basis of man's despair and of his hope for the future." Halstead conducted a second major study, confirming the value of his original empirical findings but offering only weak support for the theoretical notions (Shure & Halstead, 1958).

Time has shown that most of Halstead's original conclusions were incorrect concerning the battery's differential sensitivity to frontal-lobe function and his theoretical structure has never caught on. But what Halstead left has turned out to be a tribute to empiricism! Despite the weakness of the original standardization group and the comparison groups, the tests have been shown to consistently differentiate persons with brain dsyfunction from those without, over a wide range of disorders (Boll, 1981; Reitan & Davison, 1974). Among the many workers in the field, it has been Ralph Reitan and his colleagues that provided the kind of consistent evidence that has called worldwide attention to the contributions of a standard or fixed battery approach.

1955–1975: The Reitan Years

Starting with the 1955 publication of the corroboration of the discriminating effectiveness of Halstead's tests there was an outpouring of research from Reitan and his colleagues (e.g., Halgrim Kløve, Homer Reed, the Fitzhughs). They reported differences on the Halstead Tests between patients with left or right hemisphere EEG defined abnormalities, left and right visual field cuts, left and right sensorimotor deficits and left and right neurologically defined brain-lesioned patients. Group differences on the battery were found between controls and samples of various brain-damaged or dysfunctional groups; epileptics, alcoholics, multiple sclerotics, cerebrovascular disorders, trauma, neoplastic disease, endocrine conditions, degenerative disorders, etc. (Boll, 1978, 1981; Dodrill, 1981; Heaton & Crowley, 1981; Parsons & Farr, 1981; Parsons & Hart, 1984; Reitan & Davison, 1974).

During this time Reitan reduced Halstead's tests to the following: the Category test, the Tactual Form Board Test-Time, Memory and Location scores, the Speech-Sounds Perception test, the Seashore Rhythm test, and the Finger Oscillation test. However, he added the Wechsler Scales, Lateral Dominance test, the Hand Dynamometer test, the Halstead-Wepman Screening test for Aphasia, the Sensory Perceptual Examination, the Tactile Form Recognition test, and the Trail Making test. Several children's batteries based on the HRB were developed, one for the 9–14 year-old ages and one for the 5–8 year age span. Gradually the Halstead tests became known as the Halstead-Reitan Battery.

During this period, Reitan began giving his workshops, which by the 1970s had built up an enthusiastic following, a following that has continued to the present. There were several by-products of these workshops. First, the emphasis on careful standardization of test administration was uniform. Individuals giving the test were given elaborate protocols on how to administer them and Reitan never ceased cautioning on the importance of this variable. Thus, there has been considerable interlaboratory replicability of findings. Second, the diffusion of the battery to new settings became widespread. Although this had a decided stimulating effect, unfortunately, as with all such endeavors, some individuals who took a workshop thought that they could interpret the battery without the knowledge of neuroanatomy, neuropathology, and psychopathology that are necessary for such a process. Reitan and his colleagues have ceaselessly pointed out the importance of such background knowledge before neuropsychological data are interpreted clinically.

Although the number of publications emanating from Reitan's laboratory have diminished in the late 70s and the 80s, several stimulating articles have appeared. Selz and Reitan (1979) provide rules of a quantitative nature for distinguishing brain-damaged from learning-disorder and control children. Hom and Reitan (1982) examined the effects of lateralized cerebral damage on contralateral and ipsilateral performance in groups of patients with well validated brain lateralization. They found, contrary to some previous reports, that the right-hemisphere-lesioned group had greater deficits both contralaterally and ipsilaterally than their left-hemisphere-lesion group. However, during this period other developments with the HRB were taking place that had implications for clinical application.

1975–1984: VARIATIONS AND MODIFICATIONS OF THE BATTERY

I chose the date of 1975 because the impact of the publication of the Russell, Neuringer, and Goldstein (RNG) (1970) book, *Assessment of Brain Damage* became noticeable around this time. It was the first systematic revision of the HRB to be published. These authors presented a taxonomic approach, the neuropsychological key, that culminated in the development of a computer program for neuropsychological diagnosis. They added five tests to the Impairment Index of the HRB. These tests are the Digit-Symbol test from the Wechsler Adult Intelligence Scale (WAIS), an Aphasia score from the Aphasia Screening Test, a Spatial Relations score based on ratings of the Greek Cross from the Aphasia Test, the Trail Making test and a Perceptual Disorders Rating. They also presented a new "Impairment Rating" based on work by Phillip Rennick (another doctoral student of Halstead's). Rather than relying on a dichotomous above-or-below-the-mean score, each measure is rated on a 6-point scale (0–5) according to norms established by Rennick (Russell, 1984). Using these RNG ratings, the computer program first decides whether the patient is brain damaged versus non-brain damaged. If brain damage is present, the second decision is whether it was lateralizable or diffuse. A third decision is whether it was lateralized to left or right; a fourth decision is whether it was strongly or weakly lateralized, and a fifth decision is whether it is acute, static, or congenital.

Why was this development important? For several reasons. First, it was a meaningful extension of the HRB in terms of quantification of five additional measures covering basic and important functions for example, coding, perceptual-motor sequencing speed, language, spatial-relations, and sensory-perceptual functions. Second, if the key approach were successful it could provide rich clinical and experimental data that would be of at least heuristic if not practical value. Third, by using the RNG rating scale rather than just number of tests in the impaired range, (the latter being more a measure of consistency than severity per se), a more representative measure of severity could be obtained, that is, an average rating of 4 or more is definitely in the severely impaired range on all tests. In contrast, one could have an HRB Impairment Index of .9 but have all tests fall in the mildly impaired range.

During this time research with the HRB and its modifications has continued unabated. Reliability and validity of the HRB have been examined in a number of studies (Boll, 1981) with consistent demonstration of satisfactory levels in each category. Of particular interest as regards validity are papers by Kløve (1974), Filskov and Goldstein (1974), Schreiber, Goldman, Kleinman, Goldfader, and Snow (1976), and Tshushima and Wedding (1979). The effects of using different orders of test administration have been shown to be negligible (Neuger, et al., 1981). Reitan's quantative methods for capturing his methods of inference have been presented (Selz & Reitan, 1979). Dodrill (1981) has modified the battery for epileptics, Matthews (1981) for adults, Rourke (1981) and Boll (1981) for children, etc. Other authors have published on neuropsychological findings with the HRB in schizophrenia, affective, and other psychiatric disorders (Heaton, Baade, & Johnson, 1978). Heaton, Smith, Lehmann, and Vogt (1978) have described the problems of interpreting battery data on patients in whom there is litigation surrounding the possible brain-injury. A recent study of the effects of chronic pulmonary disease on performance on the HRB indicates mild to moderate impairment as a function of mild to severe pulmonary disease (Prigatano, Parsons, Levine, Hawryluk, & Wright, 1983). The role of the HRB in neuropsychological assessment of the elderly has been demonstrated by Price, Fein, and Feinberg (1980). It is being used to test for the effects of rehabilitation procedures, such as cognitive retraining, supportive therapy, etc. (Goldstein & Ruthven, 1983). The list goes on and on.

In summary, from my vantage point there have been three identifiable, if overlapping phases of the developments and application of the HRB. From Halstead's original tests, Reitan not only modified the battery but provided the necessary scientific evidence of brain–behavior deficits as measured by the HRB to convince fellow psychologists and other professionals of the worthwhileness of the endeavors. The extension of the battery and the development of the neuropsychological key by Russell, Neuringer, and Goldstein (1970) have provided an additional stimulus to the field as have the modifications by others. The HRB continues to be widely used for both clinical and research purposes. It remains an empirical instrument of unusual sensitivity to brain perturbations.

How is the HRB used clinically? In most laboratories it used as a fixed rather than a flexible battery (see Chapters 4 & 5 in this volume) and is frequently administered by a technician for the major portion of the tests. The clinical neuropsychologist will typically see the patient for an interview and/or specialized testing and, of course, write the report. In the following section the approach used in our laboratory is detailed. Although there are many variations of the HRB, our version and our method of making inferences as to neuropsychological functioning are probably representative of the current state of the art. Examples of other approaches are provided by Matthews (1981).

CLINICAL APPLICATION OF THE HALSTEAD-REITAN BATTERY

The HRB is designed to be given by trained individuals who understand the necessity of adhering to a prescribed protocol of administration. Reitan (1959a,b) in much circulated but never officially published documents describes the HRB, its administration, and his early approach to interpretation. Subsequent versions have been presented in his workshops. Other authors such as Boll (1981), Russell, Neuringer, and Goldstein (1970), and Swiercinski (1978) provide additional information relevant to administration.

Persons who are being trained should undergo observation by experienced users of the battery or should be videotaped so that corrective procedures can be instituted. The number of practice patients or subjects depends on the qualifications of the tester. A minimum of five practice subjects for experienced psychology graduate students is necessary; for technician-level persons 15 practice subjects is necessary. Persons administering the tests should be trained to recognize and note aspects of behavior that could affect the results, for example, motivation, competitiveness, emotional reactions, eccentric behavior, and the like. Perhaps most importantly, testers should have a supportive and encouraging attitude (Parsons & Prigatano, 1978). The battery can take up to six or more hours; the tester must be able to establish sufficient rapport to motivate the patients when the inevitable valleys in performance appear and know when to give appropriate breaks that can refresh the patient.

Interpretation of the results should be done by a clinical neuropsychologist who has training in psychometric measurement, a thorough understanding of the research literature on the HRB, and a sufficient background in neurology, neuroanatomy, neuropathology, and psychopathology. Preferably the interpreter would have more than a superficial knowledge of biomedically relevant variables, such as psychopharmacology, computerized tomography scans, EEG findings, etc. Of course these attributes are applicable to all clinical neuropsychologists regardless of the type of psychological measurement employed. Finally, there is no substitute for experience with a variety of cases to hone the clinician's interpretative skills.

The neuropsychological report can take many forms. In part the report depends on the setting and the type of neuropsychological referrals. However, there seems to be a growing concensus or at least operational similarity in what such reports should cover. Recently, as noted earlier, I have had the opportunity to read over 50 neuropsychological case reports from practicing neuropsychologists throughout the United States. Although there are indeed variations, as there should be,

I was struck by the common elements in the reports. Based on this experience, I will describe the clinical report form used in our laboratory, confident that it is representative of the content, if not the form, of many settings. The outline of the report together with the various tests that contribute to the analysis of neuropsychological functioning is presented in Table 1. (Please understand that even within a fixed or standard battery some variation occurs; for particular reasons some tests are dropped and others added.) In the subsequent paragraphs I will discuss each of the sections of the report.

IDENTIFYING DATA

Identifying data are very important. The age, education, occupation, sex, and marital status of a client can critically influence interpretation. Age and education are not adjusted for in the usual HRB scoring. But as Parsons and Prigatano (1978) have pointed out, there is ample evidence that these variables are significantly related to performance. HRB clinicians must gain a working knowledge of these effects so that they can adjust their interpretations accordingly. Resource material helpful here includes Fromm-Auch & Yeudall's (1983) age norms for ages 15–40, Pauker's (1977) norms for different ages and levels of intelligence, Harley, Leuthold, Matthews, and Berg's (1980) tables for age and HRB performance in older male VA samples. The role of sex as a variable in neuropsychological test performance has not been systematically studied. However, the results presented by Inglis and Lawson (1981) as regards differences in male and female brain-damaged patients on the WAIS and our recent findings of gender differences in verbal versus visual-spatial learning (Fabian, Parsons, & Shelton, 1984) suggests that we should be sensitive to possible differences on the HRB. We also have found that male and female alcoholics and controls differ somewhat on the Tactual Performance Test of the HRB (Fabian, Jenkins, & Parsons 1981). Fromm-Auch & Yeudall (1983) found sex differences in finger tapping speed both on overall level of performance and between hands. As regards marital status, the presence of an intact marriage may have decided positive implications for recovery from brain damage; contrariwise, divorced status may mean a lack of psychological support for the patient and hence poorer recovery. The patient's occupation certainly affects interpretation; a bank president who after a head injury has difficulty with arithmetical calculations and problem solving is much more likely to be less effective at his work than if he were engaged in manual labor; a carpenter with a right-parietal syndrome may be grossly impaired in construction tasks but able to handle a selling job with only minor difficulties.

Table 1. Outline of Neuropsychological Report

1. Identifying data
2. Reason for referral
3. Summary and recommendations
4. Extended report (see below)
   4.1. Background information
   4.2. Tests administered and behavior
   4.3. Analysis of results
   4.4. Neurological implications
   4.5. Psychological implications

4. Extended report
   4.1. Background information
      a. Information of relevance from head injury questionnaire, e.g., head trauma, hospitalizations, illnesses, drugs, and information from medical records.
      b. Record any information regarding peripheral injuries or deficits, e.g., blindness, deafness, peripheral motor injuries, etc.
      c. History of any emotional disturbances, psychiatric hospitalizations, etc.
      d. Subjective reports of problems in functioning
   4.2. Tests administered and behavior of patient
      a. List tests
      b. Note time taken for examination, patient's attitudes, interest, motivation, perseverance. Any unusual behavior should be noted.
   4.3. Analysis of results
      a. General considerations for interpretation
         1. Age, sex, and education
         2. Lateral Dominance—eyedness and handedness
         3. Any special limitations, e.g., peripheral injuries to hands or arms, sensory acuity deficit
         4. Attitudes and emotional reactions
         5. Any special cautions
      b. General intelligence and achievement
         *1. WAIS-R F.S.I.Q., verbal IQ, performance IQ
         2. WRAT reading, spelling, arithmetic
      c. Sensory-perceptual and sensory-motor functioning
         1. Sensory-Perceptual Exam findings, tactile dysstereognosia (HRB)
         2. Sensory-motor performance—finger tapping, peg board, grip strength (HRB)
      d. Attention, concentration, memory
         1. Digit Span (WAIS-R)
         2. Rhythm (HRB)
         3. Verbal Memory—Immediate & Delayed (WMS)
         4. Figural Memory—Immediate & Delayed (WMS)
         5. Tactual Performance Test Memory (HRB)
         6. Tactual Performance Test Location (HRB)
         7. Overall Memory (WMS)
         8. Luria Memory for words—immediate vs. delayed
         9. Benton Visual Retention Test
      e. Language and communication skills

(continued)

Table 1 (Continued)

   1. Vocabulary (WAIS)
   2. Aphasia Screening Test results—receptive, expressive, dysnomia, dysgraphia (HRB)
   3. Speech perception (HRB)
   4. Reading and spelling (WRAT)
   5. Patient's verbal-communicative behavior
  f. Calculational skills
   1. Arithmetic (WAIS-R)
   2. Arithmetic (WRAT)
   3. Problems on aphasia screening (HRB)
  g. Spatial relations and construction dyspraxia
   1. Block Design, Object Assembly (WAIS)
   2. Greek Cross, Key (HRB)
  h. Learning, abstracting, and problem solving
   1. Learning
    Information (WAIS-R)
    Paired Associates (WMS)
    Digit Symbol (WAIS-R)
    Improvement over trials—TPT (HRB)
    Improvement over Trials—Luria Words
   2. Abstracting
    Verbal—similarities (WAIS-R)
    Nonverbal—Category Test (HRB)
   3. Problem solving, set flexibility
    Category Test (Perseveration-qualitative aspects) (HRB)
  TPT (Time) (HRB)
  Block Design (WAIS-R)
  Trails B (HRB)
  i. Personality
   1. MMPI
 4.4. Neurological implications
  Level of functioning considering Impairment Indexes, variability, right-left difference, pathognomic signs. If impaired, acute vs. static; lateralized vs. diffuse, localization where possible. (Consider age, education, and socioeconomic status when interpreting impairment statements.)
 4.5. Psychological implications
  Specific suggestions for patient management.
  Role of personality and neuropsychological factors in life situations.
  Implications for cognitive retraining, psychologically oriented therapy, and life adaptation.

*Abbreviations used: WAIS-R—Wechsler Adult Intelligence Scale-Revised; HRB—Halstead-Reitan Battery; WMS—Wechsler Memory Scale; and WRAT—Wide Range Achievement Tests.

REASON FOR REFERRAL

The reason for referral should be stated as clearly and succinctly as possible. Is the referral one which seeks to gain information that would help distinguish brain dysfunction from a psychiatric functional disorder? These referrals still constitute a good percentage of the referrals to our laboratory, despite the advances in modern biomedical diagnostic techniques, such as the CAT scan. Is it a referral that asks for delineation of level and kind of deficits in presence of a known state of brain dysfunction? Is it a referral that wishes information relative to strengths or areas of intact functioning on which rehabilitation programs can be instituted? Obviously each of these questions and variations of them could result in different emphases in reports.

SUMMARY AND RECOMMENDATIONS

The first page of the report is probably the most important one. Referring mental health personnel, physicians, lawyers and social agency representatives are often busy professionals who wish to have a concise presentation of findings to feed into their decision-making processes regarding diagnosis or intervention. By having the summary and recommendations on the first page, the essential information is transmitted. For documentation and specifics the extended report can be consulted. Obviously in this section the referral question should be answered with as much specificity as possible. The important neuropsychological findings are summarized with their implications for the functional status of the brain and for the life adaptation of the patient. Specific recommendations for remediation, cognitive retraining, or other therapeutic approaches should be made when appropriate.

EXTENDED REPORT

Background Information

The Head Injury Questionnaire (Reitan, 1959a) is an excellent lead in for the neuropsychological evaluation. It provides systematic information on various conditions that could have resulted in brain dysfunction, for example, illness of childhood and adulthood, trauma, high fever, oxygen deprivation from near drowning or high altitudes, etc. Hospitalizations, alcohol and drug usage are noted. This questionnaire has proved useful in our laboratories time after time; frequently information is obtained that does not appear in the medical records. Of course, medical records that are available on the patient should be consulted whenever possible.

Information as to sensory deficits, for example, blindness, deafness, or greatly reduced vision or hearing, should be noted. Peripheral injuries are of particular importance to be inquired about, particularly to hands or arms. Differences in sensorimotor functions are important contributors to inferences about lateralization of brain dysfunctions; unrecognized peripheral injuries could result in erroneous inferences.

History of emotional disturbances and psychiatric hospitalizations are useful items for interpretation both in terms of the level of functioning and of the recovery of adaptive functioning. An elderly person with a history of recurrent depressions who now has memory problems resulting in impaired functioning may be suffering from pseudodementia rather than dementia of Alzheimer's type. Persons who have lived chronically in a state of tension and anxiety may well show patterns of impaired functioning resembling brain damage (Chapman & Wolff, 1959).

Finally, the patient's subjective complaints should be noted. What are the problems from his or her point of view? In what way does the patient see him or herself changed from previous level of functioning? The changes should be inquired about in intellectual, emotional, motivational, interpersonal, and social areas.

Tests Administered and Behavior of the Patient

A complete listing of all tests should be given, including the individual components of the HRB. This provides a record that has archival functions as well as current documentation for insurance companies, legal questions, etc. It also provides information of value to other psychologists should the patient be seen in other contexts (and many patients are!). The behavioral note on the patient should comment on the patient's test-taking attitudes, interests, motivation, perseverance. Interpreting the data for a person who was reluctant to be tested, gave up easily on problems, and was suspicious of the intent of the examiner is much different than a patient who was cooperative, trying very hard to do well, very interested in the tests and their performance. Perseverance and effort are important ingredients in many of the tests given in the HRB. Finally any unusual or atypical behavior should be noted, for example, inappropriate social behavior, temper outbursts, sullenness, uncommunicativeness, silliness.

Analysis of the Results

General Consideration for Interpretation

In this section note is made of those variables that provide a framework for interpretation of the results. The important variables of age, education, and sex are noted together with the presence of any peripheral injuries. Lateral dominance of eyedness and handedness results are noted. Of the two, handedness is by far the most important. Left handedness has profound implications for interpreting the results, especially with reference to laterality. As discussed earlier, attitudes and emotional reactions that may impact interpretation should be noted. Finally, any other special cautions or limitations of interpretation should be noted, for example, the patient's drug regime may cause slowing down and sleepiness.

GENERAL INTELLIGENCE AND ACHIEVEMENT

In this section general level of intellectual functioning is reported based on the WAIS-R. Full Scale, Verbal, and Performance IQs and percentile scores are reported. The general level of functioning intelligence is examined with reference to educational level and discrepancies are noted. A person with a master's degree from a university whose IQ is 85 is obviously suffering a generalized decrement in functions. Significant differences between Verbal and Performance IQs are noted with attendant implications for lateralization. Similarly, variability in pattern of subtests scores is noted. The Wide Range Achievement Test (WRAT) scores (Jastak & Jastak, 1965) on Reading, Spelling, and Arithmetic are frequently helpful in interpreting the current level of performance on these important functions that the patient is capable of at this time. Scores are reported in terms of grade equivalents or, as we prefer, standard scores.

SENSORY-PERCEPTUAL AND SENSORY-MOTOR FUNCTIONING

The sensory-perceptual examination gives rise to important data. Basic sensory receptive functions are tested and through double simultaneous stimulation, suppressions in the visual, auditory, and tactual modalities can be noted. Finger agnosia and disturbances in finger tip writing recognition are recorded. The results for the tactile test for dystereognosia are included here. Sensory-motor performance is measured by Finger Tapping, Peg Board (Purdue or Grooved) and Grip Strength. The tests in this section provide information relevant to lateralization and localization questions. Impaired sensory motor functions imply anterior lesions whereas sensory perceptual difficulties imply posterior lesions; differential impairment of the left or right side of the body gives lateralization information. Suppressions suggest acute or widespread depression of functions as opposed to chronic or specific localization of brain dysfunction.

ATTENTION, CONCENTRATION, AND MEMORY

The number of tests used here reflects the importance of measuring these functions in persons with suspected or demonstrated brain dysfunction. Digit Span from the WAIS-R provides a measure of attention and short-term memory. The HRB Rhythm test calls for keeping one rhythm in mind while listening to a second rhythm. This can be viewed as an attention and short-term memory task. Incidental memory is measured by the Tactual Performance Test (TPT)-Memory and TPT Location. The latter is typically more sensitive to brain dysfunction than the former. The Wechsler Memory Scale, despite its many limitations, provides several different scores of interest. First, there is the overall Memory Quotient that should be close to the Full Scale I.Q. In the amnestic disorders depression of the Memory Quotient (by 10 or 15 points) compared to the IQ is commonly found. Specific types of memory are obtained on the Verbal or Semantic Memory or Figural Memory scores under immediate and delayed conditions. Comparisons of RNG severity ratings contribute to decisions about laterality of dysfunction. Another type of verbal memory is obtained with the Luria words (Luria, 1980). Here a list of 10 words is presented 10 times and the number of words recalled or the trial on which all 10 are remembered is noted (most normals attain 10 words on the fourth trial). Both immediate and delayed conditions are used. Unpublished research in our laboratories by Dr. Nancy Adams and Dr. Russell Adams clearly indicates that brain-damaged psychiatric patients perform more poorly than non-brain-damaged psychiatric patients. The Benton Visual Retention test gives a somewhat different measure of figural memory. It appears to be sensitive to both generalized and right-hemisphere dysfunctions.

LANGUAGE AND COMMUNICATION SKILLS

The measurement of language skills and possible dysfunction therein are made by using the Vocabulary subtest from the WAIS, the Reading and Spelling tests from the WRAT and the Aphasia Screening Test. The WAIS and WRAT tests give a good indication of the level of verbal skills of the patient. Given the fact that patients with a history of learning disabilities in childhood frequently do poorly on measures of verbal-linguistic skills, information gathered about the patients educational experience must be considered here. The Speech–Perception test of the HRB also measures phonemic skills that may be impaired with temporal lobe dysfunctions (but performance may also be impaired because of the patient's slowness in responding or confusion in answering). The Aphasia Screening Test is useful in revealing receptive and expressive aphasic difficulties. It is scored according to Reitan's (1959a) criteria and enters into the RNG Impairment Rating. The patient's nontest related verbal communicative behavior is also informative. Word-finding difficulties, paraphasias, and loss of prosody may be present in the patient's conversation.

CALCULATIONAL SKILLS

The Arithmetic subtest of the WAIS-R is one of the major tests in this area. The Arithmetic test from the WRAT gives some indication of school-like achievement. The WAIS-R arithmetic subtest is presented auditorily so that the problems must be solved by the patient without recourse to pencil and paper. Thus the patient must conduct several operations keeping in memory the important facts. On the WRAT arithmetic the patient has problems visually displayed in front of him and uses pencil and paper to calculate his answers. Left-hemisphere lesion patients may do poorly because of difficulties with symbols whereas right-hemisphere lesion cases may do poorly because of visual-spatial problems. The Aphasia Screening Test has two arithmetic problems— one presented orally and solved by the patient in his or her head and one where the patient must copy down a written problem and then solve it.

SPATIAL RELATIONSHIPS AND CONSTRUCTION DYSPRAXIA

The Block Design test of the WAIS-R is one of the more sensitive tests to brain dysfunction. Reproducing designs may be impaired because of general obtunding of intellect. Left- or right-unilateral lesions may also result in impairment, although for different reasons, as shown by Kaplan (1979). Object Assembly also reflects difficulty in visual-spatial construction. Parsons, Vega, and Burn (1969) have shown that right-hemisphere compared to left-hemisphere lesion patients perform significantly poorer on both the Block Design and Object Assembly.

The Greek cross from the Aphasia Screening Test is scored according to RNG criteria and provides a measure of drawing dyspraxia. The score of two crosses is averaged and an RNG rating of severity is applied. The score becomes one of the 12 RNG tests comprising their Impairment Rating. The drawing of the key from the Aphasia Screening Test is not scored formally. However it can be described in terms of oversimplification and lack of detail or disturbances in reproducing the gestalt of the figure, for example, instead of notches in the shaft of the key the patient puts bumps on. The former appears to be more characteristic of left-hemisphere cases; the latter of right-hemisphere cases.

LEARNING, ABSTRACTING, AND PROBLEM SOLVING

Learning

General information from the WAIS-R provides a good measure of the patient's past ability to acquire information from the environment and retain it. Of course aphasic difficulties can interfere with performance but the scale seems to be resistant to all but the most severe brain dysfunctional states. High scores give a good indication of premorbid attainment; low scores can result from lack of attainment, learning disabilities as a child, poor educational opportunities, and brain dysfunction effects on memory and communication.

The paired associates of the Wechsler Memory Scale provides a measure of learning both new (difficult) and old (easy) paired associates. This test is particularly useful in elderly patients, especially when there is a question of dementia. The WAIS Digit-Symbol test also may involve learning to make new associations between symbols and digits but it is influenced by many other factors, such as psychomotor speed, perceptual shifting, etc. Systematic improvement in performance in the TPT may also reflect learning. Improvement or lack of it over trials in the Luria words also gives valuable information as to the patient's ability to acquire new information.

Abstracting

Verbal abstracting ability is inferred from the WAIS similarities; in cases with left-hemisphere or generalized brain-dysfunction performance on this subtest may be impaired, whereas right-hemisphere dysfunction may have little or no effect. Nonverbal abstracting ability is measured by the Category Test. This test is one of the more sensitive tests of the battery and is depressed with all types of brain dysfunctions, that is, patients with lateralized or nonlateralized lesions, or, anterior versus posterior lesions.

Problem Solving and Set Flexibility

Many of the tests noted previously have problem-solving components. It is useful to consider the TPT as an unusual or unfamiliar task to be solved and the time taken by the patient as measuring problem-solving skills. It also offers the opportunity to observe the approach and strategy of the patient in solving the problem, for example, systematic versus unsystematic, perseveration of responses, etc. Similarly the Category test has a strong problem solving component and offers the opportunity to examine set flexibility as well as perseveration. The Block Design has problem-solving components in addition to measuring construction dyspraxia. The Trails B is an excellent measure of set flexibility. It is noteworthy that these four tests frequently load on the same factor in studies of brain damaged or other groups (Fabian, Jenkins, & Parsons, 1981). Together with the Digit Symbol test and the TPT-Location, they are the tests that Reitan (Reitan & Darison, 1974) has indicated are most sensitive to brain damage and to aging.

PERSONALITY

The MMPI provides useful material relevant to diagnosis and treatment. For example, patients are frequently referred with the question of functional versus organic or raise the question of the degree to which emotional perturbations may be adding to the symptom pattern. Whereas it is appropriate for brain-damaged patients to respond with mild elevations on the Depression scale, large elevations or the absence of any depressive feelings signal an unusual reaction. Feelings of alienation, strange experiences, thought disorder, suspiciousness, heightened tension and anxiety imply different treatment approaches than the presence of denial, repressive, and avoidant behaviors.

NEUROLOGICAL IMPLICATIONS

In this section the Halstead Impairment Index and the RNG Impairment ratings are considered in light of the age and education of the patient. A statement is made as to the general level of impairment such as "not impaired, borderline, mild, moderate, or severe." In cases of disagreement between the two indexes there should be a resolution and decision as to which score more clearly reflects the disorder in the opinion of the writer. Other measures of disturbed functioning are to be noted, for example, unusual variability, pathognomic signs, right-left differences. If the decision is "impaired" or "borderline impairment," some attempt should be made to comment on whether the brain dysfunctional state is acute versus static or lateralized versus diffuse. Statements as to localization, when possible, should be offered. Finally, if the patient has a known brain dysfunctional state (e.g., neurosurgical removal of a left temporal lobe, trauma, or aneurysm in right posterior parietal region) the degree of consistency or inconsistency of the findings with the neuropsychological examination should be noted.

PSYCHOLOGICAL IMPLICATIONS

Specific recommendations for patient management should be made when possible. Implications of the findings for cognitive retraining, rehabilitation, psychologically oriented therapy, and life adaptation are given where appropriate. The possible interaction of personality and neuropsychological factors in life situations is considered.

EXAMPLES OF NEUROPSYCHOLOGICAL REPORTS

In presenting these case reports, I have selected one (Mr. X.) that is fairly typical of a large number of patients referred to a clinical neuropsychologist, that is, a patient with a psychiatric history including alcoholism, disorganized behavior, and history of possible brain injury. The second (Mr. Y.) was selected because it demonstrates the occasional unique and critically important contributions of the neuropsychological examination. The first report is based on my examination; the second report is based on a case examined by Dr. Kathleen Sullivan under the supervision of Dr. Russell Adams of our department. I am appreciative of their permission to modify the report in certain stylistic ways in the interest of consistency of presentation.

Our procedure in scoring is to assign RNG severity ratings to all tests for which they exist. For those tests where they do not exist, we use available norms and infer impairment or less efficient performance by considering means and standard deviations from similar age groups of controls. In view of space constraints I have summarized only the RNG Indexes, the Halstead Impairment Index and Wechsler data from these two cases (Table 2).

REPORT 1: NEUROPSYCHOLOGICAL EVALUATION

Name: Mr. X.
Age: 58
Education: 16 Years
Marital Status: Married
Occupation: Retired (Accountant, Auditor)

Referral Information

This 58-year-old retired male was referred for neuropsychological evaluation because of periods of disorganization and psychotic behavior. He has a history of head injury and chronic alcoholism. Evaluation of the patient's strengths and areas of possible impairment is requested.

Summary and Recommendations

This patient is functioning at the *average* level of intelligence (WAIS Full Scale IQ of 99) with Verbal IQ of 101 and a Performance IQ of 96, indicating little difference in the abilities measured by these two scales.

Table 2. HRB Data Using the RNG Tests and other Selected Test Results for the Two Case Reports

	Mr. X.		Mr. Y.	
	Score	RNG rating	Score	RNG rating
Category	104	4	116	4
TPT-time	30'	5	30'	5
TPT-memory	4	2	3	3
TPT-location	0	4	1	3
Seashore rhythm	23	2	27	1
Speech-sounds percep	11	2	15	3
Tapping	48	2	52	1
Trails B	170"	3	268"	4
WAIS dig symbol	6	3	2	4
Aphasia screening	3	1	10	2
Spatial relations	4	2	8	4
Perceptual disord.	6	1	18	2
Average impairmant rating	2.58	3	3.00	4
Halstead imp. index	1.0	—	.7	—
WAIS full scale IQ	99	—	98	—
WAIS verbal IQ	101	—	125	—
WAIS perform IQ	96	—	68	—
Wechsler mem. scale MQ	89	—	109	—

[a]RNG ratings from 0, above average performance, to 5, severely impaired, with a rating of 2 being in the mildly impaired range.

Considering his educational and work background, this level of functioning is seen as lowered compared with a previous estimated bright average level of intelligence. The Halstead-Reitan Impairment Index and more extended average Impairment rating scores suggest *moderate to severe impairment* of higher cortical functions, especially in memory, new learning, abstracting, and problem solving. This condition is seen as a product of a static diffuse brain dysfunctional state plus the effects of a loss of a sense of direction in life and lowered self-esteem. The following is recommended:

1. Complete abstinence from all alcoholic beverages
2. A structured work (other than auditing) or hobby experience
3. Counseling or psychotherapy for at least six months
4. Retesting in one year to determine progress

EXTENDED REPORT

Background Information

At the time of the neuropsychological examination, Mr. X. was hospitalized on a psychiatric service. He graduated from college with a Bachelor's Degree in Accounting and worked in that area primarily as an auditor up until 1975. In 1969, according to the history given by the wife, he started drinking heavily. This continued up until 1975 when he was hit with apparently the butt of a gun, or a similar instrument, while sitting in his car, and was examined by a neurosurgeon for possible brain damage. According to the wife, there was a hole in the skull where the bone fragments were removed, but she was told that there was no indication of any brain damage. This was in the left-temporal-parietal region. No plate was put into the skull and there has been no problem since that time, although, according to the wife, the patient is very protective of his head and anything that might harm him in that area. In March of 1975, he had a marked state of physical deterioration due to a gastric fistula and underwent surgery from which he gradually recovered. Subsequent to that operation, the wife noticed that he could not cope with the demands of the job and he was retired on physical disability from an auditing position. In July, 1979, while drunk, he fell back hitting his head and cracked his skull. As described by his wife, this crack ran from the base (occiput) of the skull to the right ear drum. After several weeks, his behavior became grossly disorganized. He was hospitalized on neurology in September, 1979, but left the ward saying that they were not doing anything for him. Disorganization of behavior continued and finally culminated in hospitalization. According to the wife's account, except for a period in 1978 when he went through an alcohol treatment program, Mr. X. has been a steady consumer of alcohol, especially beer.

Mrs. X. regarded him, previous to 1969, as a very intelligent individual who was very competent in his work and had an excellent reputation for his productivity. She now believes that he is still intelligent, but seems to have lost his ability to reason and has noticed certain personality changes that she attributes to the drinking.

Tests Administered and Behavior

The patient and his wife arrived on time for the scheduled testing. Mr. X. is a small, slight man who was quite verbal. He was very cooperative and tried very hard to complete all the tests to the best of his ability despite obviously tiring as time went on throughout the day. He kept up a fairly continuous verbal conversation during the most of the testing. He did complain about a mild headache and the fact that he had not eaten any breakfast, but nonetheless appeared to be able to mobilize sufficiently in response to the demands of the situation so that we may consider the results to be representative of his present level of functioning. His memory for events and dates was poor, rendering him an unreliable historian.

Tests Administered

Halstead-Reitan Neuropsychological Battery
Wechsler Adult Intelligence Scale

Wechsler Memory Scale
Benton Visual Retention Test
Trail Making Test
Luria Word Memory Test
Purdue Pegboard
Minnesota Multiphasic Personality Inventory

Analysis of Results

General Considerations for Interpretation. This patient is a white man of 58 years with a Bachelor's Degree in Accounting. He is right handed and right eyed. There were no peripheral impairments that would affect the results of the evaluation. His cooperation and motivation are considered sufficiently adequate. The results are considered representative of his present level of functioning.

General Intelligence and Achievement. Mr. X. achieved a Full Scale WAIS IQ of 99 placing him in the *normal* range of intelligence. His Verbal IQ of 101 and Performance of 96 suggests a relatively even balance in his abilities. Given this man's history of education and previous attainment, it would appear that there is a general lowering of both verbal and perceptual-motor skills.

Sensory-Perceptual and Sensory-Motor Functioning. Sensory-perceptual functions were generally intact and there were no suppressions. However, on the test for tactile dysterognosis, his scores with the right hand were considerably longer than the left hand. This appeared to be due to two unusually long delays in recognition time of two different shapes. Otherwise, his responses were generally slowed in tactile recognition. In sensorimotor performance, his grip strength was poor in both hands, although there was little difference in general level of performance between them. Finger tapping was within normal limits for a person of his age.

Attention, Concentration, and Memory. Mr. X. obtained five digits forward and four digits backwards in Digit Span, which was among his lower scores on the WAIS. There seems to be some difficulties in immediate recall, as evidenced by performance on the Seashore Rhythm test, which is mildly impaired, and his Wechsler Memory Scale Quotient, which is some 10 points below (89) the expected score for his intelligence. His immediate recall for both semantic and figural material was in the moderately impaired range, as was his delayed recall. His incidental memory for tactually perceived shapes was impaired and his memory for their location was markedly impaired. He had difficulty in recalling who was president before Carter, and could not name the Governor of the state of Oklahoma or the Mayor of Oklahoma City. Perhaps the most striking aspect of his delayed recall was the tendency to confabulate. For example, in his delayed recall of a story, he told of two ships colliding and a Swedish rescue helicopter coming to rescue the passengers and recover all of them. This was not mentioned in his first recall (and, of course, is not part of the story presented). His delayed memory for words on the Luria Memory test was mildly impaired. His performance on the Benton Visual Retention Test, that is, memory for geometric figures, was mildly impaired. Memory for dates in the interview was poor; confirmation had to be sought from his wife. In conclusion, a mild to moderate memory deficit is present.

Language and Communication Skills. Mr. X.'s vocabulary is in the average to slightly above average range and was marked by a great deal of variability. He would miss some relatively easy words and get some relatively difficult words. This type of scatter suggests a person of previously higher level of functioning intelligence. There was no suggestion of either receptive or expressive aphasia, as measured on the Aphasia Screening Test. The patient's communicative verbal behavior in the testing situation was good; in fact, he was quite verbal and talked a great deal.

Calculational Skills. For a person with a background in accounting, he did surprisingly poorly on the Arithmetic test of the WAIS. This, and Digit Span, were his lowest scores on the Verbal scale, and certainly suggests difficulties in concentration and calculation. Contrary to expectancy, he was able to do two verbally presented problems on the Aphasia Screening Test without error.

Spatial Relationships and Construction Dyspraxia. This patient had a Block Design score on the WAIS that was about average for his present general level but lower than would be expected for his premorbid level of functioning. Performance on the Object Assembly test was somewhat more impaired. His performance on the Greek cross was mildly impaired and there were some aspects of his drawing of the key that suggested perceptual reproductive difficulties. In conclusion, the patient manifests mild impairment in spatial relationships and constructional praxis.

Learning, Abstracting, and Problem Solving. His fund of general information was slightly below average level, suggesting a restricted interest pattern. His ability to learn new materials as evidenced by the Paired Associate Learning test is very poor. He was able to learn only 1 of the hard associates out of 12 tries, and only 5 of the 6 easy associates. There were indications of perseveration from examples that were given prior to the actual test. Similarly his learning of the 10 Luria words was impaired; it took him 10 trials to learn the words when the expectancy for a person of his age is 5 or 6. His learning during the Tactual Performance test was minimal; he was only able to get 2 blocks correctly placed in 10 minutes with his right hand and the same very poor performance with his left hand. Perhaps one of the greatest areas of impairment lies in his nonverbal abstracting performance. In contrast with his verbal abstracting ability, which is in the average range, his performance on the Category test is grossly impaired. Other measures of problem solving and set flexibility suggests a person who has many difficulties in dealing with new situations where old established verbal habits cannot carry him through. Low performance on the Category test, Tactual Performance test, Block Design, and the Trails B are in accord with this interpretation.

Personality. During the testing situation there were no signs of thought disturbance but many indications of lowered self-esteem and marital discord. The latter was chiefly centered about the patient's drinking. The MMPI was taken validly and, surprisingly, the profile was within normal limits. The pattern of the profile suggests a person who feels somewhat alienated from self and who experiences some difficulties in thinking but who is typically outgoing and likely to mildly act out difficulties. It seems likely that he is denying his lowered self-esteem and depressive feelings. Alcohol may have served to help him avoid coming to grips with his lessened effectiveness in life, loss of sense of direction, and lowered self-esteem. Should he abstain from alcohol, he may well manifest a depressive state; therefore, continued therapeutic contact over at least six months is recommended.

Neurological Implications

The patient's Impairment Index on the Halstead-Reitan Battery, according to national norms, is 1.0 (severe impairment), and on the Oklahoma norms .9 (severe impairment). The Average Impairment Index of 2.58 gives a rating of 3 for moderate impairment. Taking age into consideration, the impairment indexes indicate a person with moderate to severe impairment of higher cortical functions. There is a fairly wide variability of his performance, both within test and between tests, that suggests inefficiency of functioning due to the presence of brain dysfunction. There are no pronounced right/left differences, although the patient's performance on the test of tactual dystereognosis strongly suggests that there is some inefficiency of the right hand. The data are consistent with a static, diffuse brain dysfunctional condition that is heightened by an underlying depressive reaction.

Psychological Implications

The history given by this patient and his wife, and the results of the psychological testing suggest that recovery from the present condition will be difficult unless an active, structured program is implemented for this patient. Such a program would involve the abstinence from alcohol and some type of work situation to enable him to regain some feelings of lost self-esteem. This work situation, however, should not be auditing (given his poor arithmetic scores and problem-solving skills) unless it is of a routinized nature. Continued contact for the next six months or so in a psychotherapeutic or counseling situation is recommended for reasons noted above. Given these conditions, we would strongly suggest that he be seen again in one year to determine whether improvement at the neuropsychological level has occurred.

REPORT 2: NEUROPSYCHOLOGICAL EVALUATION

Name: Mr. Y.
Date of Birth: 8-29-26

Occupation: Retired Air Force Pilot - USAF
Education: 18 Years (Master's Degree)

Referral Information

Mr. Y. is a 57-year-old white male who is retired from the United States Air Force. For the past several years he has noticed a decline in his perception, mechanical abilities, reaction time, driving abilities, and ability to read and produce cursive handwriting. He reports that the symptoms have been especially pronounced over the past two to three years. Dr. —— requested a neuropsychological evaluation in order to evaluate the presence and/or extent of organic impairment.

Summary and Recommendations

1. Neuropsychological assessment reveals severe impairment in several areas of functioning, such as severe constructional difficulties and problems on all tasks requiring nonverbal and visual-spatial abilities, including memory for visual information. These areas of deficit point to a pathological process in the right parietal-occipital area of the brain. The possibility of major structural damage in this area should be investigated thoroughly.
2. Mr. Y.'s verbal abilities as well as verbal memory have been preserved and are, in general, in the superior range.
3. Personality assessment reveals that Mr. Y. is suffering from a moderate depression accompanied by worry and rumination. This appears to be of a rather long standing nature rather than being an acute emotional reaction.
4. The patient's neuropsychological deficits cannot be attributed to his depression.
5. In light of our findings a careful review of the patient's capabilities both at work and at home should be made with him and his wife.
6. As this information is worked through, supportive psychotherapy would be helpful.
7. A current neurological reevaluation is recommended.

EXTENDED REPORT

Background Information

Mr. Y. retired from the United States Air Force as a Colonel after 30 years of service as a pilot. Up until this time he had no major illnesses or history of psychiatric problems. Around the time of his retirement, however, he began to notice that he was unable to perform certain tasks as well as he used to. For instance, he began to notice his handwriting deteriorating as well as his ability to read the handwriting of others. He can, however, read and comprehend material that is printed. He also began to notice changes in his perception in that he often could not recognize immediately objects that he was looking for. He noticed an increased difficulty in comprehending and problem solving in terms of mechanical abilities. He often has difficulty driving, in that he reacts too slowly, often missing the street he was looking for as well as having difficulty following curved roads. Mr. Y. reports that he has difficulty reading clocks. He feels that his memory is fine although he sometimes has difficulty recalling people's names. He feels that his speech is also unimpaired. He has no complaints of hearing loss or other sensory disturbances. He does report feeling depressed but relates this to sadness and concern over his current difficulties.

Mr. Y. has consulted with his family physician as well as a neurologist. He reports that a CT Scan, EEG, and other tests performed one to two years ago produced no pathological findings. He was told that he "just needed to shape up and buckle down, and not worry so much." Because he continued to experience these difficulties, he had a reevaluation with his neurologist eight months ago. The CT and EEG were not repeated although a neurological exam was performed. Mr. Y. reports that he was again told that "there is nothing wrong" with him.

Mr. Y. is married and resides with his wife. He reports that she has some medical problems that affect her well-being. He reports the sexual relationship being nearly nonexistant due to lack of interest on the part of both parties. Family history is negative for psychiatric problems. Mr. Y. reports that his mother was quite senile at the time of her death but knows of no other neurological disorders in the family. Mr. Y. uses alcohol very rarely. He is currently on no medications.

Tests Administered and Behavior of Patient

Mr. Y. is a tall, slender, well-groomed man who was cooperative and friendly in the testing situation. He showed very good motivation throughout the testing despite the fact that he was aware of his poor performance on several of the tests. He quickly established rapport with the examiners. The results of the current evaluation are considered to be a valid index of Mr. Y.'s current level of functioning.

Order of Tests:

Halstead-Reitan Battery
Trail Making Test
Purdue Pegboard
Wechsler Adult Intelligence Scale-Revised
Wechsler Memory Scale—Immediate and Delayed Luria Memory
  Words
Bender-Gestalt Designs and Background Interference Procedure
Minnesota Multiphasic Personality Inventory
Beck Depression Inventory

Line Bisection and Line Oriention Tasks
Drawing Tasks

Analysis of Results

General Considerations for Interpretation. The patient has a master's degree and was a pilot. Thus his expected level of intelligence should be at least high average and balanced verbal and performance skills. There are no indications of peripheral injuries; he is right handed and right eyed.

General Intelligence and Achievement. On the WAIS-R Mr. Y. achieved a Verbal IQ of 125 and a Performance IQ of 68. This difference of 57 points between verbal and Performance tests is extremely rare and is viewed as highly clinically significant. In terms of his verbal abilities, Mr. Y. performs in the superior range. He has a good fund of general information as well as an accurate and precise ability to define vocabulary words. His verbal abstraction ability and verbal problem solving abilities are also very good.

In terms of nonverbal abilities, however, Mr. Y. had extreme difficulty on most subtests requiring visual-spatial analysis. His ability to discriminate essential from nonessential visual detail, to manipulate part–whole relationships, and to perform a perceptual-motor coding task was extremely impaired. His best score on the Performance scale was on a test requiring the arrangement of pictures into a logical sequence, although this also fell into the impaired range.

Sensory-Perceptual and Sensorimotor functioning. Mr. Y. is right hand, foot, and eye dominant. On a test of fine motor speed, his performance was within normal limits for the right hand but mildly impaired for the left hand. Grip strength was good bilaterally but a greater difference between the left and right hand (with the left being weaker) was seen than is usually expected. Fine motor dexterity was within normal limits bilaterally.

There were no suppressions in the visual, auditory, or tactile modes. He made several errors in finger identification (finger agnosia) on the left hand whereas he made no errors on the right hand.

Mr. Y. was unable to imitate even very simple hand postures that were demonstrated by the examiner. His right-left orientation was within normal limits. He had difficulty with alternately touching the fingers on his left hand. In summary, both sensory and motor functions in the left hand are impaired compared to the right hand, thus implicating the right hemisphere.

Attention, Concentration, and Memory. Mr. Y. was oriented to person, place, date, and situation. His attention and concentration were very good. On the Wechsler Memory Scale, Mr. Y. obtained a Memory Quotient of 109. His ability to recall verbal information such as in stories, both on immediate and long term recall, was within normal limits. Memory for Luria words was good even after a delay of one hour. His recall of visual-spatial information was severely impaired, however, both at immediate and delayed recall tests. Again, the data suggest right-hemisphere dysfunction, particularly posterior.

Language and Communication Skills. On the Aphasia Screening exam, Mr. Y.'s performance was mildly impaired. His difficulties were in naming, such as finding the name of "square" and "baby." His comprehension, articulation, and repetition of language were all within normal limits.

On a test of receptive language, where the patient is instructed to pick the correct spelling of a spoken word, his performance was moderately impaired. Rather than difficulty with receptive language, however, it was felt that his problems in letter recognition prevented him from identifying the correct spelling of the words. Interestingly, was able to read letters adequately if they were printed plainly. However, cursive writing or unusual, fancy script was difficult for him to identify.

Calculational Skills. The patient's score on the Arithmetic test was one of the lowest of his Verbal subtests but still within the normal range. He solved the two problems on the Aphasia Screening Test easily.

Spatial Relationships and Construction Dyspraxia. Mr. Y. displayed very severe difficulty in copying Bender-Gestalt designs under standard administration. When an interfering background was added, his performance further deteriorated. Difficulties with the reproduction of designs and putting together jigsaw-like puzzles have been mentioned previously. He was unable to copy a Greek cross, even after several attempts. Furthermore, he had difficulty drawing a clock accurately, placing the number "1" at the top and leaving off the number "12" completely. On the left-hand side of the clock, his numbers were outside of the outer boundary of the clock face. He was able to draw a rough approximation of a map of the United States. He was also able to label the points of a compass. His ability to bisect a series of straight lines was poor and he appeared to neglect several of the stimulus lines on the page. On a test of judgment of line orientation, however, his performance was within normal limits. The severe construction dyspraxia suggests right posterior parietal-occipital dysfunction.

Learning, Abstracting, and Problem-Solving Skills. This patient has a superior fund of general knowledge. His learning of paired associates was slower than expectancy. His learning of the Luria words was erratic, getting all 10 on the 5th trial, 9 on the 6th and 7th trial, suggesting mild retrieval difficulties, possibly due to his depression.

Whereas his verbal abstracting abilities as measured by WAIS-R similarities were in the high-average ranges, his nonverbal abstracting performance (Category Test) was severely impaired (116 errors). Similarly, his time performance to replace blocks on a form board (while blind folded) was severely impaired; after two previous trials he could only get one form correctly placed in 10 minutes (expected time for him would be 2 to

4 minutes for 10 forms). He showed no learning going from right to left hand on this task. Set shifting appeared difficult for him; his performance on an alpha-numeric sequencing task, alternating connecting numbers and letters was also severely impaired.

In summary, verbal learning is within normal limits but nonverbal learning, abstracting, and problem solving are severely impaired. The data suggest relative intactness of left hemisphere functioning but severe impairment in the right.

Neurological Implications. Mr. Y. obtained a Halstead Impairment Index of .7 (moderate impairment), an Oklahoma norm impairment of .7 (moderate impairment) and an average impairment rating of 4.00 (severe impairment). However, his verbal abilities are retained at the superior level. In contrast, almost all other tests involving visual-spatial-motor or tactual-spatial-motor analysis, organization and construction are severely impaired. These results suggest a severe brain dysfunctional state in the right parietal-occipital areas of the brain. Despite the previous negative neurological results in the past, we recommend a thorough neurological reevaluation.

Psychological Implications. The profound differences in this patient's functioning should be carefully explained to both him and his wife. As long as his work situation involves verbal cognitive abilities he is likely to do well. If nonverbal problem solving, memory, or analysis is required he will have grave difficulties. There are also some suggestions of left visual-field neglect. This caused errors on some tests and could affect his work situation. If it worsens, accidents are possible. A careful review of what he can and cannot do in relation to work and home is necessary. Supportive psychotherapy during the period of adjustment to this information is highly recommended.

FUTURE DEVELOPMENTS OF THE HALSTEAD-REITAN BATTERY

In this section four topics will be discussed. The first is concerned with methodological improvements in the HRB; the second considers the prospects for improving the clinical sensitivity of the battery; the third topic discusses the use of the HRB in rehabilitation and cognitive retraining, and the fourth topic is an attempt to place the HRB in the perspective of the directions of today's clinical neuropsychology.

METHODOLOGICAL IMPROVEMENTS IN THE HRB

A long standing concern of many neuropsychologists has been the lack of uniform age and educational norms for the HRB. Of the two, age is probably the more important. To give one example of the potency of the age variable, Prigatano and Parsons (1976) in two separate studies found correlations between the Halstead Impairment Index and age of .57 and .64 in the non-brain-damaged groups and .33 and .44 in the brain-damaged groups. All correlations were significant. As noted earlier, norms gathered by various investigators can be used. However, there has been no systematic attempt to gather the results from many laboratories and arrive at an age adjustment that is the equivalent in soundness and breadth to that used in the WAIS or WAIS-R. Given the importance of many neuropsychological examinations for questions of diagnosis, treatment, compensation, competency, and the like, it seems imperative that this development occur. Education is probably of lesser importance than age on the HRB but nevertheless contributes significantly to the variance on a number of HRB scores. Prigatano and Parsons (1976) found significant correlations between education and six Halstead measures in a non-brain-damaged nonpsychiatric group. These and many other similar findings led Golden and his Colleagues (Golden, 1981) to apply age and education corrections to scoring on the Luria-Nebraska Battery. These adjustments certainly relieve the clinician of subjectively weighing the potential influence of such variables as happens frequently with the HRB.

A final methodological improvement in my opinion would be the development of T scores. In the Vega and Parsons cross-validation study of the HRB (1967) we found that the use of mean T scores gave rise to better discrimination between the brain-damaged and controls than the Impairment Index. The advantages of converting to a common metric are obvious: scores on different tests could be directly compared for degree of impairment. I had the opportunity recently to make blind judgments concerning the presence or absence of brain damage and if present, laterality of damage on some 90 patients using both the HRB and the Luria-Nebraska findings. (See Chapter 9 in this volume). There is no question but that the format of T score profiles on the Luria-Nebraska made the process of visual inspection and ready comparison of various subtest scores much easier than on the HRB. Various laboratories including our own have advocated and used T scores based on their own norms (Harley et al., 1980; Kiernan & Matthews, 1976; Vega & Parsons, 1976).

In summary, the HRB could be greatly improved by having a national sampling much like the Wechsler scales, converting all scores to T scores and introducing age and education adjustments within the T scores.

IMPROVING THE CLINICAL SENSITIVITY OF THE HRB

Can the clinical sensitivity of the HRB be improved? This question actually has two aspects to it. The first has to do with the objective actuarial discriminating power of the HRB; the second has to do with the interpreter of the HRB test findings, the clinician. As regards the first aspect, the objective analysis of the HRB is probably best captured by the computer programs for identification, lateralization, and localization of brain damage. The original computer program for the neuropsychological Key devised by Russell, Neuringer, and Goldstein (1970) seemed to have considerable promise. Another program for the HRB entitled "Brain 1" was devised by Finklestein (1976). Anthony, Heaton, and Lehman (1980) compared the two programs using 150 verified brain-damaged patients and 100 normal controls. Both programs were able to identify brain damaged and controls with about 80% to 85% accuracy. As regards lateralization, the key correctly and significantly lateralized 69.4% of the right-hemisphere lesions and 60.6% of the left-hemisphere lesions but only 32.5% of the diffuse lesions. In contrast, the Brain 1 lateralization predictions were not significant, correctly identifying only 41% of the right-hemisphere, 31% of the left-hemisphere and 36% of the diffuse-damaged patients. Both sets of results are decidedly lower than those of the initial studies. Anthony et al. (1980) concluded that "both of these programs in their present form are of limited clinical value." Goldstein and Shelly (1982) reanalyzed the Anthony et al. (1980) data and added a new group of patients of their own. Although their reanalysis suggested that the Anthony et al. (1980) data were more comparable to the original Russell, Neuringer, and Goldstein (1970) results, Goldstein and Shelly's new cross-validation data suggested a lower percent of correct identification and lateralization than either of the other two studies.

It seems apparent that computer programs of actuarial prediction contain at best only a mild promise for the future. The limiting factors of reliability of the criteria for diagnosis against which tests are validated; the interaction of psychological and psychiatric reactions to brain damage; the interaction of the examiner and patient; and the reliability of the instruments themselves place an upper limit on the degree of successful predictions obtained by actuarial methods.

Turning to the second aspect of improving the clinical sensitivity of the HRB—the clinician, one can only reiterate what has been noted in the literature (Meier, 1981). Knowledge, experience, and wisdom are ingredients that make for clinical sensitivity. In my opinion the clinical neuropsychologist should be a person who has had training in clinical psychology, a general internship and preferably postdoctoral training in neuropsychology. The clinician should have a background in psychometric theory, psychological tests and measurement, psychopathology, cognition, perception and learning, psychotherapy, neuroanatomy, neurophysiology, clinical neurology, and neuropathology.

Postdoctoral experience should include supervised child and adult neuropsychological cases from a variety of referring disciplines and include exposure to a variety of referral questions. Experience with rehabilitation and cognitive retraining is not only useful but humbling for the fledging clinical neuropsychologist. Given these academic and experiential accomplishments and a few years as a professional handling a high percentage of neuropsychological cases, the accrued wisdom of the HRB clinical neuropsychologist will be the final determinant of how successfully the HRB can be applied. I do not expect more than modest improvement in actuarial prediction by improved quantitative methods but do expect decided improvement in clinical predictions as clinicians become more appropriately educated and experienced.

The HRB and Rehabilitation

Will the HRB be used to a greater extent in rehabilitation and cognitive retraining in the future? Perhaps! Let us first consider an area that is related to the question, namely, does the HRB predict life adjustment? Heaton, Chelune, and Lehman (1978) found that HRB scores were significantly different for full-time employed patients versus part-time versus nonemployed patients. As expected, the nonemployed had the most neuropsychological deficits, followed by the part-time employed, with the full-time employed having the least. In a prospective study, Dodrill and Clemmons (1984) found that the HRB given 3 to 11 years previously to high school epileptics predicted vocational adjustment and independent living of that group as young adults. Heaton and Pendleton (1981) review many other studies demonstrating the relevancy of the HRB and Wechsler Scales for predictions of adaptation in everyday living.

Given these studies, it would appear that the HRB might well be important in several aspects of rehabilitation. First, the degree of the deficits might suggest limits within which progress in rehabilitation might be expected. Persons who are more intact on the HRB will be more likely to achieve higher levels of adaptation than persons who manifest more widespread and severe HRB deficits. Second, the HRB can be used to monitor improvement in the functioning of the patient. Goldstein and Ruthven (1983) used the HRB to compare 19 cases of adult brain-damaged patients given systematic rehabilitation training with 22 similar patients who had no systematic rehabilitation training. The experimental group showed a significant improvement in their RNG average impairment rating whereas the control (no training) group did not. The authors concluded that the HRB tests were more sensitive to changes in functioning than the WAIS tests. (For an interesting study along these lines, see Heaton, Grant, Anthony and Lehman, 1981.)

A third way in which the HRB may contribute to rehabilitation is from an analysis of the patterns of deficit. Goldstein and Ruthven (1983) give examples of how a thorough analysis of the Russell, Neuringer, and Goldstein (1970) version of the HRB can lead to individualized programs of remedial training for left hemisphere and right hemisphere patients. Incidentally, this book contains a wealth of interesting observations, practical hints, and theoretical considerations that will amply repay the clinical neuropsychologist interested in rehabilitation.

In summary, I suggest that the HRB will receive more attention in the future as a predictor of future life adjustment for the brain-damaged patient, as a baseline of functioning that can be then used to reflect improvement and, finally, as a method for determining patterns of deficit on which rehabilitative and cognitive retraining methods can be based.

The HRB and the Future

There is little doubt that the HRB will continue to play a major role in many neuropsychology laboratories for the next decade. There are many reasons for this. There is a body of knowledge based on the HRB that is extensive (hundreds of studies) and contains some of the more interesting and convincing demonstrations of brain–behavior relationships in the literature. The standardized quantitative empirical methods of the HRB are appealing to many neuropsychologists and to other health disciplines. The fixed-battery approach, although more lengthy and perhaps on occasion less definitive than the flexible approach, nevertheless has the advantages of broad and comprehensive coverage of many areas of functioning. The validity of the HRB is unquestioned at this point. It will also be of utility in measuring change, either the natural process of recovery from brain damage or the rehabilitative process.

Will the HRB stimulate new directions of research, new models of specific functions, new cognitive retraining techniques, new theories of brain–behavior relationships? I doubt it! The HRB is largely atheoretical and empirical. As such it serves us well in many aspects of clinical neuropsychology. It will have a continuing contribution toward the growth and expansion of clinical neuropsychology. But innovative steps are likely to come from theoretical models of psychological functioning coupled with neuropsychological and brain functioning theories and empirical findings. As I see such a development, it will be largely what I term mini-models. Overall theories of brain–behavior functioning, as in systems theory or Luria's (1980) theory, are probably too abstract and too general to be useful in a predictive fashion, although they are useful as frames of reference. Models such as those emerging from the neuropsychology of memory, for example, semantic versus procedural memory and their corresponding neuroanatomical structures (Butters, 1984), or models of information processing and strategies associated with the two hemispheres (Parsons, 1984) may lead to a clearer formulation of a limited but testable hypothesis of how the brain is organized to handle such functions. Incorporating the results of viable mini-models into an overarching theory is a task for future generations of neuropsychologists.

REFERENCES

Anthony, W. Z., Heaton, R. K., & Lehman, R. A. (1980). An attempt to cross-validate two actuarial systems for neuropsychological test interpretation. *Journal of Consulting and Clinical Psychology, 48,* 317–326.

Boll, T. J. (1978). Diagnosing brain impairment. In B. B. Wolman (Ed.), *Clinical diagnosis of mental disorders.* New York: Plenum Press.

Boll, T. J. (1981). The Halstead-Reitan neuropsychological battery. In S. B. Filskov & T. J. Boll (Eds.), *Handbook of clinical neuropsychology.* New York: Wiley.

Butters, N. (1984, May). *Is there a continuity of neuropsychological deficits from social drinkers to Korsakoff patients?* Paper presented at a NIAAA Conference on Clinical Implications of Recent Neuropsychological Findings in Alcoholics, Boston, MA.

Chapman, L. F., & Wolff, H. G. (1959). The cerebral hemispheres and the highest integrative functions of man. *Archives of Neurology, 1,* 19–35.

Dodrill, C. B. (1981). Neuropsychology of epilepsy. In S. B. Filskov & T. J. Boll (Eds.), *Handbook of clinical neuropsychology.* New York: Wiley.

Dodrill, C. B., & Clemmons, D. (1984). Use of neuropsychological tests to identify high school students with epilepsy who later demonstrate inadequate performances in life. *Journal of Consulting and Clinical Psychology, 52,* 520–527.

Fabian, M. S., Jenkins, R. L., & Parsons, O. A. (1981). Gender, alcoholism and neuropsychological functioning. *Journal of Consulting and Clinical Psychology, 49,* 139–141.

Fabian, M. S., Parsons, O. A., & Shelton, M. D. (1984). Effects of gender and alcoholism on verbal and visual-spatial learning. *Journal of Nervous and Mental Disease, 172,* 16–20.

Filskov, S. B., & Goldstein, S. G. (1974). Diagnostic validity of the Halstead-Reitan neuropsychological battery. *Journal of Consulting and Clinical Psychology, 42,* 419–423.

Finkelstein, J. N. (1977). Brain: A computer program for interpretation of the Halstead-Reitan Neuropsychological Test Battery (Doctoral dissertation, Columbia University, 1976). *Dissertation Abstracts International, 37,* 5349B. (University Microfilms No. 77-8, 8864).

Fromm-Auch, D., & Yeudall, L. T. (1981). Normative data for the Halstead-Reitan neuropsychological tests. *Journal of Clinical Neuropsychology, 5,* 221–238.

Golden, C. J. (1981). A standardized version of Luria's neuropsychological tests: A quan-

titative and qualitative approach to neuropsychological evaluation. In S. Filskov & T. J. Boll (Eds.) *Handbook of clinical neuropsychology*. New York: Wiley.
Goldstein, G., & Ruthven, L. (1983). *Rehabilitation of the brain-damaged adult*. New York: Plenum Press.
Goldstein, G., & Shelly, C. (1982). A further attempt to cross-validate the Russell, Neuringer & Goldstein neuropsychological keys. *Journal of Consulting and Clinical Psychology, 50*, 721–726.
Goldstein, K. (1959). Functional disturbances in brain damage. In S. Arieti (Ed.), *American handbook of psychiatry*. New York: Basic Books.
Halstead, W. C. (1947). *Brain and Intelligence*. Chicago, IL: University of Chicago Press.
Harley, J. P., Leuthold, C. A., Matthews, C. G., & Bergs, L. E. (1980). *Wisconsin neuropsychology test battery T-score norm for older Veterans Administration Medical Center patients*, Unpublished monograph, University of Wisconsin, Neuropsychology Laboratory, Madison, WI.
Heaton, R. K., & Crowley, T. J. (1981). Effects of psychiatric disorders and their somatic treatments on neuropsychological test results. In S. B. Filskov & T. J. Boll (Eds.), *Handbook of clinical neuropsychology*. New York: Wiley.
Heaton, R. K., & Pendleton, M. G. (1981). Use of neuropsychological tests to predict adult patients' everyday functioning. *Journal of Consulting and Clinical Psychology, 49*, 807–821.
Heaton, R. K., Baade, L. E., & Johnson, K. L. (1978). Neuropsychological test results associated with psychiatric disorders in adults. *Psychological Bulletin, 85*, 141–162.
Heaton, R. K., Chelune, G. J., & Lehman, R. A. W. (1978). Using neuropsychological and personality tests to assess the likelihood of patient employment. *Journal of Nervous & Mental Diseases, 166*, 408–416.
Heaton, R. K., Smith, H. H., Lehman, R. A., & Vogt, A. T. (1978). The prospects for faking believable deficits on neuropsychological testing. *Journal of Consulting and Clinical Psychology, 46*, 892–900.
Heaton, R. K., Grant, I., Anthony, W. Z., & Lehman, R. A. W. (1981). A comparison of clinical and automated interpretation of the Halstead-Reitan battery. *Journal of Clinical Neuropsychology, 3*, 121–141.
Hom, J., & Reitan, R. M. (1982). Effect of lateralized cerebral damage upon contralateral and ipsilateral sensorimotor performance. *Journal of Clinical Neuropsychology, 4*, 249–268.
Inglis, J., & Lawson, J. (1981). Sex differences in the effects of unilateral brain damage on intelligence. *Science, 212*, 693–695.
Jastak, J. F., & Jastak, S. P. (1965). *The Wide Range Achievement Test: Manual of instructions*. Wilmington, DE: Guidance Associates.
Kaplan, E. (1979). Discussion. In N. Sollee (Chair), *Clinical Neuropsychological services in a pediatric hospital*. Presented at the 87th annual meeting of the American Psychological Association, New York City, NY.
Kiernan, R. J., & Matthews, C. G. (1976). Impairment index vs. T-score averaging in neuropsychological assessment. *Journal of Consulting and Clinical Psychology, 44*, 951–957.
Kløve, H. (1974). Validation studies in adult clinical neuropsychology. In R. M. Reitan & L. A. Davison (Eds.), *Clinical neuropsychology: Current status and applications*. Washington, DC: V. H. Winston.
Luria, A. R. (1980). *Higher Cortical Functions in Man*. New York: Basic Books.
Matthews, C. G. (1981). Neuropsychology practice in a hospital setting. In S. B. Filskov & T. J. Boll (Eds.), *Handbook of clinical neuropsychology*. New York: John Wiley & Sons.

Meier, M. (1981). Education for competency assurance in human neuropsychology: Antecedants, models and directions. In S. B. Filskov & T. J. Boll (Eds.). *Handbook of clinical neuropsychology*. New York: Wiley.

Neuger, G. J., O'Leary, D. S., Fishburne, F. J., Barth, J. T., Berent, S., Groidani, B., & Boll, T. J. (1981). Order effects on the Halstead-Reitan neuropsychological battery and allied procedures. *Journal of Consulting and Clinical Psychology*, 49, 722–730.

Parsons, O. A. (1984). Recent developments in clinical neuropsychology. In G. Goldstein (Ed.), *Advances in clinical neuropsychology* (Vol. 1). New York: Plenum Press.

Parsons, O. A., & Prigatano, G. P. (1978). Methodological considerations in clinical neuropsychological research. *Journal of Consulting and Clinical Psychology*, 46, 608–619.

Parsons, O. A., & Farr, S. P. (1981). Neuropsychology of alcohol and drug use. In S. B. Filskov & T. J. Boll (Eds.), *Handbook of clinical neuropsychology*. New York: Wiley.

Parsons, O. A., & Hart, R. L. (1984). Behavior disorders associated with cerebral nervous system dysfunction. In H. E. Adams & P. B. Sutker (Eds.), *Comprehensive handbook of psychopathology*. New York: Plenum Press.

Parsons, O. A., Vega, A., & Burn, J. (1969). Different psychological effects of lateralized brain damage. *Journal of Consulting and Clinical Psychology*, 33, 551–557.

Pauker, J. D. (1977, February). *Adult norms for the Halstead-Reitan neuropsychological battery-preliminary data*. Paper presented at the International Neuropsychological Society Meetings, Albuquerque, NM.

Price, L. J., Fein, G., & Feinberg, J. (1980) In L. W. Poon (Ed.), *Aging in the 1980's*. Washington, DC: American Psychological Association.

Prigatano, G. P., & Parsons, O. A. (1976). Relationship of age and education to Halstead Test performance in different patient populations. *Journal of Consulting and Clinical Psychology*, 44, 527–533.

Reitan, R. M. (1955). An investigation of Halstead's measures of biological intelligence. *Archives of Neurology and Psychiatry*, 73, 28–35.

Reitan, R. M. (1959a). *Manual for administration of neuropsychological test batteries for adults and children*. Unpublished manual, Indiana University Medical Center, Neuropsychology Laboratory, Indianapolis, IN.

Reitan, R. M. (1959b). *The effects of brain lesions on adaptive abilities in human beings*. Unpublished monograph, Indiana University Medical Center, Neuropsychology Laboratory, Indianapolis, IN.

Reitan, R. M., & Davison, L. A. (Eds.) (1974). *Clinical neuropsychology: Current status and applications*. Washington, DC: V. H. Winston.

Rourke, B. P. (1981). Neuropsychological assessment of children with learning disabilities. In S. B. Filskov & T. J. Boll (Eds.), *Handbook of clinical neuropsychology*. New York: Wiley.

Russell, E. W. (1984). Theory and development of pattern analysis methods related to the Halstead-Reitan battery. In P. E. Logue & T. M. Schear (Eds.), *Clinical neuropsychology*. Springfield, IL: Charles C. Thomas.

Russell, E. W., Neuringer, C., & Goldstein, G. (1970). *Assessment of Brain Damage*. New York: Wiley.

Schreiber, D. J., Goldman, H., Kleinman, K. M., Goldfader, P. R., & Snow, M. Y. (1976). The relationship between independent neuropsychological and neurological detection and localization of cerebral impairment. *Journal of Nervous and Mental Disease*, 162, 360–365.

Selz, M., & Reitan, R. M. (1979). Rules for neuropsychological diagnosis: Classification of brain function in older children. *Journal of Consulting and Clinical Psychology*, 47, 258–264.

Shure, G. H., & Halstead, W. C. (1958). Cerebral localization of intellectual process. *Psychological Monographs, 72,* (12, Whole No. 465).

Swiercinsky, D. (1978). *Manual for the adult neuropsychological evaluation.* Springfield, IL: Charles C Thomas.

Tshushima, W. T., & Wedding, D. (1979). A comparison of the Halstead-Reitan neuropsychological Battery and computerized tomography in the identification of brain disorder. *Journal of Nervous and Mental Disease, 167,* 704–707.

Vega, A., & Parsons, O. A. (1967). Cross-validation of the Halstead-Reitan tests for brain damage. *Journal of Consulting Psychology, 31,* 619–625.

7

The Luria-Nebraska Neuropsychological Battery

CHARLES J. GOLDEN and MARK MARUISH

Based on Luria's (1966, 1973) functional systems theory of brain organization, the Luria-Nebraska Neuropsychological Battery (LNNB) (Golden, Purisch, & Hammeke, 1985) consists of 269 items representing approximately 700 test procedures. These comprise the battery's 11 ability scales, which are constructed to assess skills in 11 broad ability areas. In addition, two sensorimotor scales, a pathognomonic scale, and sets of localization and factor scales are also derived from the 269 items. Test results are interpreted by analysis of the patterns on all sets of scales, performance on individual items, and qualitative data.

Development of the LNNB represented an attempt by Golden and his associates to standardize and quantify methods used by the acclaimed Russian neuropsychologist A. R. Luria in the evaluation of brain injured patients. Although it is reported that Luria was opposed to the standardization, it was believed that efforts in this direction would add to the acceptance of these procedures. The items contained within the LNNB essentially represent the carrying of Christensen's (1975a,b) standardization of Luria's evaluation procedures a step further to include greater standardization in the administration and scoring of these procedures, as well as providing statistical norms against which performance can be judged. Whereas it provides a quantitative

---

CHARLES J. GOLDEN • University of Nebraska Medical Center, 602 South 45th, Omaha, NE 68105.   MARK MARUISH • National Computer Systems, 5605 Green Circle Drive, Minnetonka, MN 55343

approach to assessment, the LNNB also allows for the integration of qualitative data (gleaned from behavioral observations and testing-the-limits strategies) in generating hypotheses regarding the nature of the patient's deficits.

The popularity that the battery has gained among clinicians appears to be the result of several factors. These include a relatively short administration time (2–3 hours), the low price of the equipment necessary for the administration of the test items, and the portability of the test materials. Also, use of objective rules for interpretation (based on empirical data and clinical experience) allow the clinician to make reasonably accurate general statements about the neuropsychological status of the patient, whereas the more sophisticated clinician has available extensive data for more complex interpretations.

THE LURIA-NEBRASKA SCALES

Clinical Scales

The 269 items presently contained on the LNNB are grouped into 11 ability scales, representing very broad skill areas outlined by Luria (1966). Items contained within each scale primarily focus on the ability suggested by the scale title. However, these items deliberately vary within scales to test the interaction of the central skill with a variety of secondary input, output, or integrational abilities. Empirically derived scale scores for each item within a given scale are summed. The total score is subsequently converted to a T-score, based on normative data. Given the fact that no items in any of the scales are measures of a pure ability, a particularly high T-score (which would be indicative of impairment) could result from deficits other than those implied by the scale title.

The 11 basic ability scales are as follow:

Motor. Items on this scale are designed to measure bilateral and bimanual motor coordination and speed, kinesthetic-based movement, oral-motor movement, and drawing ability and speed. The ability to copy hand and arm movements from a model, to perform complex forms of praxis, and to use verbal commands to guide behavior is also measured.

Rhythm. The patient's ability to attend to, discriminate, and produce both verbal and nonverbal rhythmic stimuli is measured on this scale.

Tactile. Tasks on this scale require the patient to identify the location and direction of tactile stimuli, discriminate both hard and soft touches and painful stimuli, and identify numbers, letters, and shapes traced on the wrists. Position sense and the ability to identify objects placed in the hands are also assessed.

Visual. The patient is presented with objects and asked to identify them. Pictures of objects that are either clear, unfocused, or of high contrast quality, along with those of drawings of overlapping objects, must also be identified. Progressive matrix, clock reading, clock setting, and directional orientation tasks, as well as tasks measuring intellectual operations in space (e.g., counting the blocks in a pictured three-dimensional stack) are also included.

Receptive Speech. The items here require the patient to discriminate between phonemes, and measure the patient's ability to comprehend simple words, phrases, and sentences. Understanding of complex and inverted grammatical structures is also assessed.

Expressive Speech. Items on this scale assess the patient's ability to articulate speech sounds, words, and sentences presented either orally or visually. The ability to identify pictured or described objects, fluency and automatization of speech, and the ability to construct grammatically correct sentences are also measured.

Writing. In addition to assessing motor writing abilities in general, phonetic analysis and the ability to copy and write from dictation letters, sounds, words, and phrases is assessed. A spontaneous writing task is also included.

Reading. Items on this scale require the reading of letters, unfamiliar syllables, simple and complex words and sentences, and a simple paragraph. Two items requiring the synthesis of letters into sounds and words are also included.

Arithmetic. Designed to measure basic arithmetic skills, the Arithmetic scale's items include tasks requiring the reading and writing of single and multi-digit numbers and Roman numerals, the performing of simple and complex computations, the comprehension of mathematical signs and number structure, and the performing of serial subtractions.

Memory. Short-term memory for verbal and nonverbal stimuli under both interference and noninterference conditions is assessed here. Paired-associate learning and the ability to recall the gist of a paragraph are among the types of skills assessed here.

Intellectual Processes (Intelligence). This scale includes items that measure comprehension of thematic pictures and texts, concept formation (in the form of definitions, analogies, opposites, and elaboration of similarities, differences, and relationships between objects), and the ability to solve complex discursive mathematical problems.

Three other clinical scales are derived from the LNNB items.

Pathognomonic. The Pathognomic scale consists of items drawn from 10 of the ability scales. These items have been found to be highly sensitive to the presence of brain dysfunction. In addition to being the best indicator of brain integrity (among the clinical scales), it also provides an estimate of the degree to which the brain has compensated for sustained injuries.

Left Hemisphere. Items measuring right-hand performance on the Motor and Tactile scales are included here. This scale provides specific information regarding the integrity of the left hemisphere sensorimotor strip (Broadman areas 1–4).

Right Hemisphere. Included here are Motor and Tactile items measuring left-hand performance, thus rendering an indication of the intactness of the right-hemisphere sensorimotor strip.

LOCALIZATION SCALES

Recognizing the potential effectiveness of utilizing specific items for the localization of lesions, McKay and Golden (1979a) selected sets of items that were more statistically sensitive to localized damage in one of eight areas of the brain (i.e., the frontal, sensorimotor, parietal-occipital, and temporal areas in the left and right hemispheres). Item selection was based on the performance of 53 brain-injured patients with lesions localized primarily to one of these eight areas. In Golden, Moses, Fishburne et al.'s (1981) cross-validation of these scales, accurate localization for 74% of their 87 patients with localized lesions occurred when either of the two highest scales was the criterion for classification. Also notable is the fact that the highest scale allowed for accurate lateralization in 92% of the cases.

FACTOR SCALES

From studies that factor analyzed each of the ability scales (Golden, Hammeke et al., 1981; Golden, Osmon et al., 1980; Golden, Purisch et al., 1980; Golden, Sweet et al., 1980) a total of 34 factors was derived for these scales. A comparison of the performance of patients with localized lesions from the Lewis, Golden, Moses et al. (1979) and Golden, Moses, Fishburne et al. (1981) studies to that of a group of normal patients on scales designed to measure these factors was subsequently undertaken (McKay & Golden, 1981a). This resulted in the establishment of normative data for these scales. Although limitations in this set of scales (e.g., the number of items on each scale, findings of relatively low reliability for some, differential ceilings) warrant caution in its use, the factor scales have been found to be beneficial in testing hypotheses drawn from examination of the clinical scales.

RESEARCH ON THE LURIA-NEBRASKA

Reliability

During the initial development of the LNNB, Golden, Hammeke, and Purisch (1980) sought to determine the degree to which the battery could be reliably scored. By using a different set of two raters for each of 5 patients who were administered the then 282 items, they obtained a 95% agreement rate (on the 1410 pairs of scores). Correlations between the two sets of raw scores for each patient ranged from .97 to .99. Although Bach, Harowski, Kirby, Peterson, and Schulein (1981) found somewhat lower interrater agreement when both normal and marginal responding was demonstrated by confederates, the results supported the initial finding of high interrater reliability.

In investigating the test–retest reliability with a small group of brain-impaired psychiatric patients, Golden, Berg, and Graber (1982) obtained an average correlation coefficient of .88 between the T-scores of the 11 ability scales and 2 sensorimotor scales from one testing to another (mean interval = 167 days). With a similar group tested during a longer mean interval of time (8.1 months), Plaisted and Golden (1982) obtained a mean correlation of .89 for the 11 ability scales, 2 sensorimotor scales, and the Pathognomonic scale. A mean correlation of .89 was also achieved for the localization scales. The reliability of the 30 factor scales was found to be much more variable, with the mean correlation equalling .75. Two of the factor scale correlations did not achieve significance at the .01 level. Golden, Fross, and Graber (1981) evaluated the split-half reliability of the 11 ability scales using the results of 388 normal, neurologic, and psychiatric patients. The results of the three groups combined yielded correlations ranging from .89 to .93.

Alpha coefficients (which yield the mean reliability coefficients derived from all possible ways in which items can be grouped into halves) have been computed in several studies. Mikula (1981) obtained alphas ranging from .82 to .94 on the 14 clinical scales for his combined group of medical control and neurologic patients. From the LNNB results of 285 alcoholic, brain-damaged, and schizophrenic patients, Moses, Johnson, and Lewis (1983a) obtained alphas ranging from .78 to .88 for the 14 summary scales; from .73 to .88 for the localization scales; and from .04 to .87 for the factor scales. Four of the 6 factor scales that were found not to meet the minimum standards for internal consistency (i.e., .5 or greater) were those derived from the Receptive Speech scale. This led the authors to question the stability of this scale's factor structure.

Moses, Johnson, and Lewis (1983b, 1983c) further investigated the

LNNB internal consistency by computing separate alphas for each of three diagnostic groups. Comparison of each group's alphas for the summary and localization scales to those of the previous study led the authors to conclude that adequate reliability is present to justify the clinical use of the 14 summary scales and 8 localization scales with brain-damaged patients. Maruish, Sawicki, Franzen, and Golden (1985) also computed separate alphas for the 14 summary scales for large groups of brain-injured, schizophrenic, and mixed psychiatric patients, as well as for a group of normals. Those for the normal group ranged from .40 to .78, whereas the alphas for the other three groups ranged from the lower .80s to the lower .90s.

Recently, an alternate form of the LNNB (Form II) has been developed (Golden, Purisch, & Hammeke, 1984). The first 269 of the 279 items of this form are essentially the same as those on the original form. Only the specific content varies in the majority of these items. The last 10 items comprise the new Intermediate Memory scale. Ariel and Golden (1981) reported significant correlations from .79 to .87.

VALIDITY

Golden, Hammeke, and Purisch (1978) selected 285 items from a pool of items constructed from Christensen's (1975a,b) procedures. They were subsequently administered to 50 neurologic and 50 medical control patients and scored on a 0, 1, 2 system (with "0" indicating normal performance, "2" impaired performance, and "1" borderline performance). Two hundred eighty-five t tests, comparing the performance of each group on all items, showed the controls to have performed significantly better on 253 of the items, and better on nearly all of the remaining items, far exceeding the 5% chance differences expected statistically. Hammeke, Golden, and Purisch (1978) then administered the same items to 50 normals and 50 brain-impaired patients, and summed the raw scores of all items in each of the clinical scales. The controls were found to have performed better on all summary measures. Moses and Golden (1979), in cross-validation, administered the LNNB in its present form to a like group of subjects and achieved comparable results using Hammeke, Golden, and Purisch's (1978) cutoff scores and discriminant function. Duffala (1979) found her group of 20 head trauma patients to perform significantly worse than 20 controls on the Pathognomonic and all ability scales except the Visual.

Several studies have demonstrated the LNNB's ability to identify brain-impaired subjects. Bach (1983) found all but the Reading and Memory scores to discriminate between orthopedic, Parkinson's disease, and alcohol dementia patients. In light of the effects of age and education on LNNB performance found by Marvel, Golden, Hammeke, Purisch, and Osmon (1979) and seen again in their sample of normals, Golden, Moses, Graber, and Berg (1981) generated a regression equation to predict the average T score of patients from their age and educational level. When the 60 normals were combined with a group of 60 brain-damaged patients, and two or more scores (excluding the Writing, Arithmetic, and the two sensorimotor scales) above the critical level (predicted average T score plus 10) were used as the criterion for brain impairment, 84% of the total sample was accurately classified. Using this same criterion, similar total hit rates were achieved when normals were combined with groups of primary and secondary epileptics (85%, Berg & Golden, 1981), patients with relatively mild brain impairment (78%, Malloy & Webster, 1981), patients with various neurological disorders (77%, Sawicki & Golden, 1984a), patients with relatively discrete brain lesions (86%, Golden, Moses, Fishburne et al., 1981) and elderly patients with either confirmed or suspected brain dysfunction (88%, MacInnes, Gillen et al., 1983; 93% Spitzform, 1982). Moses (1984d) also found that this rule accurately classified 83% of his group of 165 brain-damaged patients.

The discriminant ability of the LNNB has undergone testing in a number of studies designed to evaluate the degree to which schizophrenics can be differentiated from brain-damaged patients, and the degree to which brain-damaged schizophrenics can be differentiated from those without brain impairment. Purisch, Golden, and Hammeke (1978) found accurate classification of 88% of 50 schizophrenic and 50 brain-damaged patients. The schizophrenics were also found to have significantly better performance on all scales but the Rhythm, Receptive Speech, Memory, and Intelligence scales. These findings (including the pattern of significant differences found between the two groups on the ability scales) were essentially the same as those obtained by Moses and Golden (1980) in a cross-validation study. Shelly and Goldstein (1983a) partially replicated the Purisch, Golden, and Hammeke findings. The lack of complete replication was attributed to differences between their and the original brain-damaged groups.

Puente, Heidelberg-Sanders, and Lund (1982b) found schizophrenics without any neurological signs to have performed significantly better than brain-damaged patients on all 14 LNNB summary scales. In another study by the same authors (1982a), the schizophrenic group performed significantly better on all but the Expressive Speech and Pathognomonic scales and those four scales initially found by Purisch, Golden, and Hammeke to be nondiscriminating for these two populations. Moses, Cardellino, and Thompson (1983) combined two sets of samples of schizophrenic and neurologic patients and achieved a 74% hit rate for the total group using a discriminant function derived from the results of one of the sets of samples. This represented a drop from the previous hit rate (81%) obtained for the one sample from which the function was derived. Respective hit rates of 81% and 78% occurred when both the 14 individual scale T scores and the sum of the 14 T scores for each subject of the combined sets were submitted to discriminant analysis.

Noting that schizophrenics identified by the LNNB in previous studies as being brain-damaged might actually have been brain-impaired, Golden, Moses et al. (1980) investigated the relationship between LNNB summary scores and ventricular size (a measure of cerebral atrophy). Eight of the 14 scales were found to significantly correlate with ventricular size. When submitted to multiple regression, a multiple correlation of .72 ($p < .001$) between ventricular size and the 8 LNNB scales was obtained. A 90% classification rate of subjects was achieved based on a set of rules derived in this sample. Use of a revision of these rules by Golden, Graber, Moses, and Zatz (1980) lowered the hit rate to 81% for the identification of enlarged ventricles, but allowed for the accurate classification of 90% of schizophrenics in terms of whether or not they had visible sulci (another measure of cerebral atrophy). Using these same rules with a somewhat younger group of schizophrenics with nearly twice the number of hospitalizations as the previous sample, Golden, MacInnes et al. (1982) achieved a 77% hit rate. A significant relationship ($p < .01$) between the 14 LNNB variables and ventricular size was again found, but the scales originally found to correlate individually with ventricular size were somewhat different than those found previously.

The ability of the LNNB to aid in the lateralization and localization of brain impairment was first investigated by Osmon, Golden, Purisch, Hammeke, and Blume (1979). McKay and Golden (1979b), using the direction of the difference between the scores from empirically derived left and right hemisphere scales, were able to accurately lateralize the site of dysfunction for all of their initial subjects, and for 83% of their cross-validation sample. Lewis et al. (1979) found their eight groups of subjects with localized lesions (left and right frontal, temporal, sensorimotor, and parietal-occipital) to differ from each other significantly.

Using patients determined to have localized lesions in one of eight areas (according to the Lewis et al. criteria), McKay and Golden (1979a) developed a set of 8 localization scales. The site of the dysfunction of these patients was subsequently identified with 89% accuracy when the highest localization scale T score was used as the criterion. In cross-validation, Golden, Moses, Fishburne et al. (1981) accurately identified the locus of the lesion in 74% of their 87 patients. The study also revealed the highest localization scale to achieve a better hit rate for lateralization than the direction of difference between the empirically derived lateralization scales (92% vs. 79%).

Further validation of the LNNB can be found in studies investigating its relationship to other neuropsychological, psychological, and physiological measures. The most important of these are those that have sought to determine its relationship to the Halstead-Reitan Neuropsychological Battery (HRNB). Vicente et al. (1981) and Golden, Kane et al. (1981) found significant multiple Rs between the LNNB scales and each of the HRNB measures investigated. They, along with Shelly and Goldstein (1982b), also noted significant correlations between many or all of the variables from both batteries. Comparability in the two batteries' discriminant ability was found in studies that utilized raters expert in the use of one or the other battery (Kane, Sweet, Golden, Parsons, & Moses, 1981), and objective rules (Johnson, Moses, & Bryant, 1984) as a means of classifying brain-impaired and nonimpaired subjects. Correlations of .60 or better were found between indices of global impairment for the two batteries (Johnson, Moses, & Bryant, 1984; Shelly & Goldstein, 1982b). After submitting the LNNB scales and HRNB measures together to factor analysis, Shelly and Goldstein (1982b, 1983b) concluded that both batteries assess similar abilities (i.e., language, nonverbal cognitive, perceptual-motor) and that laterality measures from the two batteries were tapping similar skills.

Investigations into the relationship between the LNNB and Wechsler Adult Intelligence Scale (WAIS) have revealed significant correlations between the LNNB ability scales and the WAIS subtests (McKay, Golden, Moses, Fishburne, & Wisniewski, 1981; Shelly & Goldstein, 1982a). Of particular note are the significant and high correlations that were found between the LNNB Intelligence scale and the three WAIS IQs (McKay, Golden, Moses et al., 1981; Picker & Schlottmann, 1982; Prifitera & Ryan, 1981). Similar correlations were found between the Intelligence scale and WAIS-R IQs (Dill & Golden, 1983). It should be noted that to some extent, as has been asserted by some (e.g., Chelune, 1982), the relationship between the LNNB and the HRNB is due to the relationship of both to psychometric intelligence, as well as the relationship of both to brain dysfunction.

Research has also demonstrated relationships between LNNB scales and other psychological measures. Both Ryan and Prifitera (1982) and McKay and Ramsey (1983) found significant correlations ($p < .001$) between the LNNB Memory scale and the Memory Quotient derived from the Wechsler Memory Scale. Correlations between each of the LNNB summary and localization scales and the score of each of the five subtests of the Peabody Individual Achievement Test were significant (Gillen, Ginn, Strider, Kreuch, & Golden, 1983). Also, the summary, factor, and localization scales that correlated the highest with the PIAT subtests were those that intuitively would be expected to do so. A similar pattern of correlations was obtained by Shelly and Goldstein (1982a) when the relationship between the 11 ability scales and the 3 subtests of the Wide Range Achievement Test was investigated.

In studies of multiple sclerosis patients, Kaimann (1981) corroborated previous HRNB findings of sensory and motor impairment in this population. With essentially the same sample, Kaimann, Knippa, Schima, and Golden (1983) found a factor analysis-derived density factor to correlate significantly with 10 of the LNNB ability scales, and sulcal width and cerebral distance factors to each correlate significantly with five of the ability scales. Performance on some of the scales was later reported to also be related to demographic and illness variables (Kaimann, in press).

Support for the use of the LNNB has come from studies investigating populations that one would intuitively expect to have neurological and neuropsychological impairment. On the LNNB, 83% of Berg and Golden's (1981) epileptics were classified as impaired, and both of these groups differed from the normal controls on 11 of the summary scales. LNNB results indicated expected progressive cognitive deterioration over time in Huntington's disease patients in various states of the illness (Moses, Golden, Berger, & Wisniewski, 1981). The effects of alcoholism and/or treatment of the same on LNNB performance were found to be as one might expect in studies by Chmielewski and Golden (1980), Gechter, Griffith, and Newell (1983), De Obaldia, Leber, and Parsons (1981), and Teem (1981). Normal controls were found to render better performances on the LNNB than groups of diabetics (Strider, 1982), dyslexics (Grey, 1982), and learning disabled adolescents (Parolini, 1983), whereas greater neuropsychological dysfunction was indicated for violent or assaultive male schizophrenics or criminals than for their nonassaultive counterparts (Bryant, 1982; Bryant, Scott, Golden, & Tori, 1984; McKay, 1981; Scott, Cole, McKay, Leark, & Golden, 1982; Scott, Martin, & Liggett, 1982; West, 1982).

Other results have included the finding that substance abusers diagnosed as schizophrenics perform better than schizophrenics without an abuse history on all LNNB summary scales (Scott, Cole, McKay, Golden, & MacInnes, 1982), and that of psychiatric patients experiencing visual hallucinations (often an indication of acute organic dysfunction) performing consistently better than counterparts experiencing auditory or no hallucinations on all ability scales (after acute episodes were over) (McKay, Golden, & Scott, 1981).

In that the LNNB is initially based on Luria's (1966, 1973) theory of the functional organization of the brain, its characteristics should reflect this theory. Also, results obtained from the battery should accurately reflect the status of abilities associated with the different brain areas as would be predicted by the theory. Support for the functional systems theory being reflected in the LNNB comes from studies in which the majority of the 269 items were found to correlate highest with the score for the scale in which it is contained (Golden, Fross, & Graber, 1981), as well as from the Golden and Berg (1980a,b,c,d, 1981a,b,c, 1982a,b, 1983a,b,c) findings of significant ($p < .0001$) correlations between nearly all items and items from other scales. Several factor analytic investigations of the summary scales (Golden, Hammeke et al., 1981; Golden, Osmon et al., 1980; Golden, Purisch et al., 1980; Golden, Sweet et al., 1980; McKay and Golden, 1981b; Moses, 1983, 1984a, 1984b, 1984c; Moses & Johnson, 1983; Sawicki & Golden, 1984b), and of the battery as a whole (McKay, Golden, Wolf, & Perrine, 1983) have generally resulted in the emergence of factors representing component abilities that would be predicted by Luria's theory. For the most part, deficit patterns of eight groups of patients with localized lesions have been found by Lewis et al., (1979) and Golden, Moses, Fishburne et al. (1981) to be consistent with those that one might expect, based on Luria's work. Also supportive of the LNNB's construct validity is the robustness of the LNNB variables on a language factor derived from a factor analysis of LNNB and HRNB measures (Shelly & Goldstein, 1982a).

The literature, in general, supports the contention that the LNNB is a reliable and valid instrument for the assessment of neuropsychological functioning. It has been shown to produce a relatively stable set of scores over time and to discriminate brain-damaged from normal patients at a level approximately equal to that found for the Halstead-Reitan battery. It has also been found to correlate highly with other neuropsychological, psychological, and physiological tests and measures, and to display sensitivity to disorders that would be expected to cause disruption in cognitive processing. In addition, the construction of the LNNB is such that it adequately reflects Luria's theory of brain functioning.

Among current issues of importance are preliminary findings suggesting poor internal consistency for the factor and localization scales for some diagnostic groups (Moses, Johnson & Lewis, 1983b, 1983c), a relationship of clinical profile elevations to various demographic and treatment variables (e.g., Sawicki & Golden, 1984a), differences between different subtypes of schizophrenics (Langell, Purisch, & Golden, 1985; Purisch & Langell, 1983), and effects of depression on LNNB scores (Sweet, 1983).

INTERPRETATION OF THE LURIA-NEBRASKA

The empirical data gathered since the initial stages of development of the LNNB and the experience gained from its clinical use with various diagnostic populations have allowed for the formulation of general strategies of interpretation. These strategies are presented by Moses, Golden, Ariel, and Gustavson (1983) and Golden, Hammeke, and Purisch (1980) and are summarized below. They permit the user to make some fairly accurate general statements regarding the patient's neuropsychological status. However, knowledge of these guidelines in no way makes one an expert in the use of the LNNB. Such expertise comes only with close familiarity with the anatomy and functioning of the nervous system, knowledge of the neuropsychological literature (including the works of Luria), and effective supervised clinical experience.

As with any neuropsychological instrument, even the most general interpretation of the results begins with the assumptions that the patient was sufficiently motivated to render the best performance, and that the battery has been properly administered and scored. Inability to elicit maximum performance is not uncommon and subsequently must be taken into consideration when test data is analyzed. Proper administration and scoring of the LNNB can only come with understanding of the standardized test procedures, including the degree to which one can be flexible in these approaches. In addition, adequate background information obtained from interview data and medical records must be analyzed to determine whether variables that may influence test performance (e.g., medical, psychological, educational) are present.

The accuracy of the interpretation of the LNNB results at any level is dependent on the user's ability to integrate all patient data. Demands for increased sophistication in interpretation will call not only for greater knowledge of brain functioning but also for more sophistication in integrating test and other data with such knowledge.

It is also important to note that use of the LNNB does not necessarily supplant or eliminate the need for the administration of other instruments. Occasionally, the examiner may want to further investigate the presence of deficits suggested by LNNB results, or assess skills not measured by the LNNB (e.g., reading comprehension, remote memory). Administration of supplemental tests can only aid in increasing the knowledge that one has about the patient's strengths and deficits, and is recommended when time and financial factors are of no great consideration.

With these considerations in mind, methods of interpreting the results from the LNNB will be presented. Subsequently, application of these methods will be demonstrated with clinical cases.

Identification of Brain Damage

Classification of a patient as having some form of brain damage using the LNNB is essentially determined by comparing the patient's performance on the scales to that which would be predicted for a person of the same age and education. Education (an estimate of premorbid intellectual functioning) and age have been found to be significantly related to the average of the summary T scores of 60 normal subjects (Golden, Moses, Graber, & Berg, 1981). From this data, a regression equation was developed to predict a mean T score (or baseline) of the patient's performance. Adding 10 T score points (one standard deviation) to this predicted mean establishes a critical level, any T scores above which are considered abnormal. The formula for determining the critical level is:

$$\text{critical level} = 68.8 + (.213 \times \text{age}) - (1.47 \times \text{education})$$

Age in the formula refers to actual age in years, except when the patient is 13 to 24 years of age, or older than 70. In the former instance, the age of 25 is used, whereas the age of 70 is used in the latter.

Education in the above formula refers to years of education, up to 20 years. Thus, a high school graduate is given credit for 12 years of education, and an individual who had dropped out of school after completing the eighth grade is given credit for 8 years of education. Regardless of the number of years needed to complete their degrees, those with bachelor's degrees are credited with 16 years of education, those with master's degrees are given 18 years, and those with doctorate degrees are credited with 20 years. At times, the examiner may need to adjust the number of years of education for which the patient is given credit, based on other considerations regarding intellectual abilities, as detailed in Moses, Golden et al. (1982).

In most cases, the determination of an appropriate critical level will present no particular difficulty. Use of the prescribed formulas and guides will allow for a fairly accurate estimate of this cut-off value. However, there are summary scale profile patterns that should lead one to question whether the critical level is accurate and, subsequently, needs to be adjusted. Inaccurate critical levels may be due to poor estimates of educational achievement, or to other factors that are not considered and that may have a significant effect on premorbid functioning (e.g., geographical residence, socioeconomic status, occupation of parents).

Because all scores should fall within the range of the critical level and 20 points below this cutoff value, any score(s) falling more than 20 points below the critical level suggests that the critical level is too high and needs to be adjusted. Adding 25 points to the lowest scale T-score will establish a new, more conservative critical level. This should then replace the original critical level if it is found to be lower. Given the statistical properties of the battery, adjustment of the critical level should also be considered in cases where two-thirds or more of the summary scales fall between the critical level and 10 points below it.

With the establishment of the critical level, determination of the statistical probability of brain damage becomes relatively easy. In general, brain damage is indicated when three or more of the scores from the 11 ability scales and the Pathognomonic scale exceed the critical level, whereas only one or none of these scores above the critical level is more consistent with a normal profile. Given an accurate critical level, 85%–90% of patients will be accurately classified. In cases where two scales are significantly elevated (above the critical level), the profile is likely to be abnormal if neither of the scales is the Arithmetic nor Writing scale. If either or both of the two scales is the Arithmetic or Writing scale, one must determine the reason for the elevation(s). Elevations due solely to problems in spelling or calculations that are consistent with the patients's history can be ignored. Elevations due to motor writing problems (for the Writing scale), or to an inability to read or recognize numbers (for the Arithmetic scale) should be considered significant and counted as a significant elevation.

Caution is warranted in the classification of profiles where elevations are due solely to educational deficits. This is usually indicated by elevations on only the Reading, Writing, and Arithmetic scales, and occasionally, owing solely to poor performance on those items requiring reading skills, the Expressive Speech scale is also elevated. This profile is commonly seen in individuals with poor school histories. In cases where the individual has had adequate exposure to this material but was unable to learn, the presence of brain damage is highly suggested. However, when no other deficits are noted on the LNNB and the individual appears to have not been exposed to the material for any number of reasons (including emotional disturbance unrelated to academic performance), the presence of brain damage is more questionable.

Alertness to the presence of two other profile types may allow for the correct identification of some of the 10%–15% of the brain-damaged patients misidentified by the use of the above rules. One of these has already been discussed in regard to considerations having to do with the determination of the critical level. This is the profile where two-thirds or more of the scores fall between the baseline and the critical level. If, on consideration, the critical level is deemed accurate, such a profile may indicate subtle yet real brain dysfunction. Further examination of the patient's performance (by a qualified neuropsychologist) is thus warranted. Also, a profile in which the difference between the highest and lowest scale T scores exceeds 30 should be considered indicative of brain damage. However, when the highest scores(s) is due solely to spelling and/or arithmetic calculation problems, this score should be ignored and the range recalculated using the next highest score.

Support for the initial hypothesis regarding the patient's brain status may come from examination of the localization scales. Here, two or more scales elevated above the critical level is associated with brain damage. The accuracy rate of this classification procedure is similar to that seen with the clinical scales. A difference of 30 or more points between the highest and the lowest scale T scores is also highly suggestive of brain damage. If both the clinical and localization scale patterns indicate the presence of brain damage, the probability that the patient is actually brain damaged is increased. If the hypotheses suggested by the two sets of scales are not in concordance, consideration of subtle brain dysfunction or an interfering psychiatric disorder is warranted. In regard to using the localization scales to support initial hypotheses, it must be noted that performance on the left parietal-occipital scale (LPO) is highly related to academic achievement. A poor academic history may thus result in an elevation on this scale. In this case, one must exercise caution (given the above discussion on elevations on the academic scales) in using the performance on this scale to support or disconfirm previously generated hypotheses.

SCALE AND PATTERN ANALYSIS

The previously described methods of interpretation only permit the examiner to make very basic tentative statements regarding the status of the patient's brain. Frequently, this is all that is requested by referral sources. However, the results of the LNNB permit one trained in its use to go beyond making elementary statements to provide a description of the patient's neuropsychological functioning. This is accomplished by an overall analysis of the clinical, localization, and factor scales and individual items, combined with an analysis of qualitative data. It is beyond the scope of this chapter to go into any great depth in regard to the interpreting of LNNB results. To be discussed are ways of looking at the data that will allow the formulation of hypotheses regarding the integrity of the patient's brain.

CLINICAL SCALE ELEVATIONS

In analyzing the results of the LNNB clinical scales, little emphasis is placed on interpreting scale elevations in and of themselves. Because all scales are composed of heterogeneous items, an elevated scale score may reflect one or more of a number of deficits possibly caused by an injury in any part of the brain. However, when viewed in relationship to the other scales, elevations on individual scales allow one to begin generating hypotheses about the patient's neuropsychological status. The experienced clinician will, of course, arrive at these hypotheses only after other factors that affect scale elevations (e.g., peripheral or brain stem injuries, type and extensiveness of the neurological dysfunction, time since the injury, expressive or receptive language problems, premorbid functioning, psychiatric problems) have been taken into consideration.

Because we use a language referring to specific brain areas in the rest of this chapter, the reader must be cautioned from overinterpreting the statements. A "right parietal" focus behaviorally refers to a certain pattern of deficits seen in such injuries, not to the injury itself. Physiologically, although the brain is somewhat localized, there is extensive intraindividual variation and a variety of disconnection syndromes and learning history problems that force one to regard the localization as reflecting not a real brain but a behavioral model of the brain. With sufficient knowledge, localization to this model is possible and useful in understanding a patient. However, the model does not predict to physiology on a one-to-one basis, nor would we expect it to, given current theories of brain function and development. In light of this the reader should keep in mind these limitations, and avoid localization statements that suggest a specific physiological (as opposed to behavioral) etiology for specific symptoms until they are well supported.

Motor

Interpretation of Motor scale elevations is best made by comparing this scale to Tactile and Left and Right Hemisphere scales. When these scales are not elevated, the Motor elevation suggests problems in performing complex motor tasks. Generalized impairment of the sensory and motor areas (often in the context of diffuse deficits) is suggested when all four scales are highly elevated. An anterior lesion is suggested when the Motor scale is much higher than the Tactile scale and one of the hemisphere scales is at least 10 points higher than the other (with the meaningfulness of this result increasing with the amount of difference between the two hemisphere scales).

Significant involvement of one sensorimotor area or related subcortical structures is indicated when the Motor, Tactile, and one of the hemisphere scales are elevated, particularly if the difference between the two hemisphere scales is 20 points or more. This pattern is frequently seen in patients with unilateral middle cerebral artery strokes involving the sensorimotor area. Comparison of the raw scores of the first four items may be found useful in the absence of clear sensorimotor deficits. Here, left-hand scores should be 90%–110% of those of the right-hand scores, such that a left hand score that is more than 110% of the right-hand score may be a sign of subtle left-hemisphere loss. A left-hand score less than 90% of the right-hand score would suggest subtle right-hemisphere dysfunction. This scale is neither sensitive to lower limb (leg) motor deficits, nor particularly sensitive to subcortical disorders that do not affect voluntary motor movements.

Rhythm

Right-hemisphere injuries that are usually in the more anterior areas (frontal and temporal lobes) or subcortical injuries in either hemisphere are most likely to be present when the Rhythm scale elevation is the highest. This is particularly the case when the highest elevations occur on the Rhythm, Memory, Intelligence, and Arithmetic scales. This pattern can occur in left-hemisphere disorders, but it is accompanied by at least subtle verbal deficits. When the Rhythm and Visual scales are elevated, either anterior or posterior lesions may be present. Here, the probability that the lesion is posterior increases with greater elevations on the Visual scale. With the Right-Hemisphere scale greatly elevated in relation to the Left-Hemisphere scale, the possibility of a lesion crossing the sensorimotor area and involving posterior and anterior areas of the hemisphere must strongly be considered. Rhythm elevations are not uncommon in left-hemisphere injuries; however, they are generally below those of the other scales. In general, very low elevations on the Rhythm scale are inconsistent with right-hemisphere lesions not involving the sensorimotor area.

Tactile

When the highest, the Tactile score must be interpreted with equal regard to the Motor and two hemisphere scales. Generally, when the Tactile scale is equal to or much greater than Motor, and one hemisphere scale is significantly elevated over the other, a posterior lesion in the hemisphere indicated is likely to be present. When both hemisphere scales are elevated, one must consider either a severe left-hemisphere or bilateral injury. It is important to note that elevations on this scale may be due either to deficits in the patient's ability to integrate and identify all stimuli, or to tactile/spatial deficits. The former will likely be revealed on other items requiring naming and identification and should lead one to consider the presence of a left parietal lesion.

When the Visual scale is also elevated, a right parietal-occipital involvement is likely to be present. Concentration problems will also cause errors on the Tactile scale for many patients.

Visual

A right-hemisphere or left occipital lesion is suggested when the Visual scale is the highest. Although it can be elevated in other left-hemisphere injuries, it likely will not be the highest scale. Right anterior or mild parietal disorders are suggested when deficits are noted on only the more complex tasks of the Visual scale. An elevation on the Motor scale suggests that a right-hemisphere disorder is also present. Severe peripheral visual problems and subcortical lesions that impede visual processing may also result in patterns suggesting right-hemisphere dysfunction. Problems due solely to naming should be interpreted as reflecting dominant hemisphere dysfunction.

Receptive Speech

A left-hemisphere injury is usually indicated when the Receptive scale is the highest and is at least 15 points above the critical level. A less significant elevation (which remains the highest) resulting from problems with the more complex items of the scale may be seen with a lesion to the right anterior portion of the brain. This may particularly be the case when mild elevations on the Receptive scale are combined with those on either the Memory, Rhythm, Visual, and/or Intelligence and Arithmetic scales. However, when the Receptive scale is more highly elevated, this will generally indicate a left-hemisphere disorder.

Expressive Speech

As with the Receptive scale, a left-hemisphere disorder is indicated when the Expressive scale is significantly elevated above the critical level. Mild elevations (resulting mostly from errors on the last, more complex items of the scale) may be associated with disorders of the right hemisphere.

Writing

Elevations on the Writing scale may be the result of spelling, motor writing and/or spatial deficits. Motor writing deficiencies suggest disorders in the hemisphere contralateral to the hand used for writing, whereas spatial deficits (when seen in other portions of the battery) should lead one to consider a right-hemisphere lesion. Spelling deficits, which had not been present premorbidly, can result from injuries to either hemisphere.

Reading

Left posterior disruption is almost always indicated with Reading scale deficits in individuals who had good reading skills premorbidly. However, left frontal lesions are suggested when highly educated individuals experience mild problems with complex words. Also, deficits due to spatial disruption (inability to follow a line of print) or neglect of the left side (for which the examiner should correct the patient) suggest right-hemisphere disruption.

Arithmetic

Performance on the Arithmetic scale possibly may be influenced by a lesion in any part of the brain, or by preexisting deficits in some normals who are achieving at a level well below that which would be expected (based on years of education). Left-hemisphere injury is suggested by an inability to read or write numbers. When difficulty is experienced only when dealing with the spatial aspects of numbers, a right-hemisphere disorder is suggested. However, this might also be indicative of left hemisphere involvement.

Memory

Very high elevations on the Memory scale suggest left hemisphere dysfunction whereas lower elevations may reflect dysfunction in either hemisphere. Along with Rhythm, the Memory scale is very sensitive to dysfunction of a subcortical nature. In fact, impaired performance on this scale often accompanies subtle subcortical dysfunction, particularly that of the temporal lobes.

Intelligence

Although affected by injuries to both the anterior and posterior regions, the Intelligence scale is generally more susceptible to posterior injuries. This scale may also reflect injuries to either hemisphere. A highly elevated Intelligence scale combined with relatively elevated scores on the Rhythm, Visual, Memory, and Arithmetic scales indicates disturbance in the right hemisphere, whereas combinations with the Expressive, Reading, and Writing scales suggest left-hemisphere dysfunction.

Pathognomonic

The Pathognomonic scale not only discriminates the best between normal and brain-damaged patients, but also allows for a determination of the degree to which the patient has compensated for the brain injury. It is important to note here that the degree of compensation attained is generally affected by several factors, including the time since the injury, severity of permanent brain damage, premorbid functioning, and the nature of the disorder (progressive vs. static). When the Pathognomonic scale is quite high, that is, usually 20 points above the critical level, and is the highest scale, an uncompensated injury is likely. Most frequently, this is seen with severe, acute injuries; however, it may also occur in patients with severe chronic dysfunction for which total compensation has not occurred. When quite high but approximately equal to the mean of the scales, compensation has probably occurred and some recovery of function has been demonstrated by the patient. A chronic injury that has generally recovered to the maximum expected level is suggested when the Pathognomonic scale is high but is the lowest of the scales. With this same pattern, but with the Pathognomonic scale not as elevated, a less serious state of affairs is likely.

If the Pathognomonic scale is not extremely elevated (less than 20 points above the critical level) but remains the most highly elevated scale, the injury may either be a recent one that is just starting to recover, or one that is more long standing where general recovery has taken place but one or more areas of function have not been compensated for. This would then likely reflect a rather limited area of dysfunction. An elevated but not high score that is at the same level as the other scores suggests partial recovery with some (generally diffuse) dysfunction remaining.

When the Pathognomonic scale is below the critical level, compensation has likely taken place. In this case and when it is the highest scale, the injury is likely to be compensated for but an area of dysfunction remains. It might also indicate a small, slow-growing lesion causing little disruption, or the beginning of a very small, fast-growing lesion. When all scores fall below the critical level and Pathognomonic is either at or below the level of the other scales, a normal profile or a generally recovered injury is indicated.

Left- and Right-Hemisphere Scales

Left and Right Hemisphere scales basically reflect sensorimotor functions and are interpreted with respect to the Motor and Tactile scales. The lateralization suggested by the difference between the two scales should be given serious consideration if it is 10 points or more. A difference of more than 20 points generally suggests sensorimotor strip involvement if no peripheral problems are present.

PATTERN ANALYSIS OF THE CLINICAL SCALES

After initial hypotheses are generated from examination of the relationships of the clinical scales to each other, one may wish to proceed to a more global examination of the summary profile. From the work of Lewis et al. (1979) and Golden, Moses, Fishburne et al. (1981), LNNB summary profiles have been obtained for small groups of patients with localized lesions in either the left or right frontal, temporal, sensorimotor, or parietal-occipital areas. Comparison of the patient's profile to these average localized patterns may serve as a means of developing initial localization hypotheses.

Left Frontal

For the left frontal group of patients, the highest scores are found on the Expressive, Arithmetic, Receptive, and Pathognomonic scales. This combination of scales and the next highest combination of scales (Writing, Reading, Motor, and Memory) point to left-hemisphere difficulties. The equally mild elevations on the two hemisphere scales suggest an injury lying outside of the sensorimotor area. A lower score on the Tactile scale (as compared to Motor), the lack of visual deficits, and mild elevations on the Reading and Writing scales (which are sensitive to posterior injuries) relative to the first combination of scales, all point to an anterior locus of dysfunction. The elevation on the Receptive scale and its relationship to the Expressive scale are also consistent with a left-anterior locus.

Further support for a conclusion of a left-frontal focus can come from an examination of the types of errors made during the course of the testing. On the Expressive scale, the primary deficits are found on the items requiring spontaneous and complex speech, and possibly on the sequencing items. Clear expressive speech deficits, if present, will generally be found on the most complex items unless the lesion extends into the sensorimotor area. In this case, lateralized deficits will be noted on the Motor and Left-Hemisphere scales. Conduction aphasia (generally representing a subcortical temporal lesion) should be considered if repetition errors are present and reading errors are not, or if such errors exist when the Receptive scale is less than or equal to the Expressive scale. Naming deficits on the Expressive scale can be attributed to parietal injuries when these are the only errors noted.

Most frequently, Arithmetic errors are limited to the more complex calculation problems, particularly if these are performed mentally. Problems in complex number reading and writing errors, and errors due to impulsive responding may also be seen. On the Receptive, Reading, and Writing scales, difficulty is encountered on the more compelx items. Slow response rate, as well as deficits on the complex speech/motor items and those requiring bilateral motor movements or complex motor sequences may be seen on the Motor scale. Indications of severe, lateralized motor impairment suggests involvement of the posterior frontal areas. Rhythm deficits are seen on the expressive and complex receptive items for this scale.

Right Frontal

The highest elevations for the right-frontal groups are found on the Motor, Rhythm, Arithmetic, and Pathognomonic scales. This pattern may be seen in various right-anterior injuries. However, the high elevation on the Arithmetic scale and the elevations of the Right-Hemisphere over the Left-Hemisphere and the Motor over the Rhythm scales suggest a frontal rather than a temporal locus of dysfunction. In addition, the relatively low elevations on the Receptive, Memory and Intelligence scales (which are more sensitive to temporal lesions) are consistent with a frontal focus.

Examination of individual items reveals a particular pattern of deficits. On the Motor scale, these include slowness in performing left-body-side movement (not as pronounced as in sensorimotor lesions) and drawing speed, and problems on tasks requiring sequential processing. Rhythm scale errors tend to be worse on the expressive rhythm items. When present, Tactile and Visual scale errors are found only on the complex spatial problems. Problems on the Arithmetic scale are similar to those seen in the left-frontal injuries, with the possible added problem of working with sequencing numbers. Except in cases of mixed or reversed dominance, true dysgraphia or dyslexia will likely not be seen. If present, it is more like those seen in left-frontal patients. Expressive deficits may be seen on items requiring sequencing (days of the week and counting). Spontaneous speech and the sequencing of sentences may be impaired due more to slowness than to impoverished speech production. Generally, Receptive errors are limited to items that are syntactically more complex and inverted.

Left Sensorimotor

Elevations on the Motor, Expressive, Left-Hemisphere, Arithmetic, Intelligence, and Pathognomonic scales characterize left-sensorimotor patients. Particularly characteristic is the Left Hemisphere scale being greatly elevated over the Right Hemisphere. Although this may be due solely to motor or tactile deficits of the right hand, it is generally due to both.

Deficits on the Expressive scale are seen on both repetition and reading items, as well as on the more complex items. This is due to losses in basic motor speech. Although most characteristics of Broca's aphasia may be present, it is also possible that expressive speech skills will remain intact if the lesion is small and distant from the speech zone. Motor and Tactile scale errors generally reflect errors on items measuring right-body-side sensimotor skills, bilateral coordination, and drawing skills. Due to either sensory or motor deficits, motor writing problems may occur. However, true construction dyspraxia is rare. Arithmetic deficits are similar to those found with the frontal patients, but the severity of the problems may increase if the lesion extends into the parietal area. Concrete thinking and problems with the discursive arithmetic problems (particularly noted on the most difficult problems) are seen with elevations on the Intelligence scale.

Deficits on the Reading, Visual, Rhythm, Receptive, and Memory scales are likely to be absent if care is taken to correct for any speech deficits that may be present. If such has not occurred, all scales may be elevated. If there is significant subcortical involvement, memory deficits may be present. These will likely be seen on the verbal items of the scale. It is important to note that with stroke patients, it is not unusual to find a left-sensorimotor pattern combined with patterns suggestive of involvement of one or more of the other areas.

Right Sensorimotor

The pattern of elevations on the Motor, Right Hemisphere, and Tactile scales seen with right-sensorimotor patients is characteristic of right-hemisphere disorders. Additional elevations on the Rhythm, Visual, Memory, and/or Arithmetic scales might also be present if the lesion extends beyond the sensorimotor area. Motor deficits here are likely to represent mild constructional problems and left-body-side sensory and motor problems. Rhythm deficits are similar but less severe than those seen with right-frontal injuries, whereas tactile deficits are generally limited to the left-body side. Bilateral dysgraphesthesia and astereognosis may also be noted. However, loss of spatial analysis rather than bilateral perception problems is primarily responsible for this.

Left Parietal-Occipital

Elevations on the Expressive, Reading, Writing, Arithmetic, Intelligence, and Receptive scales, as well as impairment of the verbal memory items and the more complex spatial visual items, attest to the severity of the disruption caused by lesions in this area. Visual deficits will increase and language deficits will decrease the more the occipital lobe is involved. If both areas are involved, deficits in both areas will be noted.

Patterns of deficits will allow for further localization. With occipital-parietal lesions, arithmetic deficits will be noted. These may include loss of the ability to read and write numbers, or to do simple calculations or comprehend the meaning of arithmetic signs (dyscalculia). Temporal-occipital lesions will result in reading deficits, including an inability to associate letters and sounds in the presence of the ability to process auditory and visual information (dyslexia). Lesions more posterior to this area may result in an inability to recognize letters (visual dyslexia) and problems with other visual stimuli. Writing scale deficits, in the form of letter substitution and inability to write to dictation or spell phonemically, result from temporal-parietal lesions. Loss of all these skills, and naming problems (dysnomia) across all scales may occur as a result of injury to the integrative temporal-parietal-occipital area.

Right Parietal-Occipital

Right parietal-occipital patients show elevations on the Motor, Tactile, Visual, and Right Hemisphere scales. The elevation on the Left Hemisphere scale tends to be higher than what is seen in other right-sided injuries. The difference between the two hemisphere scales is often much less than is seen with right-sensorimotor lesions, and the mildness or absence of verbal scale elevations is quite different from what is present in right-temporal profiles. Left-body-side deficits are noted on both the Motor and Tactile scales. Deficits on items requiring visual feedback bilaterally, other bilateral items, and constructional items may also be present on the Motor scale, whereas bilateral dysgraphesthesia and astereognosis may be seen on the Tactile scale. On the Visual scale, complex spatial items, spatial orientation items, and, in severe injuries, complex recognition items are performed poorly. Severe spatial disruption may also result in borrowing and carrying problems and difficulty in the alignment of numbers when performing arithmetic problems. Difficulty in comprehending visual material and solving discursive arithmetic problems may occur on the Intelligence scale.

Left Temporal

An extreme elevation on the Receptive scale, accompanied by lesser elevations on all of the verbal scales, is most characteristic of lesions in the posterior half of the temporal lobe. Deficits on the Memory scale or a pattern similar to one found with frontal lesions (especially with extensive subcortical involvement) may characterize more anterior lesions. With the more posterior lesions, the profile begins to resemble the parietal-occipital profile. The extent of the disorder can be determined by examination of the Receptive scale. With an increase in basic errors will come an increase in the overall profile elevation. Deficits in comprehension of phonemes will be seen in the most severe cases, whereas less severe cases will result in difficulties in understanding all but the simplest language. Problems in phonemic analysis or association of phonemes to written letters will result in reading deficits, whereas poor phonemic skills (except for overlearned material) will lead to writing deficits. The patient's understanding of grammar and syntax will be impaired. Arithmetic problems presented auditorily will present more problems than those presented visually. Problems in comprehending questions and thinking abstractly may impair the Intelligence scale performance. Repetition will be more impaired than reading on the Expressive scale, reflecting the comprehension problems. Deficits in both comprehension and repetition likely indicates a cortical (Wernicke's) aphasia. With repetition deficits in the presence of unimpaired comprehension and fluent speech, conduction aphasia (involving a subcortical temporal lesion) would be suspected. In either case, one will find spontaneous speech to be less impaired than what is seen in frontal or sensorimotor lesions.

Right Temporal

Patients with lesions in the right-temporal area show elevations on the Motor, Rhythm, Visual, Receptive, Tactile, Intelligence, and Memory scales. Unlike other right-hemisphere lesions, the Left-Hemisphere scale is usually slightly elevated over the Right-Hemisphere scale (when peripheral or subcortical deficits are absent). This area's role in analyzing unfamiliar stimuli is seen in an improvement that occurs with repeated administration of graphesthesia items on the Tactile scale and similarly, on visually presented Motor scale items. Slowing is noted on constructional items, and the more complex visual discrimination and visual-spatial items are performed poorly. The ability to analyze nonverbal auditory stimuli is more impaired than in the right frontal patient. Problems in discriminating closely related phonemes and comprehending inverted and complex grammatical structures may be present on the Receptive scale. Spontaneous speech, sequencing, and, occasionally, visual identification on the Expressive scale may be impaired. Deficits on the Memory scale usually reflect difficulties with the complex, interference, and visual items. Impaired interpretation of visual thematic and some verbal material, along with difficulty on the discursive arithmetic problems usually appear on the intelligence scale.

Subcortical

Very little data has been gathered on lesions that affect only the subcortical regions of the brain. Preliminary work suggests that subcortical involvement should be considered when a right-hemisphere focus is suggested by the clinical scale profile pattern but not by an analysis of the cognitive symptoms, or when evidence of coexisting left- and right-hemisphere lesions is present. Also, discrepancies between patterns found on the clinical scales and those on the localization and factor scales (see below) should also alert one to the possibility of a subcortical disorder.

Mixed Lesions

Mixed lesions will combine aspects of localized patterns and, in general, are much more common than the classical patterns. Unless one is experienced in localizing injuries, and is specifically asked to do so, the examiner should focus on describing deficits, generate hypotheses about the patient's performance from hypotheses regarding localization, and determine the degree to which the three sets of scales are comparable.

PATTERN ANALYSIS OF THE LOCALIZATION SCALES

The empirically derived localization scales can provide the examiner with assistance in generating or confirming hypotheses regarding the nature of the cortical injury. The eight scales are most sensitive to injuries in different brain areas, but not exclusively so. (These scales will subsequently be referred to as L1, L2, L3, L4, L5, L6, L7, L8.) As with the clinical scales, elevations on any of this set of scales are considered only within the context of their relationship to the other scales.

The scales have been found useful in formulating both general and specific hypotheses regarding the status of the patient's brain. The use of the scales for the determination of the presence of brain damage has already been discussed. If determined to be present, the scales can then assist in lateralizing the injury. In general, lateralization is most simply determined by looking at the highest scale. If the highest scale is a left hemisphere (L1 to L4), then the injury is likely to be to the left hemisphere; if the highest scale is a right-hemisphere scale (L5 to L8), the injury is likely to be located in the right hemisphere. One will find this method to work best if the highest scale is at least 5 points above the next highest scale. If this is not the case, classification of the two highest scales (in order) may be helpful. In those cases where the two highest scales represent the same hemisphere (LL or RR), the hypothesis is clear. When the two highest scales are found to represent each of the hemispheres (LR or RL), lateralization is likely to be primarily in the hemisphere of the highest scale. Because the probability of error increases here, caution is advised in cases with these profiles.

LOCALIZATION

Accuracy in localization is best attained when the two highest localization scales are used. When these two scales represent adjacent areas of the brain, an overlapping lesion is likely to be involved. If the three highest scales are within 5 points of each other, then this profile should be used. If the three scales are consistent with a single, large lesion (e.g., elevations on L5, L6, L8), this then becomes the most likely site of dysfunction. If the three scales are greatly disparate in their localization (e.g., L1, L3, L5), one must consider either a subcortical or diffuse disorder.

Initial hypotheses generated by the localization scales are, of course, investigated for accuracy through comparison with the other sets of scales, item patterns, and qualitative data. Consistency between all sources of information increases the likelihood of the initial localization hypothesis being valid. When only some of the deficits are accounted for by this method, the presence of a second area of dysfunction, suggested by secondary elevations, needs to be considered. When the initial hypothesis is generally not confirmed by comparison with the other data, a subcortical lesion must be considered. Here, subcortical interconnections between the areas represented by the two highest scales are first investigated. If this is not substantiated, the possibility of involvement of many areas (suggested by the lower elevations) needs to be considered. Because these scales were derived from items designed to measure cortical functioning, it is not surprising that they may not be adequate in localizing or describing a subcortical lesion. In these cases, other data obtained from history, observations, or neurological examination may be helpful in understanding the deficit.

Left Frontal (L1)

Although most effective in identifying left-frontal lesions, the Left-Frontal scale is also sensitive to subcortical and left-temporal lesions, as well as severe left-parietal problems. When elevated 10 points above all other scales, a lesion in the left-frontal cortical or subcortical area, or at least the left-anterior portion, is usually indicated. Frequently, L3 or L7 will be elevated with L1 when frontal lesions are present. When combined with an elevation on L2, involvement of the entire frontal lobe is likely. Motor and speech deficits, as well as general problems in carrying out tasks, will frequently be noted. When combined with L3 elevations that are at about the same level but are lower than L1, a residual disorder involving most of the hemisphere, or a lesion in the tracts connecting the frontal and more posterior areas may be present. Approximately equal elevations on L1 and L4 may represent a lesion overlapping the two areas represented. With both L1/L3 and L1/L4 combinations, limbic-system involvement should strongly be considered.

Except for L7, it is rare to find an L1 elevation combined with a Right Hemisphere scale elevation. Although the L1/L7 combination usually indicates a left-frontal focus, a bilateral injury should be considered when severe construction dyspraxia is present. Combinations with either L5 or L8 may be indicative of bilateral frontal involvement. They are both also seen frequently in patients with a variety of psychiatric disorders, being associated with poor emotional control, insight, and understanding. The L1/L6 combination is quite rare. Although its interpretation is vague, one may speculate that it suggests bilateral motor involvement resulting from lesions in deep brain structures.

Right Frontal (L5)

When highly elevated over the other scales, L5 suggests very limited cortical or subcortical sesions in the right-frontal area. Right-frontal injuries are suggested by both L7/L5 and L5/L7 combinations. With the L5/L7 profile, right-anterior hemisphere and subcortical disorders must be considered. From examination of the clinical and factor scales, one may be able to determine which area represented by the former combination is the locus of the lesion. L5/L8 likely indicates an anterior right-hemisphere lesion, although a subcortical lesion must also be considered. The L5/L6 profile indicates a lesion within the whole right-frontal lobe that, when L7 is also elevated, may be part of a more general right-hemisphere disorder. Such is frequently seen with right-middle cerebral artery disorders. Combinations of L5 with left hemisphere scale elevations are rare and generally indicate bilateral dysfunction. These are most often seen with traumatic head injury cases and those where the left-hemisphere injury is present before the right-hemisphere dysfunction.

Left Sensorimotor (L2)

L2 is usually elevated with either L3 or L1, reflecting a lesion that overlaps into another area. Less frequently, L4 will be elevated with L2. This likely reflects either an overlapping lesion, or one that extends from the temporal lobe to the internal capsule or to the tracts leading to the internal capsule. Combinations of L2 and right-hemisphere scale elevations are rare. One might speculate that they represent bilateral or subcortical disorders. Evidence regarding the presence of cortical left- or right-hemisphere involvement possibly would aid in determining this.

Right Sensorimotor (L6)

Combinations of L6 with other right hemisphere scales are generally indicative of an overlapping lesion. Typically, a right sensorimotor area lesion will show an L7 elevation accompanying the L6 elevation. When L6 is the highest scale and a right-sensorimotor problem is not suggested by the clinical scales, a lesion in the more anterior right hemisphere needs to be considered. Bilateral injuries are suggested when a left-hemisphere scale elevation accompanies an L6 elevation. The most common among these profiles is the L6/L3, which may represent L3 deficits that predate the L6 deficits.

Left Parietal-Occipital (L3)

Along with those on L7, L3 elevations are among those most commonly seen. This is due to both scales' association with middle-cerebral-artery strokes and academic deficiencies—conditions that often result in referrals to neuropsychologists. Lesions in this area result in a marked L3 elevation, which is accompanied by a lesser L1 elevation. However, these lesions may also be reflected in L7/L3 profiles. Such profiles thus need to be carefully evaluated. Both L3/L2 and L3/L4 combinations generally reflect overlapping lesions. A combination of L3 with any right hemisphere scale at a comparable level suggests a bilateral disorder. However, an accompanying L7 elevation may only reflect extensive visual deficits resulting from a lesion in the posterior left parietal/occipital area.

Right Parietal-Occipital (L7)

Elevations on L7 can be associated with lesions in any area of the right hemisphere. Generally, the locus will be in the right-parietal area when L7 is elevated well above the other scales. When any of the other right-hemisphere scales are elevated to a level similar to L7, one must strongly consider an overlapping lesion or one outside of the parietal area. The L7/L5 combination may indicate a frontal lesion or one located between the two areas. Although seen in left-frontal injuries, the pattern is usually L1/L7 instead of L7/L1. Bilateral injuries are strongly suggested when L7 is the highest scale and one of the left-hemisphere scales is about equally elevated. The importance of such a combination disappears when L7 is considerably higher than the left-hemisphere scale.

Left Temporal (L4)

When L4 is the highest elevation, this will most likely reflect middle- or posterior-temporal area dysfunction, as well as that of adjacent parietal areas. However, more anterior-temporal injuries may show an L1 elevation, whereas more posterior-temporal injuries may show an L3 elevation. Generally, when L4 is the highest scale, temporal-lobe involvement is indicated. If a secondary scale is equally elevated, this other represented area may additionally be involved. Particularly in the case of subcortical or anterior temporal lesions, left-temporal involvement may present without L4 being notably high.

Combinations with other left-hemisphere scales (L4/L1, L4/LZ, L4/L3) likely represent overlapping lesions. Because verbal deficits associated with L4 lesions may result in difficulty in comprehending instructions and thus lead to L7, L8, and L5 elevations, caution in the interpretation of combinations of L4 with right-hemisphere scales is warranted. Elevations of 100 or more on L4 may represent comprehension deficits so severe that the ability to understand instructions during the entire battery may be impaired. Associated rhythm deficits may also be present with L4 elevations and affect right-hemisphere elevations. Thus, the nature of the deficits seen on the right-hemisphere elevations that accompany the L4 elevation must be considered before concluding the presence of a bilateral disorder.

Right Temporal (L8)

Because of the loading of items that are unfamiliar or novel on this scale, repeated administration (and thus practice) has the effect of improving (lowering) the L8 score. This scale is thus the most unreliable of the localization scales. Elevations on L7 tend to overlap with L8 elevations such that L7/L8 combinations, where the two scales are about equal, likely indicate overlapping lesions. However, they may also reflect a pure right-temporal focus in some cases. It is best to consider L8/L5 and L8/L6 combinations as indicating right-anterior dysfunction without trying to localize further. L8/L3 may reflect either a bilateral disorder or a preexisting L3 lesion. Both L8/L1 and L8/L4 may be seen in cases of psychomotor epilepsy or psychiatric disorders. These profiles possibly represent bilateral involvement of subcortical areas with or without cortical involvement.

PATTERN ANALYSIS OF THE FACTOR SCALES

Further assistance in determining the nature of the deficits being investigated is provided by an analysis of the empirically derived factor scales. These scales include Kinesthetic-Based Movement, Drawing Speed, Fine, Motor Speed, Spatial-Based Movement, and Oral-Motor Movements for the Motor scale; Rhythm and Pitch Perception for the Rhythm scale; Simple Tactile Sensation and Stereognosis for the Tactile scale; and Visual Acuity and Naming and Visual-Spatial Organization for the Visual scale. Receptive Speech scale factors include Phonemic Discrimination, Relational Concepts, Concept Recognition, Verbal-Spatial Relations, Word Comprehension, and Logical Grammatical Relations. Simple Phonetic Reading, Word Recognition, and Reading Polysyllabic Words comprise the Expressive Speech scale factors, whereas Reading Complex Material and Reading Simple Material comprise the Reading scale factors. Writing scale factors include Spelling and Motor Writing Skills. Factor scales for Arithmetic are Arithmetic Calculations and Number Reading. Memory scale factors include Verbal Memory and Visual and Complex Memory, while General Verbal Intelligence, Complex Verbal Arithmetic, and Simple Verbal Arithmetic comprise the scales derived for the Intelligence scale. The title of each of the factor scales may mislead those who are unfamiliar with these scales. Some titles provide a good description of the content of the items contained within the scales. Other scales are more general, making the titles somewhat misleading and thus require closer scrutiny of the items.

Below are presented factor-scale profile patterns found by McKay and Golden (1981a) for patients with localized lesions. These provide the examiner with information regarding patterns of deficits that might be seen with localized areas of dysfunction. This additional information can be useful in attempting to confirm hypotheses formulated earlier.

Left Frontal

A generalized elevation of nearly all scales is seen with a major left-frontal lesion. This may be due as much to the general problem of behavioral control as it is to specific deficits represented by the scales. The only prominent deficit is seen on the Number Reading scale. However, the items missed here may, on another occasion, be performed correctly. Other notable deficits include those seen on the Kinesthetic-Based Movement, Simple Tactile Sensation (likely reflecting attentional difficulties), and on Visual-Spatial Organization scales (representing problems in integrating disparate stimuli). For this same reason, Relational Concepts, Logical Grammatical Relations, and Verbal-Spatial Relations are also somewhat impaired. Generalized mild verbal deficits, particularly seen on the more complex verbal items, and mild problems in abstracting, as indicated on the General Verbal Intelligence scale, are also noted.

Right Frontal

With the exception of the deficits on Relational Concepts, elevations on the verbal scales are much less than those seen with left-frontal lesions. Motor deficits are indicated by elevations on Kinesthetic-Based Movement, Drawing Speed, and, to a lesser extent, Oral-Motor Movement. Impaired performance on the former as well as on the Simple Tactile Sensation scale likely represent the spatial integrative deficits which are not uncommonly seen with these patients.

Left Sensorimotor

Expected impairment in motor and sensory abilities are reflected in elevations on the Kinesthetic-Based Movement, Fine Motor Speed, Simple Tactile Sensation, and Stereognosis scales. Approximately half of the items on these scales measure right-body-side performance. The motor deficits indicated on the Number Reading and Motor Writing and, to a lesser extent on the Drawing Speed scales, are also consistent with the locus of the lesion. However, the fact that the dominant hand of many of the patients studied was so impaired that the nondominant hand had to be used for writing and drawing suggests that the elevations on these latter three scales may reflect the poorer nondominant hand performance. The lack of elevation on the three Expressive scales is explained by the absence of patients with severe aphasic problems who could not take the test from any of the localized groups.

Right Sensorimotor

Right-Sensorimotor scale elevations are similar to those for the Left-Sensorimotor group, but with the sensory and motor deficits lateralized to the right hemisphere. This similarity likely will decrease as motor speech deficits increase. Also, Spatial-Based Movement is somewhat more elevated.

Left Parietal-Occipital

Left-Parietal lesions result in elevations on the academic achievement scales, with particular deficits noted on Number Reading, Arithmetic Calculations, and Motor Writing. Minor elevations seen on the General Verbal Intelligence, Kinesthetic-Based Movement, and the Receptive scales measuring more complex understanding abilities (Relational Concepts, Verbal-Spatial Relations, and Logical Grammatical Relations). The elevation on the Visual Acuity and Naming scales may result from visual recognition or severe naming deficits.

Right Parietal-Occipital

Lesions in this area are reflected in elevations on Kinesthetic-Based Movement, Simple Tactile Sensation, and Stereognosis (indicative of left body side tactile deficits), and on Visual Acuity and Naming and Visual-Spatial Organization (indicating visual deficits). Typically, single errors resulting in elevations on Motor Writing and Reading Simple Material will occur. While Drawing Speed is normal, the quality of the drawings will likely not be.

Left Temporal

As one might expect, several verbal deficits are present. In addition to deficits on the Reading and Writing scales, elevations are seen on a number of scales sensitive to deficits in both verbal auditory comprehension (Phonemic Discrimination, Number Reading, Word Repetition) and some higher level comprehension and abstraction tasks (Relational Concepts, Verbal-Spatial Relations, Logical Grammatical Relations, General Verbal Intelligence). As a result of dysgraphesthesia problems, Simple Tactile Sensation is elevated. A slight elevation on Spatial-Based Movement is noted. The Verbal Memory and Visual and Complex Memory scale elevations reflect difficulties in remembering verbal and complex material.

Right Temporal

The unexpected elevation on the Oral-Motor Movements scale may reflect either peculiarities in the particular sample studied, or the uncommon nature of the spatial mouth movements for individuals with lesions in this area. This area's function of organizing input from all sensory modalities is seen in deficits on the more complex items of the Simple Tactile Sensation and Stereognosis, as well as on Rhythm and Pitch Perception and Visual Acuity and Naming scales. Also notable are mild elevations on both Memory scale factors, and General Verbal Intelligence (reflecting problems in verbally interpreting thematic pictures).

Item Pattern and Qualitative Analysis

Analysis of all three sets of scales will lead one to arrive at a hypothesis that may represent the product of many revisions and modifications of the initial hypothesis. At this point, ideas regarding the nature of the disorder in question are tested against an analysis of individual items. Consistency of the item pattern (which items are missed, which items are performed adequately) strengthens one's confidence in a hypothesis. Inconsistency, on the other hand, will necessitate further revision and, subsequently, checking again for consistency of the new hypothesis with the item pattern.

It is generally recommended that one initially entertain the likelihood of a single focus of dysfunction. Lack of substantiating evidence for the simple explanation would then necessitate considering more complex hypotheses that, again, are checked against all available data. Of course, the ability to arrive at a hypothesis that adequately explains all of the results will depend on the knowledge, experience, and skill of the interpreter.

An integrated hypothesis will also take into consideration qualitative information. This data has to do with the way in which the patient performed an item rather than whether or not his performance was adequate. It is obtained either through standard testing procedures or by initiating testing-the-limits procedures. At this time, a standardized system for scoring qualitative data is being developed. No data for groups of patients with localized lesions or normals is currently available. Even without such norms, however, the LNNB user who is knowledgable of behavior associated with localized lesions can use qualitative data to verify or question a hypothesis. For example, observations of frequent impulsive or perseverative responding (both frequently seen in patients with frontal lobe disorders) would be consistent with a hypothesis of a left-frontal lesion. If the hypothesis was that of a left-parietal disorder, one would need to reconsider the hypothesis. It is also important to consider historical and other patient factors when doing any clinical case. Pattern analyses at the item level are discussed more fully in Golden, MacInnes, et al. (1982) and Moses et al. (1983).

REFERENCES

Ariel, R., & Golden, C. J. (1981). *An alternate form of the Luria-Nebraska Neuropsychological Battery: Form II.* Paper presented at the meeting of the National Academy of Neuropsychologists, Orlando, FL.

Bach, P. J. (1983). *Empirical evidence mitigating against the diagnostic utility of the Luria-Nebraska Neuropsychological Battery.* Paper presented at the meeting of the International Neuropsychological Society, Mexico City.

Bach, P. J., Harowski, K., Kirby, K., Peterson, P., & Schulein, M. (1981). The interrater reliability of the Luria-Nebraska Neuropsychological Battery. *Clinical Neuropsychology, 3* (3), 19–21.

Berg, R. A., & Golden, C. J. (1981). Identification of neuropsychological deficits in epilepsy using the Luria-Nebraska Neuropsychological Battery. *Journal of Consulting and Clinical Psychology, 49,* 745–747.

Bryant, E. T. (1982). The relationship of learning disabilities, neuropsychological deficits, and violent criminal behavior in an inmate population (Doctoral dissertation, California School of Professional Psychology, Berkeley, 1982). *Dissertation Abstracts International, 43,* 3182B.

Bryant, E. T., Scott, M. L., Golden, C. J., & Tori, C. D. (1984). *Neuropsychological deficits, learning disability, and violent behavior.* Journal of Clinical and Consulting Neuropsychology, 52 (2), 323–324.

Chelune, G. J. (1982). A re-examination of the relationship between the Luria-Nebraska and Halstead-Reitan Batteries: Overlap with the WAIS. *Journal of Consulting and Clinical Psychiatry, 50,* 578–580.

Chmielewski, C., & Golden, C. J. (1980). Alcoholism and brain damage: An investigation using the Luria-Nebraska Neuropsychological Battery. *International Journal of Neuroscience, 10,* 99–105.

Christensen, A. L. (1975a). *Luria's neuropsychological investigation.* New York: Spectrum.

Christensen, A. L. (1975b). *Luria's neuropsychological investigation: Manual.* New York: Spectrum.

De Obaldia, R., Leber, W. R., & Parsons, O. A. (1981). Assessment of neuropsychological functions in chronic alcoholics using a standardized version of Luria's neuropsychological technique. *International Journal of Neuroscience, 14,* 85–93.

Dill, R. A., & Golden, C. J. (1983). *WAIS-R and Luria-Nebraska intercorrelations.* Unpublished manuscript.

Duffala, D. (1979). Validity of the Luria-South Dakota Neuropsychological Battery for brain-injured persons. (Doctoral dissertation, California School of Professional Psychology, Berkeley, 1978). *Dissertation Abstracts International, 39*, 4439B.

Gechter, G. A., Griffith, S. R., & Newell, T. G. (1983). *Changes in neuropsychological functions among detoxifying and recovering alcoholics as measured by the Luria-Nebraska Neuropsychological Battery*. Paper presented at the meeting of the American Psychological Association, Anaheim, CA.

Gillen, R. W., Ginn, C., Strider, M. A., Kreuch, T. J., & Golden, C. J. (1983). The relationship of the Luria-Nebraska Neuropsychological Battery to the Peabody Individual Achievement Test: A correlational analysis. *International Journal of Neuroscience, 21*, 51–62.

Golden, C. J., & Berg, R. A. (1980a). Interpretation of the Luria-Nebraska Neuropsychological Battery: The Writing scale. *Clinical Neuropsychology, 2*, (1), 8–12.

Golden, C. J., & Berg, R. A. (1980b). Interpretation of the Luria-Nebraska Neuropsychological Battery by item intercorrelation: Items 1–24 for the Motor scale. *Clinical Neuropsychology, 2* (2), 66–71.

Golden, C. J., & Berg, R. A. (1980c). Interpretation of the Luria-Nebraska Neuropsychological Battery by item intercorrelation: Items 25–51 of the Motor scale. *Clinical Neuropsychology, 2* (3), 105–108.

Golden, C. J., & Berg, R. A. (1980d). Interpretation of the Luria-Nebraska Neuropsychological Battery by item intercorrelation: The Rhythm scale. *Clinical Neuropsychology, 2* (4), 153–156.

Golden, C. J., & Berg, R. A. (1981a). Interpretation of the Luria-Nebraska Neuropsychological Battery by item intercorrelation: The Tactile scale. *Clinical Neuropsychology, 3* (1), 25–29.

Golden, C. J., & Berg, R. A. (1981b). Interpretation of the Luria-Nebraska Neuropsychological Battery by item intercorrelation: VI. The Visual scale. *Clinical Neuropsychology, 3* (2), 22–26.

Golden, C. J., & Berg, R. A. (1981c). Interpretation of the Luria-Nebraska Neuropsychological Battery by item intercorrelation: VII. Receptive Language. *Clinical Neuropsychology, 3* (3), 21–27.

Golden, C. J., & Berg, R. A. (1982a). Item interpretation of the Luria-Nebraska Neuropsychological Battery: VIII. The Expressive Speech scale. *Clinical Neuropsychology, 4* (1), 8–14.

Golden, C. J., & Berg, R. A. (1982b). Interpretation of the Luria-Nebraska Neuropsychological Battery by item intercorrelation: The Reading scale. *Clinical Neuropsychology, 4* (4), 176–179.

Golden, C. J., & Berg, R. A. (1983a). Interpretation of the Luria-Nebraska Neuropsychological Battery by item intercorrelation: Intellectual Processes. *Clinical Neuropsychology, 5* (1), 23–28.

Golden, C. J., & Berg, R. A. (1983b). Interpretation of the Luria-Nebraska Neuropsychological Battery by item intercorrelation: The Memory scale. *Clinical Neuropsychology, 5* (2), 55–59.

Golden, C. J., & Berg, R. A. (1983c). Interpretation of the Luria-Nebraska Neuropsychological Battery by item intercorrelation: The Arithmetic scale. *Clinical Neuropsychology, 5* (3), 122–127.

Golden, C. J., Hammeke, T. A., & Purisch, A. D. (1978). Diagnostic validity of a standardized neuropsychological battery derived from Luria's neuropsychological tests. *Journal of Consulting and Clinical Psychology, 46*, 1258–1265.

Golden, C. J., Graber, B., Moses, J. A., & Zatz, L. M. (1980). Differentiation of chronic schizophrenics with and without ventricular enlargement by the Luria-Nebraska Neuropsychological Battery. *International Journal of Neuroscience, 11*, 131–138.

Golden, C. J., Hammeke, T. A., & Purisch, A. D. (1980). *A manual for the administration and interpretation of the Luria-Nebraska Neuropsychological Battery.* Los Angeles: Western Psychological Services.

Golden, C. J., Moses, J. A., Zelazowski, R., Graber, B., Zatz, L. M., Horvath, T. B., & Berger, P. A. (1980). Cerebral ventricular size and neuropsychological impairment in young chronic schizophrenics. *Archives of General Psychiatry, 37,* 619–623.

Golden, C. J., Osmon, D., Sweet, J., Graber, B., Purisch, A., & Hammeke, T. (1980). Factor analysis of the Luria-Nebraska Neuropsychological Battery III. Writing, Arithmetic, Memory, Left, and Right. *International Journal of Neuroscience, 11,* 309–315.

Golden, C. J., Purisch, A., Sweet, J., Graber, B., Osmon, D., & Hammeke, T. (1980). Factor analysis of the Luria-Nebraska Neuropsychological Battery II: Visual, Receptive, Expressive and Reading scales. *International Journal of Neuroscience, 11,* 227–236.

Golden, C. J., Sweet, J., Hammeke, T., Purisch, A., Graber, B., & Osmon, D. (1980). Factor analysis of the Luria-Nebraska Neuropsychological Battery: I. Motor, Rhythm, and Tactile scales. *International Journal of Neuroscience, 11,* 91–99.

Golden, C. J., Fross, K. H., & Graber, B. (1981). Split-half reliability of the Luria-Nebraska Neuropsychological Battery. *Journal of Consulting and Clinical Psychology, 49,* 304–305.

Golden, C. J., Hammeke, T., Osmon, D., Sweet, J., Purisch, A., & Graber, B. (1981). Factor analysis of the Luria-Nebraska Neuropsychological Battery: IV. Intelligence and Pathognomonic scales. *International Journal of Neuroscience, 13,* 87–92.

Golden, C. J., Kane, R., Sweet, J., Moses, J. A., Cardellino, J. P., Templeton, R., Vicente, P., & Graber, B. (1981). Relationship of the Halstead-Reitan Neuropsychological Battery to the Luria-Nebraska Neuropsychological Battery. *Journal of Consulting and Clinical Psychology, 49,* 410–417.

Golden, C. J., Moses, J. A., Fishburne, F. J., Engum, E., Lewis, G. P., Wisniewski, A. M., Conley, F. K., Berg, R. A., & Graber, B. (1981). Cross-validation of the Luria-Nebraska Neuropsychological Battery for the presence, lateralization, and localization of brain damage. *Journal of Consulting and Clinical Psychology, 49,* 491–507.

Golden, C. J., Moses, J. A., Graber, B., & Berg, R. (1981). Objective clinical rules for interpreting the Luria-Nebraska Neuropsychological Battery: Derivation, effectiveness, and validation. *Journal of Consulting and Clinical Psychology, 49,* 616–618.

Golden, C. J., Berg, R. A., & Graber, B. (1982). Test-retest reliability of the Luria-Nebraska Neuropsychological Battery in stable, chronically impaired patients. *Journal of Consulting and Clinical Psychology, 50,* 452–454.

Golden, C. J., MacInnes, W. D., Ariel, R. N., Ruedrich, S. L., Chu, C., Coffman, J. A., Graber, B., & Bloch, S. (1982). Cross-validation of the Luria-Nebraska Neuropsychological Battery to differentiate chronic schizophrenics with and without ventrivular enlargement. *Journal of Consulting and Clinical Psychology, 50,* 87–95.

Golden, C. J., Purisch, A. D., & Hammeke, T. A., (1985). *Luria-Nebraska Neuropsychological Battery: Forms I and II Manual.* Los Angeles: Western Psychological Services.

Grey, P. T. (1982). A neuropsychological study of dyslexia using the Luria-Nebraska Neuropsychological Battery. *Dissertation Abstracts International, 34,* 1236B. (University Microfilms No. 82–16, 284)

Hammeke, T. A., Golden, C. J., & Purisch, A. D. (1978). A standardized, short, and comprehensive neuropsychological test battery based on the Luria neuropsychological evaluation. *International Journal of Neuroscience, 8,* 135–141.

Johnson, G. L., Moses, J. A., & Bryant, E. (1984). *Development of an impairment index for the Luria-Nebraska Neuropsychological Battery. International Journal of Clinical Neuropsychology, 6,* 242–247.

Kaimann, C. (1981). *A neuropsychological investigation of multiple sclerosis.* Paper

presented at the meeting of the American Psychological Association, Los Angeles, CA.
Kaimann, C. R. (in press). A neuropsychological investigation of multiple sclerosis (Doctoral dissertation, University of Nebraska, Lincoln, 1983). *Dissertation Abstracts International.*
Kaimann, C., Knippa, J., Schima, E., & Golden, C. J. (1983). *Relationship of performance on the Luria-Nebraska Neuropsychological Battery to CT-scan findings in multiple sclerosis patients.* Manuscript submitted for publication.
Kane, R. L., Sweet, J. J., Golden, C. J., Parsons, O. A., & Moses, J. A. (1981). Comparative diagnostic accuracy of the Halstead-Reitan and standardized Luria-Nebraska Neuropsychological Batteries in a mixed psychiatric and brain-damaged population. *Journal of Consulting and Clinical Psychology, 49,* 484–485.
Langell, M. E., Purisch, A. D., Green, & Golden, C. J. (1985). Left frontal functioning in paranoid and nonparanoid schizophrenics. In A. Yozawitz, (Chair.), *Cognitive changes with psychiatric disorders.* Paper session conducted at the meeting of the American Psychological Association, Los Angeles, CA.
Lewis, G. P., Golden, C. J., Moses, J. A., Osmon, D. C., Purisch, A. D., & Hammeke, T. A. (1979). Localization of cerebral dysfunction with a standardized version of Luria's neuropsychological battery. *Journal of Consulting and Clinical Psychology, 47,* 1003–1019.
Luria, A. R. (1966). *Higher cortical functions in man.* New York: Basic Books.
Luria, A. R. (1973). *The working brain.* New York: Basic Books.
MacInnes, W. D. (1981). *Aging and its relationship to neuropsychological and neurological measures.* Paper presented at the meeting of the National Academy of Neuropsychologists, Orlando, FL.
MacInnes, W. D., Cole, J. K., & Kaimann, C. (1981). *Aging and the Luria-Nebraska Neuropsychological Battery.* Paper presented at the meeting of the American Psychological Association, Los Angeles, CA.
MacInnes, W. D., Golden, C. J., Sawicki, R. F., Gillen, R. W. Quaife, M., Graber, B., Uhl, H. S., & Greenhouse, A. J. (1982). *Aging, neuropsychological functioning, and regional cerebral blood flow: Interrelationships.* Unpublished manuscript.
MacInnes, W. D., Franzen, M. D., Sawicki, R. F., Golden, C. J., Mahoney, P., McGill, J., & Uhl, H. S. (1983). *Aging, neuropsychological functioning, and brain density: Interrelationships.* Paper presented at the meeting of the American Psychological Association, Anaheim, CA.
MacInnes, W. D., Gillen, R. W., Golden, C. J., Graber, B., Cole, J. K., Uhl, H. S., & Greenhouse, A. H. (1983). Aging and performance on the Luria-Nebraska Neuropsychological Battery. *International Journal of Neuroscience, 19,* 179–190.
Malloy, P. F., & Webster, J. S. (1981). Detecting mild brain impairment using the Luria-Nebraska Neuropsychological Battery. *Journal of Consulting and Clinical Psychology, 49,* 768–770.
Maruish, M. E., Sawicki, R. F., Franzen, M. D., & Golden, C. J. (1985). Alpha coefficient reliabilities for the Luria-Nebraska Neuropsychological Battery summary and localization scales by diagnostic category. *International Journal of Clinical Neuropsychology, 7,* 10–12.
Marvel, G. A., Golden, C. J., Hammeke, T., Purisch, A., & Osmon, D. (1979). Relationship of age and education to performance on a standardized version of Luria's neuropsychological tests in different patient populations. *International Journal of Neuroscience, 9,* 63–70.
McKay, S. (1981). The neuropsychological test performance of an assaultive psychiatric population (Doctoral dissertation, University of Nebraska, Lincoln, 1980). *Dissertation abstracts International, 41,* 4269B.

McKay, S., & Golden, C. J. (1979a). Empirical derivation of experimental scales for localizing brain lesions using the Luria-Nebraska Neuropsychological Battery. *Clinical Neuropsychology, 1* (2), 19–23.

McKay, S., & Golden, C. J. (1979b). Empirical derivation of neuropsychological scales for the lateralization of brain damage using the Luria-Nebraska Neuropsychological Test Battery. *Clinical Neuropsychology, 1* (2), 1–5.

McKay, S. E., & Golden, C. J. (1981a). The assessment of specific neuropsychological skills using scales derived from factor analysis of the Luria-Nebraska Neuropsychological Battery. *International Journal of Neuroscience, 14,* 189–204.

McKay, S. E., & Golden, C. J. (1981b). Re-examination of the factor structure of the Receptive Language scale of the Luria-Nebraska Neuropsychological Battery. *International Journal of Neuroscience, 14,* 183–188.

McKay, S. E., Golden, C. J., Moses, J. A., Fishburne, F., & Wisniewski, A. (1981). Correlation of the Luria-Nebraska Neuropsychological Battery with the WAIS. *Journal of Consulting and Clinical Psychology, 49,* 940–946.

McKay, S. E., Golden, C. J., & Scott, M. (1981). Neuropsychological correlates of auditory and visual hallucinations. *International Journal of Neuroscience, 15,* 87–94.

McKay, S. E., Golden, C. J., Wolf, B. A., & Perrine, K. (1983). *Factor analysis of the Luria-Nebraska Neuropsychological Battery.* Unpublished manuscript.

McKay, S., & Ramsey, R. (1983). Correlation of the Wechsler Memory Scale and the Luria-Nebraska Memory scale. *Clinical Neuropsychology, 5,* 168–170.

Mikula, J. A. (1981). The development of a short form of the standardized version of Luria's neuropsychological assessment (Doctoral dissertation, Southern Illinois University, Carbondale, 1979). *Dissertation Abstracts International, 41,* 3189B.

Moses, J. A. (1983). An orthogonal factor solution of the Luria-Nebraska Neuropsychological Battery items: I. Motor, Rhythm, Tactile, and Visual scales. *Clinical Neuropsychology, 5,* 181–185.

Moses, J. A. (1984a). An orthogonal factor solution of the Luria-Nebraska Neuropsychological Battery items: II. Receptive Speech, Expressive Speech, Writing, and Reading scales. *International Journal of Clinical Neuropsychology, 6,* 24–28.

Moses, J. A. (1984b). An orthogonal factor solution of the Luria-Nebraska Neuropsychological Battery items: III. Arithmetic, Memory, and Intelligence scales. *International Journal of Clinical Neuropsychology, 6,* 103–106.

Moses, J. A. (1984c). An orthogonal factor solution of the Luria-Nebraska Neuropsychological Battery items: IV. Pathognomonic, Right Hemisphere, and Left Hemisphere scales. *International Journal of Clinical Neuropsychology, 6,* 161–165.

Moses, J. A. (1984d). The relative effects of cognitive and sensorimotor deficits on the Luria-Nebraska Neuropsychological Battery performance in a brain-damaged population. *International Journal of Clinical Neuropsychology, 6,* 8–12.

Moses, J. A., & Golden, C. J. (1979). Cross validation of the discriminative effectiveness of the standardized Luria Neuropsychological Battery. *International Journal of Neuroscience, 9,* 149–155.

Moses, J. A., & Golden, C. J. (1980). Discrimination between schizophrenic and brain-damaged patients with the Luria-Nebraska Neuropsychological Test Battery. *International Journal of Neuroscience, 10,* 121–128.

Moses, J. A., & Johnson, G. L. (1983). An orthogonal factor solution for the Receptive Speech scale of the Luria-Nebraska Neuropsychological Battery. *International Journal of Neuroscience, 20,* 183–188.

Moses, J. A., Golden, C. J., Berger, P. A., & Wisniewski, A. M. (1981). Neuropsychological deficits in early, middle, and late stages of Huntington's disease as measured by the

Luria-Nebraska Neuropsychological Battery. *International Journal of Neuroscience,* 14, 95–100.

Moses, J. A., Cardellino, J. P., & Thompson, L. L. (1983). Discrimination of brain damage from chronic psychosis by the Luria-Nebraska Neuropsychological Battery: A closer look. *Journal of Consulting and Clinical Psychology,* 51, 441–449.

Moses, J. A., Golden, C. J., Ariel, R., & Gustavson, J. L. (1983). *Interpretation of the Luria-Nebraska Neuropsychological Battery* (Vol. 1). New York: Grune & Stratton.

Moses, J. A., Johnson, G. L., & Lewis, G. P. (1983a). Reliability analyses of the Luria-Nebraska Neuropsychological Battery summary, localization, and factor scales. *International Journal of Neuroscience,* 20, 149–154.

Moses, J. A., Johnson, G. L., & Lewis, G. P. (1983b). Reliability analyses of the Luria-Nebraska Neuropsychological Battery summary, and localization scales by diagnostic group: A follow-up study. *International Journal of Neuroscience,* 21, 107–112.

Moses, J. A., Johnson, G. L., & Lewis, G. P. (1983c). Reliability analyses of the Luria-Nebraska Neuropsychological Battery factor scales by diagnostic group: A follow-up study. *International Journal of Neuroscience,* 21, 113–118.

Osmon, D. C., Golden, C. J., Purisch, A. D., Hammeke, T. A., & Blume, H. G. (1979). The use of a standardized battery of Luria's tests in the diagnosis of lateralized cerebral dysfunction. *International Journal of Neuroscience,* 9, 1–9.

Parolini, R. (1983). Reading, spelling, and arithmetic disabilities: A neuropsychological investigation using Luria's methods (Doctoral dissertation, University of Nebraska, Lincoln, 1982). *Dissertation Abstracts International,* 43, 1996B.

Picker, W. T., & Schlottmann, R. S. (1982). An investigation of the Intellectual Processes scale of the Luria-Nebraska Neuropsychological Battery. *Clinical Neuropsychology,* 4 (3), 120–124.

Plaisted, J. R., & Golden, C. J. (1982). Test-retest reliability of the clinical, factor, and localization scales of the Luria-Nebraska Neuropsychological Battery. *International Journal of Neuroscience,* 17, 163–167.

Prifitera, A., & Ryan, J. J. (1981). Validity of the Luria-Nebraska Intellectual Processes scale as a measure of adult intelligence. *Journal of Consulting and Clinical Psychology,* 49, 755–756.

Puente, A. E., Heidelberg-Sanders, C., & Lund, N. (1982a). Detection of brain damage in schizophrenics measured by the Whitaker Index of Schizophrenic Thinking and the Luria-Nebraska Neuropsychological Battery. *Perceptual and Motor Skills,* 54, 495–499.

Puente, A. E., Heidelberg-Sanders, C., & Lund, N. L. (1982b). Discrimination of schizophrenics with and without nervous system damage using the Luria-Nebraska Neuropsychological Battery. *International Journal of Neuroscience,* 16, 59–62.

Purisch, A. D., & Langell, M. E. (1985). Neuropsychological functioning of paranoid and nonparanoid schizophrenics. In A. Yozawitz (Chair.), *Cognitive changes with psychiatric disorders.* Paper session conducted at the meeting of the American Psychological Association, Los Angeles, CA.

Purisch, A. D., Golden, C. J., & Hammeke, T. A. (1978). Discrimination of schizophrenic and brain-injured patients by a standardized version of Luria's neuropsychological tests. *Journal of Consulting and Clinical Psychology,* 46, 1266–1273.

Ryan, J. J., & Prifitera, A. (1982). Concurrent validity of the Luria-Nebraska Memory scale. *Journal of Clinical Psychology,* 38, 378–379.

Sawicki, R. F., & Golden, C. J. (in press-a). Examination of two decision rules for the global interpretation of the Luria-Nebraska Neuropsychological Battery summary profile. *International Journal of Neuroscience.*

Sawicki, R. F., & Golden, C. J. (in press-b). Multivariate statistical techniques in neuropsychology: I. Comparison of orthogonal rotation methods with the Receptive scale of the Luria-Nebraska Neuropsychological Battery. *International Journal of Clinical Neuropsychology, 6,* 126–134.

Scott, M. L., Cole, J. K., McKay, S. E., Golden, C. J., & MacInnes, W. D. (1982). Neuropsychological performance in schizophrenics with histories of substance abuse. *International Journal of Neuroscience, 17,* 209–213.

Scott, M. L., Cole, J. K., McKay, S. E., Leark, R., & Golden, C. J. (1982). Neuropsychological performance of sexual assaulters and pedophiles. In J. Cole (Chair), *Psychological and neuropsychological concomitants of violent behavior.* Symposium conducted at the meeting of the American Psychological Association, Washington, DC.

Scott, M. L., Martin, R. L., & Liggett, K. R. (1982). Neuropsychological performance of persons with histories of assaultive behavior. In J. Cole (Chair), *Psychological and neuropsychological concomitants of violent behavior.* Symposium conducted at the meeting of the American Psychological Association, Washington DC.

Shelly, C., & Goldstein, G. (1982a). Intelligence, achievement, and the Luria-Nebraska Battery in a neuropsychiatric population: A factor analytic study. *Clinical Neuropsychology, 4* (4), 164–169.

Shelly, C., & Goldstein, G. (1982b). Psychometric relations between the Luria-Nebraska and Halstead-Reitan Neuropsychological Batteries in a neuropsychiatric setting. *Clinical Neuropsychology. 4,* (3), 128–133.

Shelly, C., & Goldstein, G. (1983a). Discrimination of chronic schizophrenia and brain damage with the Luria-Nebraska Battery: A partially successful replication. *Clinical Neuropsychology,* 5(2), 82–85.

Shelly, C., & Goldstein, G. (1983b). *Relationships between language skills as assessed by the Halstead-Reitan Battery and the Luria-Nebraska language factor scales in a nonaphasic patient population.* Paper presented at the meeting of the American Psychological Association, Anaheim, CA.

Spitzform, M. (1982). Normative data in the elderly on the Luria-Nebraska Neuropsychological Battery. *Clinical Neuropsychology,* 4(3), 103–105.

Strider, M. A. (1982). Neuropsychological concomitants of diabetes mellitus (Doctoral dissertation, University of Nebraska, Lincoln, 1982). *Dissertation Abstracts International, 43,* 888B.

Sweet, J. J. (1983). Confounding effects of depression on neuropsychological testing: Five illustrative cases. *Clinical Neuropsychology,* 5(3), 103–109.

Teem, C. L. (1981). Neuropsychological functions in chronic alcoholism. *Dissertation Abstracts International, 42,* 791B. (University Microfilms No. 81–14, 380)

Vicente, P., Kennelly, M. A., Golden, C. J., Kane, R., Sweet., J., Moses, J. A., Cardellino, J. P., Templeton, R., & Graber, B. (1980). The relationship of the Halstead-Reitan Neuropsychological Battery to the Luria-Nebraska Neuropsychological Battery: Preliminary report. *Clinical Neuropsychology,* 2(3), 140–141.

West, C. Y. (1982). Discrimination of violent and nonviolent inmates with the standardized Luria-Nebraska Neuropsychological Test Battery (Doctoral dissertation, California School of Professional Psychology, Berkeley, 1981). *Dissertation Abstracts International, 42,* 4218B.

Wolf, B. (1981). *Prediction of the Luria-Nebraska by changes in brain density.* Paper presented at the meeting of the American Psychological Association, Los Angeles, CA.

Zelazowski, R., Golden, C. J., Graber, B., Blose, I. L., Bloch, S., Moses, J. A., Zatz, L. M., Stahl, S. M., Osmon, D. C., & Pfefferbaum, A. (1981). Relationship of cerebral ventricular size to alcoholics' performance on the Luria-Nebraska Neuropsychological Battery. *Journal of Studies on Alcohol, 42,* 749–756.

8

An Overview of Similarities and Differences between the Halstead-Reitan and Luria-Nebraska Neuropsychological Batteries

GERALD GOLDSTEIN

INTRODUCTION

Chapters 6, 7, 8, and 9, provide a review of the relationships between the two major standard neuropsychological test batteries; the Halstead-Reitan (HRB) and the Luria-Nebraska (LNNB). Such a review can serve several purposes. Because the LNNB is a relatively new procedure and the HRB an established one, a high degree of concordance between the two procedures would aid in establishing the concurrent validity of the LNNB. The clinician wishing to do standardized neuropsychological assessment may want to know the relative advantages and disadvantages of the two procedures. Furthermore, the clinician who maintains access to both procedures may require information regarding the type of patient or assessment situation that gives the advantage to one battery over the other. Finally, if both batteries have comparable validity, reliability, and sophistication with regard to the drawing of interpretive inferences, then cost-effectiveness may become an issue, with the more economic procedure becoming the one of choice.

GERALD GOLDSTEIN • Veterans Administration Medical Center, Highland Drive, Pittsburgh, PA 15206, and Department of Psychiatry and Psychology, University of Pittsburgh, Pittsburgh, PA 15213.

Our discussion of similarities and differences between these two procedures will be divided into three parts. In this chapter, matters of a historical, practial, and theoretical nature will be reviewed. In subsequent chapters there will be presentations of empirical, psychometric findings and of case studies in which patients were administered both batteries. The four chapters combined should provide a broad overview of how the two procedures resemble and differ from each other from theoretical, empirical, and clinical perspectives. Contrasts between these two standard procedures, or fixed batteries, and flexible, or individualized, procedures will be provided in other chapters contained in this volume.

HISTORICAL CONSIDERATIONS

The two procedures to be discussed here have names associated with two of the major figures in neuroscience in general and neuropsychology in particular; Ward Halstead and Alexander Luria. Although they are both deceased, they were both alive at the same time over much of the first half of the 20th century. However, to the best of my knowledge, they did not know each other well, Luria doing his work in the Soviet Union and Halstead in the United States. Luria's name is never mentioned in Halstead's *Brain and Intelligence* (1947), whereas only one reference is made to Halstead in *The Working Brain* (1973), and only on a list of investigators interested in frontal-lobe function. Therefore, it seems apparent that the two investigators worked independently, and only perhaps shared the zeitgeist of their generation of brain scientists. It is also clear from their writings that they were both influenced to some extent by Kurt Goldstein and Martin Scheerer (Goldstein & Scheerer, 1941), but were not always in full agreement concerning the theoretical formulations of these other pioneers of neuropsychology. However, Luria and Halstead are generally viewed as being in the cognitive psychology camp and were relatively unsympathetic, each in their own ways, with the so-called diagram makers or localization-oriented behavioral neurologists of the late 19th and early 20th centuries. Thus, the standard test batteries we will be discussing are based on some form of holistic theory, but not on its extreme version, with substantially less derivation from classical localization theory.

Although the scientific histories of these two pioneer neuropsychologists are important and instructive ones, it must be stated from the outset that neither Halstead nor Luria utilized what are now described as comprehensive standard (or fixed) neuropsychological batteries, these procedures really only having developed during the 1950s. We will therefore be more concerned with Halstead's and Luria's legacies, as passed down through their students and advocates, than with their individual biographies. As we will see, the actual legacies of both Luria and Halstead did not turn out to be what these men might have predicted, and there is some debate, expressed perhaps only by cynics, as to whether one or both of them are "turning over in their graves." With regard to Luria, the knowledge of his work and of his legacy are somewhat clouded by two factors. First, his major publications, although written largely around the time of the Second World War, were not translated into English until the mid-1960s. Second, there seems to be little information concerning what developments have taken place in neuropsychology in the Soviet Union since Luria's death. Apparently, he was quite active until the end of his life, and a number of posthumous publications have recently appeared in English translation, including a revised version of the *Higher Cortical Functions* (Luria, 1980). Thus, we appear to have a relatively complete picture of Luria's work up until the time of his death, but what happened thereafter in the Soviet Union, to the best of my knowledge, is unclear. We know mainly about those aspects of Luria's work that were exported. This work includes the translated books and papers, the research done based on this work, and the contributions of individuals who personally worked or studied with Luria and his group and who pursued their interests in this work following their experiences in the Soviet Union. The names of Elkhonon Goldberg, Lawrence Majovski, J. T. Hutton, and Anne-Lise Christensen are probably the most familiar ones in this group. Of course, Karl Pribram and Hans-Lukas Teuber knew Luria, and there was a good deal of mutual influence they had on each other.

There appears to have been a division of labor among the associates of Luria, perhaps reflecting his diverse interests, not only in neuropsychological assessment, but in basic psychophysiological research and rehabilitative treatment as well. Goldberg has pursued the basic science work; Majovski has taken up Luria's work in rehabilitation and restoration of function; Hutton published a paper on aphasia with Luria (Luria & Hutton, 1977), and I believe has pursued some of his eye movement work. Christensen (1975a) took up Luria's method of neuropsychological assessment, and later also worked with some of his rehabilitation methods (Christensen, 1984). Although Luria did, in fact, publish information concerning his procedures, particularly in the *Traumatic Aphasia* (Luria, 1970) and *Higher Cortical Functions* books (Luria, 1980), in recent years his assessment techniques have become known to the English speaking public largely through the two volumes and kit prepared by (Christensen, 1975a,b,c), and several of her papers and presentations that appeared later (Christensen, 1979, 1984). It may be mentioned that although Luria is said to have never published a battery as such, Part III of *Higher Cortical Functions* contains quite a detailed description of his procedures. His language examination is also rather thoroughly described in *Traumatic Aphasia*. Nevertheless, Christensen provided us with specific, discrete items organized into sections, as well as the appropriate stimulus material. This material was modified by Majovski *et al.* (1979) into an adolescent's neuropsychological battery based on Luria's method, and by Golden and various collaborators into what is now known as the Luria-Nebraska battery.

When we focus on Luria's legacy in the area of assessment, it would appear that his work has come down to us in what may be described as three branches. There is, first of all, the type of individualized, targeted assessment advocated by Christensen (1975a, 1979, 1984), which, in essence, represents a continuation of Luria's original methods, not only in reference to the test materials, but to the mode of assessment as well. This particular mode is epitomized by Christensen in the following passage.

> The neuropsychologist must adapt a flexible approach to the examination. He/she should be a skilled observer of the patient's reactions in all situations, be able to invent small experimental situations and formulate hypotheses that can be verified or rejected by more specific examinations. Luria has compared the work of the neuropsychologist with that of a detective; it is more complex and less logical than a scientific experiment. (1984, p. 10)

Thus, the mode of assessment Christensen describes is apparently much like the kinds of diagnostic evaluations performed in the clinics and hospitals in which Luria and his associates practiced.

Another group of neuropsychologists borrowed some of Luria's procedures and certain aspects of this theory, but incorporated what was borrowed into their already established methods and conceptualizations. For example, Albert and Kaplan (1980) report the use of a test in which the patient is asked to copy 3-looped figures for 90 seconds. This test was used by Luria to examine an aspect of frontal-lobe function. It is probably fair to say that this borrowing of items and theoretical formulations has had most to do with frontal-lobe tests. Although the verification of Luria's theories of frontal lobe function has had its difficulties (e.g., Drewe, 1975; Goldberg & Tucker, 1979), most informed clinicians (e.g. Lezak, 1983; Walsh, 1978) are sympathetic with Luria's views concerning the role of the frontal lobes in the formation of intentions, and in the planning, regulation, and verification of behavior. Thus, Luria's tests based on these ideas have been accepted by many clinicians as measures of frontal-lobe function.

The third branch eventuated in the development of the Luria-Nebraska Battery, originally known as the Luria-South Dakota Battery or as A Standardized Version of Luria's Neuropsychological Tests. In this case, many of the items used for neuropsychological assessment by Luria, and particularly those items contained in the Christensen kit (1975c), were incorporated into a standardized, quantitatively oriented test battery. Thus, Luria's items were used, but in a manner quite different from the way in which he and his associates used them. Whereas efforts have been made to provide interpretations of these items to some extent within the framework of Luria's conceptualizations (e.g., Golden et al. 1982), the mode of conduct of the assessment differs sharply from Luria's procedure. In summary then, Luria's work in the area of neuropsychological assessment has come down in the form of (a) a continuation of the originally described method and theory, as practiced mainly by people who were directly associated with Luria; (b) a borrowing of selected procedures that become incorporated into other more comprehensive systems. The "frontal-lobe items" are most commonly used in this way; and (c) a quantified, standard battery based preponderantly on the test items described in *Higher Cortical Functions, Traumatic Aphasia* and the materials provided by Christensen.

Halstead's legacy took a somewhat different course from Luria's. In his case, the leadership of his work in the neuropsychological assessment area was clearly taken over by Ralph Reitan, his student and longtime associate. Philip Rennick, a younger student of Halstead, also subsequently took over some of this work following a period of study with Reitan. However, as in the case of Luria, I do not believe that it was ever Halstead's intention to establish a fixed, standard battery to be used in routine clinical assessment. Indeed, the original "Halstead Battery" was not a standard battery at all but was an extensive series of tests utilized in Halstead's Chicago laboratory in efforts made to develop discriminating measures. The Appendix to *Brain and Intelligence* (Halstead, 1947) contains 27 tests. The ones most clinical neuropsychologists are familiar with, such as the Category and Tactual Performance tests, are on that list, but so are such items as the "Halstead Color Gestalt Test," and "The Halstead Schematic Face Test." It is unlikely that most contemporary clinicians are familiar with those and several other procedures used in Halstead's laboratory.

Like Luria, Halstead also had diverse interests, including basic animal model research in physiological psychology, aging, and the neurochemistry of memory. One of his early papers was entitled "The effects of cerebellar lesions upon the habituation of post-rotational nystagmus" (Halstead, 1935), and later in life he was among the first scientists to surmise the neurochemical basis for memory (Katz & Halstead,

1950). His work in assessment appeared to focus largely around his specific interests in frontal lobe function and his theory of biological intelligence, and probably it can be fairly said that he utilized neuropsychological tests primarily to pursue those interests. Thus, tests were added, dropped, and modified in accordance with the requirements of specific research goals. If I may be permitted a personal anecdote, the only time I met Halstead was late in his life when he was quite ill. I recall describing to him a version of the Category test that I was using in which the informative feedback regarding whether the response was right or wrong was delayed, and provided after the stimulus was no longer on the screen. Halstead became very excited about this modification, and expressed the idea that it was just what he needed for his studies of high-level executives who required measures that were more challenging than the available procedures for detection of subtle impairment. He thought that the modification of the Category test described could help fulfill this need. It seems clear that this incident represented Halstead's entire scientific career, during which he was constantly exploring and changing methods of assessment and analysis in pursuit of answers to significant scientific questions. To describe Halstead as the inventor of a standard, fixed-assessment battery would be a gross distortion of his intentions and scientific achievements.

To a great extent, what Halstead passed on to Ralph Reitan was what can be described as a laboratory approach to clinical assessment. Perhaps it is this laboratory approach that really creates the major dividing line between the advocates of standard procedures such as the Halstead-Reitan and Luria-Nebraska Batteries on the one hand, and the followers of Luria's and other qualitative approaches on the other. In what at the time was a major departure from the medical model, Halstead took the patient out of the clinic and placed him or her into a laboratory designed especially for neuropsychological assessment. What was formerly done in a doctor's office with little if any instrumentation was done in a laboratory with what at the time was state-of-the-art apparatus specifically designed for study of higher brain functions. The quantitative data derived from these procedures became a significant basis for interpretive conclusions reached, and at least supplemented pure clinical opinion. Two important features of the laboratory approach were the principle that all tests had to be given to the patient in the same manner, utilizing the same instructions regardless of the level of impairment, and the procedure of having tests administered by trained technicians. These two innovations obviously had major implications for the subsequent development of neuropsychological assessment, and, in my view, were really initiated in Halstead's laboratory.

Most neuropsychologists would probably agree that Halstead's theoretical formulations were not pursued by others following his death and the eventual phasing out of the Chicago laboratory. However, it is not entirely accurate to say that all he left us were his tests. As indicated previously, he also contributed both the concept and the substance of a neuropsychological laboratory, as well as the now not uncontroversial but widely accepted and utilized practice of applying sophisticated forms of statistical analysis to neuropsychological test data. Halstead published in *Brain and Intelligence* (1947), and in a less well known work (Halstead, 1945), the first factor analyses ever done with neuropsychological test data. The factor analysis reported in *Brain and Intelligence* (Halstead, 1947) is reproduced below. It is particularly interesting to note that Halstead obtained the same four factor solution many more recent analyses obtained, despite differences in the tests included and the different solutions used. Those familiar with current versions of the Halstead-Reitan battery will no doubt recognize the Category, Tactual Performance, Speech Perception and Finger Oscillation measures, but may be less familiar with the Carl Hollow Square Test, the Flicker Fusion Procedure used by Halstead, the Time sense Test and the D.V.F. (Dynamic Visual Field) measures. We will not describe those procedures here because they are rarely if ever used, but mention them as illustrative of the point made earlier that the Halstead-Reitan battery is not really a fixed procedure, and has evolved substantially over the years. Not only were the unfamiliar tests contained in the original analysis dropped, but many new tests were added (e.g., the various Wechsler Intelligence Scales and the Trail Making Test).

The use of multivariate analysis was not taken up to any great extent by Reitan and his collaborators, with the exception of a series of discriminant function papers (Wheeler, 1964; Wheeler, Burke, & Reitan, 1963; Wheeler & Reitan, 1963), but was adopted by a number of other investigators utilizing some version of the Halstead-Reitan battery. The growing availability of high speed computers and statistical software packages will no doubt promote even greater expansion into that area. Although we will not review the extensive literature here, there are already numerous publications involving not only factor analysis but other highly sophisticated multivariate analysis methods as well, including multiple group discriminant analysis, canonical correlation, and multivariate analysis of variance.

Needless to say, Halstead did leave us with his tests, several of which eventually became the major component of what is now known as the Halstead-Reitan Battery. This battery remains the most widely used comprehensive standard neuropsychological assessment battery

Table 1. *Oblique-Factor Matrix (Thurstone) for Thirteen Neuropsychological Indicators*

		Factors			
Indicator	Description	C	A	P	D
1	Carl hollow-square (I.Q.)	.25	.45	−.07	.04
2	Category	.49	.63	.09	−.03
3	Flicker fusion	.00	.04	.54	.05
4	Tactual performance speed	.12	−.02	.04	.61
5	Tactual incidental recall	−.02	.66	.43	−.02
6	Tactual incidental localization	.19	.34	.25	.29
9	Henmon-Nelson (I.Q.)	.58	.27	−.05	.23
13	Speech perception	.49	−.06	.06	.22
14	Finger oscillation	.40	−.06	.25	.18
16	Time-sense memory	.43	.11	.08	.02
17	D.V.F. form	.41	.07	.64	−.03
18	D.V.F. color	.41	−.03	.61	−.06
19	D.V.F. peripheral	−.15	.11	−.06	.54
		Σ 3.60	2.51	2.81	2.04

Note: From *Brain and Intelligence: A Quantitative Study of the Frontal Lobes* by W. C. Halstead. Copyright 1947 by The University of Chicago Press. Reprinted by permission.

in the United States and perhaps all of North America. It has been translated into several languages, and so now has international application going beyond North America. For example, Hans Bergman and various collaborators (Bergman, 1984) have translated the battery into Swedish, and use portions of it in their extensive alcoholism research program. The battery, in its original or altered forms, is administered in whole or part by psychologists to a wide variety of patient groups in widely varying clinical and research settings. It is now generally viewed by the professional community as the state-of-the-art standardized comprehensive assessment system for neuropsychology, although, as we will see elsewhere in this chapter and volume, it is not an entirely uncontroversial procedure.

In concluding these historical remarks concerning the legacies of Halstead and Luria, certain remarkable resemblances are noteworthy, despite the fact that they worked in two different countries and had somewhat different philosophies and theoretical orientations. They both had a fascination for the frontal lobes, and made major contributions to our understanding of their role in human behavior. Although the theories were somewhat different from each other, they both stressed the executive role of the frontal lobes in the modulation and control of behavior. Had they the opportunity to discuss their theoretical formulations with each other, they may have found themselves in substantial agreement. Second, they were both very much involved with patient contact. Halstead was known to form strong associations with brain-damaged patients that allowed him to see the impact of their disorders on their personal lives outside of clinical settings. Correspondingly, Luria did many of his own examinations, and followed several patients for many years. Therefore, although both of them were scientists, neither was a detached laboratory scientist having no association with patients outside of the clinic or the laboratory. Although there were, of course, many differences, and although their intellectual descendants went in many differing directions, these resemblances are nonetheless quite striking.

SIMILARITIES AND DIFFERENCES

Having briefly traced the histories of the scientific developments that led to the construction and widespread utilization of the Halstead-Reitan and Luria-Nebraska Batteries, we can now go on to consider how they are alike and different. Before doing so, however, we would point out that there are three elements to be considered, although we will only deal with two of them. Aside from those associated with the two standard batteries, there is also the approach of Luria himself, who apparently never used or advocated the use of a standard battery. Indeed, he is generally thought of as being philosophically opposed to such usage. Currently, Christensen (1975a, 1979, 1984) appears to be the major advocate of the original Luria method, and has written extensively about it. As is well known, Luria's method of assessment is highly individualized and is accomplished by an experienced examiner, who, following a preliminary conversation with the patient, explores some preliminary hypotheses that become increasingly focused with the aim of ultimate establishment of a syndrome, or organized pattern of functional deficits. The tests utilized are selected from a large repertoire available to the examiner, and are chosen for administration on the basis of the patients' prior responses to other tests. Recently Christensen (1984) has divided the process of examination into four stages; the preliminary conversation, the preliminary investigation, the selective investigation, and the formulation of a clinical neuropsychological conclusion through grouping findings into a syndrome. There has been a great deal of ultimately nonproductive controversy regarding whether or not the Luria-Nebraska battery is in any way a representation of Luria's methods and theories, with those on the negative side suggesting that it actually bears very little if any relationship to what Luria and his associates and students actually did or do at present. This debate is strongly associated with the controversy between advocates of fixed and flexible assessment procedures.

For our present purposes, we will grant that the actual basis for application of the LNNB in clinical practice has very little to do with major aspects of Luria's neuropsychological theories, notably the concepts of functional systems and syndrome analysis. We will return to this matter later, because it reflects a particular disposition toward the LNNB that requires further elaboration. Here we will simply offer our respects to those scholars and clinicians who answer "neither" when asked which of the two systems should be generally used in clinical practice. However, contrasting the LNNB with the HRB is a different matter from contrasting standard, quantitative approaches to neuropsychological assessment with individualized, qualitative approaches. Here, we will attempt to adhere to the former matter, only touching on the latter issue when necessary to clarify some point with reference to comparisons between the two standard procedures. In other words, if one elects to take a quantitative approach to assessment, utilizing a relatively fixed, comprehensive, standardized battery, which of the two major ones is preferable under what circumstances, if there are indeed major differences between the two?

SIMILARITIES

Although the structure and content of the LNNB and HRB are quite different from each other, there are many similarities between the two procedures. As most readers know, the HRB consists of a number of discrete tests assembled into a battery, whereas the LNNB consists of 269 items divided into a number of scales. Thus, the structures, from a psychometric standpoint, are quite different from each other, and although there is some overlap in content, there are clearly different emphases. Nevertheless, both procedures constitute standardized psychometric batteries, most likely complying with the spirit if not necessarily the letter of the technical requirements for psychological tests promulgated by the American Psychological Association (APA, 1984), and necessary for publication by professionally recognized publishers of psychological tests. Although this similarity might not strike one as being particularly remarkable, it may be noted that the achievement of psychometric standardization of a test series is not viewed by some authorities as a particularly desirable accomplishment (e.g., Hamsher, 1984; Lezak, 1983), and that we might do better at assessment were we to select individual standardized tests on a patient-by-patient basis. Nevertheless, the developers of the HRB and LNNB appeared to have the common goal of developing a standard procedure that can be administered in the same way to all patients.

Both procedures are described as "fixed batteries," but as indicated previously that term is somewhat misleading for several reasons. With regard to the HRB, the procedure has changed over the years both in and outside of Halstead's and Reitan's laboratories. One often finds in the literature the phrase "modified version" of the HRB. Whereas there is a core of procedures most clinicians would agree is essential, other procedures have been added and subtracted in numerous settings. For example, one could really not call a procedure the HRB without the Category and Tactual Performance tests. The LNNB is relatively new and continues to evolve, both with regard to the items themselves and the ways in which scoring is accomplished. For example, the factor scales (McKay & Golden, 1981) are a relatively new development. Numerous changes have been made since the earliest versions appeared, and it seems that the LNNB will continue to change as critiques and new research findings appear. For example, delayed recall items have recently been added to the battery, primarily because published reviews (e.g., Russell, 1981) called the test developers' attention to the fact that the original Memory Scale neglected this crucial area. Thus, neither approach can be described as fixed, except in the following sense. The HRB or LNNB are fixed in that the adoption of one or the other of these procedures in a particular laboratory or clinic generally involves utilization of some version of the entire procedure with all patients and over a sustained period of time. It is rare, for example, to give some patients just the Category Test and certain subtests of the WAIS, whereas other patients receive just the Speech Perception and Rhythm Tests. That is, portions of these batteries are typically not administered on an individualized basis. Rather, the goal is usually that of getting every patient through the entire procedure. When the battery is changed, the change is usually made on the basis of some general consideration, such as new research findings or lack of feasibility for some particular setting or population. However, when the change is made, it is anticipated that essentially all patients will receive the entire modified version. Thus, the two batteries are both fixed in the sense that it is customary to administer the entire procedure to all patients, but not in the sense that these procedures do not change over the years on the bases of scientific and clinical input. Indeed, they have both changed substantially. This pattern of stability and change appears to reflect a strong similarity between the two procedures.

Both procedures are also described as being comprehensive, but this term also requires some clarification. The term *comprehensive* might imply that the procedure not only evaluates the entire gamut of behaviors generally viewed as belonging within the purview of neuropsychological assessment, but also does so in depth in the case of each function evaluated. It would appear on the face of it that the former definition is appropriate, but not the latter one. It seems inconceivable that a single procedure of reasonable length could possibly evaluate all neuropsychological functions to the greatest depth possible in consideration of the current state of the art. Just taking the area of memory as an example, a comprehensive evaluation of memory in itself requires administration of a memory battery in which one systematically examines in detail numerous forms of memory. Even such traditional memory batteries as the Wechsler Memory Scale (Wechsler & Stone, 1945) are not generally considered state of the art, and more sophisticated procedures such as those described by Butters and Cermak (1980) or Ferris and Crook (1983) are now thought to provide a more detailed evaluation of the neuropsychological aspects of memory. For example, if one considers such categories as short-term, long-term, and remote memory, or semantic versus episodic memory or memory for verbal as contrasted with nonverbal material, it becomes apparent that a sophisticated and comprehensive memory assessment is a lengthy and detailed procedure that would be essentially impossible to implement along with equivalently sophisticated and comprehensive assessments of other areas, such as language or motor abilities. Indeed, advocates of the flexible approach have pointed to this dilemma as a rationale for doing more individualized, targeted assessments of the patient's specific presenting difficulties, as may be gleaned from the referral for testing, the medical records, or a brief initial conversation.

It would appear that neither the HRB nor the LNNB are comprehensive in both senses described above, and we would do better by characterizing them as screening procedures. Unfortunately, the term screening has become associated with the brief "tests for brain damage" commonly used in large clinical settings to screen out the brain-damaged patients from patients with functional psychiatric disorders. Typically, a brief copying task, such as the Bender-Gestalt Test (Bender, 1938) or Graham-Kendall Memory for Designs Test (Graham & Kendall, 1960) is used for this purpose, and the presence or absence of brain damage is predicted on the basis of some established cutoff score. The term *screening*, however, need not be used in that sense, but may be used to characterize procedures that screen a number of functions. Thus, the HRB and LNNB can be seen as comprehensive screening procedures that provide highly useful but not necessarily in-depth evaluations of such areas as conceptual abilities, language, memory, and motor skills. Indeed, the batteries have a strong similarity in the sense that they cover essentially the same areas: general intelligence, visual-spatial abilities, attention, language and associated academic skills, perceptual and motor functions, and conceptual abilities. Some clinicians, in fact, utilize one or the other of these batteries to obtain cues as to those areas requiring more intensive assessment. It may be added that there has been some discussion of how comprehensive these procedures are, even when one defines the term as suggested here. The HRB has been criticized in particular because of the lack of formal memory testing, whereas the LNNB has been criticized, particularly by Spiers (1981), because it is felt that none of the functional areas assessed are treated in sufficient depth or with sufficient sophistication. Thus, problems remain with the term comprehensive even if it is used in a limited sense, but these problems are shared by both batteries.

The matter of similarities in content is not clear-cut, and there appear to be both similarities and differences. Dr. Kane's chapter will discuss empirical similarities, which are generally evaluated with correlational statistics, and so we will deal mainly with conceptual issues here. The major question would appear to involve whether or not the two batteries assess similar or different aspects of brain function. As we will see in Dr. Kane's chapter, essentially every study using correlational statistics or clinical judgment yielded results indicating that there is exceptionally high concordance between the two procedures. However, the matter is complicated somewhat by the influence of general intelligence, as has been pointed out by Chelune (1982). Thus, the apparently high correlations between the HRB and LNNB may be somewhat artifactual in the sense that the correlations may be so high because measures derived from both procedures tend to also correlate highly with IQ measures. Nevertheless, it is clear that both batteries have components devoted to the assessment of conceptual abilities, language, visual-spatial skills, basic perceptual and motor functions, and general intelligence. We know from one study at least (Goldstein & Shelly, 1984) that the language tests from each procedure are highly correlated with each other. This study utilized canonical correlation, and reported finding very high canonical correlation coefficients between a multivariate combination of language tests derived from the HRB and a similar combination derived from the LNNB.

Perhaps a useful way of looking at similarities in content might be to ask the question, "If one of the two procedures were administered to a patient, what more would be learned by administering the other procedure?" The question is raised here not because we have the answer, but because it may offer a direction in which clinical research in standardized neuropsychological assessment could profitably go. We do have some clinical hints, but it is clear that any definitive answer to the question would have to be quite specific. Clinically, we have found that when it is important to learn more about memory specifically, the LNNB can add to the HRB because it includes formal memory testing. On the other hand, when it is important to determine how well an individual responds to a highly challenging problem solving task, the LNNB does not appear to have any substitute for the Category or Tactual Performance tests from the HRB.

As a final comment on the matter of content similarities, it seems pertinent to point out that correlation matrices derived from neuropsychological data, whether derived from samples of brain-damaged or non-brain-damaged patients, generally contain preponderantly significant correlations. It is as though every neuropsychological test contains some component in common with other tests, perhaps a general impairment factor, and a specific component. In other words, our tests tend not to assess exquisitely discrete abilities, but generally appear to be more or less sensitive to general level of impairment. It is therefore probably the case that the similarities between the HRB and LNNB are not the product of something the two procedures have uniquely in common, but rather may be based on the general finding that essentially all neuropsychological tests correlate more or less among themselves. Chapman and Chapman (1973) have described a general deficit syndrome in the case of schizophrenia research, and it seems quite plausible that this syndrome may be an important consideration in regard to neuropsychological research with brain-damaged patients. It is perhaps commonplace to note that regardless of the test in question, if it has any sensitivity at all to brain dysfunction in general, the patient group will typically do worse than the normal control group. In the clinical realm, well trained clinical neuropsychologists typically think in terms of functions and abilities rather than scores on particular tests. It is therefore understandable that a competent clinician may have the capacity to review data from a variety of test batteries and come to the same conclusions as long as the basic information regarding these functions and abilities is available somewhere in the test data. Based on these empirical and clinical considerations, it is perhaps not at all surprising that we find the reported content similarities between the two batteries.

There is clearly a great deal of similarity with regard to method of interpretation. Both procedures may be administered by a technician who scores the tests and presents the clinician with both the scores and the test protocol itself. The clinician then has the option of attempting to provide an interpretation through an examination of the quantitative relationships among the various tests or scales, through a detailed qualitative examination of the test protocol itself or through a combination of both methods. The opportunity for blind interpretation is available with both batteries because the clinician need not see the patient personally or review the case history prior to interpretation. In essence, both batteries incorporate the laboratory concept originally developed by Halstead. We know of no instance in which Luria and his co-workers had technicians administer neuropsychological tests, nor did they appear to utilize the method of blind interpretation to any extent at all. Within clinical neuropsychology, the method of blind interpretation appears to have been greatly encouraged by Ralph Reitan and his collaborators, who have been strong advocates over the years of that particular method of interpretation.

The final point of similarity between the two procedures that we will discuss here has to do with the way in which they were developed. We will attempt to elaborate the view that the scientific epistemology that characterized the research associated with the development of the HRB is essentially the same as was the case for the LNNB. The LNNB can be seen as having traced the same course during the middle 1970s and subsequently as the HRB traced beginning in the 1950s with Reitan's (1955a) initial cross-validation study of some of Halstead's tests. The basic scientific model revolved around psychometrics, with a strong emphasis on the matter of empirical validity, notably concurrent validity utilizing neurological diagnostic criteria. There are, of course, some differences. Golden and his co-workers had the initial problems associated with quantification and scale development, whereas in the Reitan group's case, this work was largely already accomplished in Halstead's laboratory. Second, the LNNB really emerged during a different era in neuroscience, allowing for the use of such advanced methods as the CT scan in validation studies. Computer techniques had also advanced greatly, and complex multivariate data analyses could be accomplished much more readily. However, despite these differences, very similar scientific careers can be traced.

The first issue addressed by the developers of both batteries was the basic validity question; Is the procedure sensitive to the effects of brain damage in a general way? Thus, we have Reitan's (1955a) paper and the paper by Golden, Hammeke, and Purisch (1978) that deal with the same issue. The subsequent studies in both cases dealt with issues related to further specification of impairment in terms of lateralization, localization, and identification of specific disorders. Even the disorders themselves were more or less the same: alcoholism, epilepsy, multiple sclerosis, Huntington's disease, etc. The extensive literature concerned with these matters is reviewed elsewhere in this volume and in numerous review articles (e.g., Boll, 1981; Golden, 1981; Klove & Matthews, 1974; Reitan, 1966). We will not repeat that documentation here, but it was generally determined that both procedures are sensitive to the differential effects of brain lesions in different portions of the cerebral hemispheres, and to the neuropsychological consequences of several disease entities that may affect brain function. Aside from these matters of content, the research designs utilized by developers of the two procedures are similar. The use of relatively large samples and application of group statistics was common to both sets of investigators. The remarkable point about this is not that these battery developers utilized adequate samples and appropriate statistical methods, but that they did so in a field in which that methodology is still not fully accepted. The detailed case history, clinical judgment, and the performance of classical experiments with small groups of rather precisely defined patients remain viable epistemological alternatives to the psychometric, objective, and statistically oriented methodologies of the HRB and LNNB developers.

In order to illustrate these points, and to further demonstrate that the LNNB is scientifically and philosophically closer to the HRB than to the neuropsychology of Luria, we may compare Luria's methods with those of the HRB and LNNB developers. First, let's look at some studies that Luria never did. Luria never did a study to determine whether his tests discriminated between patients with and without brain damage. He never did a study to evaluate the reliabilities of his procedures. He never formed patients into groups with left hemisphere, right hemisphere and diffuse lesions, in order to determine what statistically significant neuropsychological differences existed among those groups. He never evaluated whether his tests could discriminate between brain-damaged and schizophrenic patients. He never looked at whether his tests were sensitive to the effects of chronic alcoholism. The point seems clear that Luria had an entirely different research agenda from one that could be approached through studies of the type just mentioned. His work in assessment based on his theoretical formulations also never addressed itself to issues of this type. As indicated earlier, his assessments were directed toward delineation of an identifiable syndrome through the use of a preliminary conversation and examination, followed by individualized sets of tests applied selectively from a wide variety of procedures he had in his repertoire. This syndrome-analysis approach is strongly tied to the functional systems concept, and aims at identification of what are described as impaired links in functional systems. It was felt that such identification has strong implications for diagnosis and rehabilitation. Examples of this approach are to be found throughout Luria's writings. The first sentence in Christensen (1975a) is the following.

> It is the main purpose of the study of the higher cortical functions in the presence of local brain lesions to attempt to explain the *syndrome* of disturbances in mental activity resulting from the fundamental defect. (p. 23)

Reitan (1976) has taken issue explicitly with the syndrome-analysis approach, and does not accept the soundness of its scientific basis. Those investigators associated with the LNNB have not disavowed syndrome analysis, but do not make it clear in their clinical writings (Golden et al., 1982; Moses, Golden, Ariel, & Gustavson, 1983) that they attempt to perform such analyses using the LNNB. It would therefore appear that criticisms of the LNNB approach to the effect that it does not reflect Luria's method or theories miss the point. Apparently, the approach does not involve an attempt to utilize the method of syndrome analysis, and if one were to attempt to describe the method the LNNB developers have adopted, it would be much closer to those methods of interpretation that have been traditionally associated with the HRB. These methods include (a) level of performance considerations; (b) pattern of performance considerations; (c) specific behavioral deficits or pathognomonic signs; and (d) comparison between performance of the two sides of the body.

In general, then, it would be my view that the development of the LNNB involves an application of the research model and strategies initially associated with the HRB, as well as an interpretive system associated with the HRB, to a series of test items utilized in a completely different way by Luria and his co-workers. This point should be apparent simply on the basis of what we know about the belief systems and backgrounds of the LNNB and HRB developers. These individuals are essentially empiricists, with strong beliefs in quantification, the use of objective tests, and interpretation based to the greatest extent possible on pertinent research findings. These findings, in turn, are felt to be best established when based on studies involving sufficiently large samples to allow for the application of appropriate statistical procedures. In this regard, there are numerous examples of instances in which interpretive rules or principles were dropped or modified on the basis of the results of replication or cross-validation attempts.

At a more personal level, Charles Golden, prior to his interest in the LNNB, had published several papers involving the HRB (Golden, 1977), and continues to maintain an interest in it (Golden, Osmon, Moses, & Berg, 1980). Furthermore, none of the developers of the LNNB were trained in the qualitative neuropsychology or behavioral neurology associated with approaches developed in Europe and exported to the United States under the leadership of such individuals as Kurt Goldstein, Martin Scheerer, and Heinz Werner. Nor did they study with members of the Russian school led by Luria. It is not being maintained that a group of scientists and clinicians cannot adopt a method, approach, and philosophy without direct contact with the founders. It is only being suggested that when one looks at the whole picture, including the nature of the extensive research done, the methods of clinical interpretation developed, and the backgrounds of the developers of these methods, there is little to suggest that the LNNB method and approach to assessment has any strong relationship to Luria's neuropsychological method and theory. However, there would appear to be such a relationship with the HRB. Elsewhere (Goldstein, 1984) we have expressed the thought that the HRB and the LNNB are essentially the same procedure in different costumes.

Differences

Given the similarities we have discussed, there are also practical and substantive differences between the HRB and LNNB. Beginning with the practical differences, the LNNB is a briefer procedure with regard to administration time, taking perhaps half the time it generally takes to administer the full HRB. A skilled examiner can administer the LNNB in two and a half to three hours, whereas the HRB, including the appropriate Wechsler scale, generally takes five to six hours. Scoring time may not differ substantially. The LNNB is portable, and less expensive to purchase and maintain than the HRB. The HRB generally requires the permanent allocation of a room of at least office size, whereas that is not the case for the LNNB. Thus, the LNNB would appear to have several crucial practical advantages in a busy clinical setting having the customary space shortage. However, these allurements should not immediately convince one to adopt the LNNB. At this point in our knowledge, it would be foolhardy to allege that the LNNB can do everything the HRB can do in less time, using less space, and costing less money.

The significance of the substantive differences between the HRB and LNNB is to some extent dependent on the particular goals of the assessment and the nature of the patient population from which referrals are accepted. If the major interest is in screening of the type commonly done in large neuropsychiatric facilities where classification into the traditional neurological and psychiatric categories is the goal, current evidence suggests that the HRB and LNNB are equally effective, and better than the WAIS (Goldstein & Shelly, 1984). That is, the HRB and LNNB appear to be equally good at identifying the brain-damaged patients in a mixed neuropsychiatric population. However, an examination of the contents of the two procedures would strongly suggest that if the objective of the assessment is a detailed analysis of specific functions, each procedure has rather clearly identifiable strengths and weaknesses. This consideration leads into an analysis of the structure and content of the two batteries from a technical, relatively detailed standpoint. Much has been written about this matter, particularly with regard to the LNNB (e.g., Crosson & Warren, 1982; Delis & Kaplan, 1982; Spiers, 1981), and we will employ some of this published material in our discussion.

The structure of the HRB is fairly well understood. Repeated factor analyses in numerous settings generally come quite close to replicating Halstead's original four factors plus a verbal or language factor when the WAIS and/or educational achievement tests are included, such as the Wide Range Achievement Test (Jastak & Jastak, 1965). The original four factors can be roughly described as involving abstraction and related conceptual abilities, perceptual and motor skills, which may or may not include language and language-related abilities, attention, and what essentially amounts to a kind of knowledge or data base that Halstead (1947) called the "central integrative field." This latter factor may be simply described as memory, particularly long term memory, but that may be an oversimplification. The thrust of the HRB, and particularly of Halstead's original tests, toward assessment of higher levels of adaptive abilities apparently motivated the development and utilization of very complex tests. Although there are simple tests, such as Finger Tapping, the bulk of the battery consists of quite complex tests. It is generally not possible to identify the particular source of whatever impairment is noted on these tests, making the HRB a less than ideal procedure for performing syndrome or information processing type analyses. However, in some cases such analyses may be possible through the combining of findings from several tests. In fact, Reitan added some simpler tests to the battery, such as an Aphasia Screening test and the Perceptual Disorders subbattery in order to make it more suitable for clinical neuropsychological diagnostic assessment in a general medical and neurological setting. It may be recalled that the subtitle of Halstead's *Brain and Intelligence* was "*A Quantitative Study of the Frontal Lobes,*" and that the original intent of the battery was largely that of studying frontal lobe function. That consideration probably largely explains the emphasis on complex, problem-solving tests. However, in addition to the postulated sensitivity to frontal lobe function, complex tests serve at least two very important roles in neuropsychological assessment. First, they tell us a great deal about real world performance (Heaton & Pendleton, 1981). It appears that performance on such procedures as the Category or Trail Making test may be quite predictive of the way in which the individual may approach situations involving problem solving in a natural setting. Thus, if the referral question revolves around how the hospitalized patient will function in the outside world, my personal preference would be to look at performance on tests of the Category and Trail Making type, as opposed to simple tasks, such as having the patient point to the ceiling. We are not belittling the use of simple tasks in general because of their crucial role in differential diagnosis of various types, but are suggesting that a battery that contains simple and complex tasks has the capability of dealing with both diagnostic and functional issues.

The second asset of complex tests is their sensitivity to subtle impairment. The abundant research literature clearly demonstrates the sensitivity of several of the HRB tests to such conditions as chronic alcoholism and general systemic diseases that have implications for brain function. It is particularly interesting to note that individuals with these more subtle types of brain disorder typically do exceptionally poorly on tests of the Category or Trail Making type, although they perform normally on tests of simpler functions, such as Finger Tapping or tests of verbal intelligence that only require access to semantic memory. Although Chimelewski and Golden (1980) demonstrated the sensitivity of the LNNB to the effects of chronic alcoholism, the abilities that separated the alcoholics from the controls involved complex functions, notably memory and the capacity to comprehend complex grammatical constructions.

The LNNB consists largely of apparently simple tests of discrete functions, thus allowing for the possibility of identifying discrete dysfunctions in patients with local brain lesions. It therefore may be the battery of choice if the goal of the evaluation is detailed neuropsychological diagnosis in a population of patients with stroke, open head injury, brain tumors, or other processes that give rise to relatively isolated syndromes. The capacity of the LNNB to support such evaluations, however, remains a matter of ongoing controversy in the field of clinical neuropsychology. Although some of the criticisms have been quite broad ranging (e.g., Spiers, 1981), it would probably be fair to say that most of them relate to the language and memory portions of the LNNB. However, the extent to which the LNNB can actually reliably identify and delineate specific syndromes in other areas, such as purposive movement and visual-spatial abilities, has not been established either. There is also a related issue that has to do with the capacity of patients with specific syndromes to produce an interpretable performance on the LNNB as a whole. The issue has been raised in the case of the aphasic patient (Crosson & Warren, 1982; Delis & Kaplan, 1982), particularly aphasic patients with severe comprehension deficits. It would appear that this issue was raised for the LNNB in particular because the procedure as a whole is highly dependent on comprehension of verbal instructions. It has been described by some as a "language loaded" procedure. It would appear that this matter has to be dealt with prior to a discussion of the content of the LNNB because the specific contents of the various scales may be irrelevant for the patient who does not understand any of the instructions. In the case of patients with language disorders, the end result of this difficulty is that patients who have a specific language related syndrome may look globally impaired, and in particular, may be impaired in abilities that can be demonstrated to be intact when an adequate channel of communication is established with the patient. Smith (1975) has provided a general solution to this problem in a set of principles he proposes for selection of neuropsychological tests. The pertinent principle is as follows.

> Tests should be selected that *permit differentiation of the sensory and motor modalities involved in perception and execution of the task* (or lower level cerebral functions) *from the mental or cognitive processes* (higher level cerebral functions) *they were designed to measure.* (p. 87)

Thus, a truly comprehensive assessment battery should have the capability of allowing the clinician to distinguish between peripheral or primary perceptual-motor impairment and impairment of higher cortical functions. In the case of language comprehension, the procedure should have the capability of supporting a related consideration having to do with distinguishing between failure because of inability to understand the instructions and failure because of inability to cope effectively with the substantive requirements of the task. In my view, the failure to understand the instructional language is a problem of the type we also have in assessing individuals with significant sensory and motor handicaps. Failure on certain neuropsychological tests may simply be a function of inability to see or hear well, or inability to control one's movements to the extent that skilled motor tasks, such as copying or handwriting, cannot be performed normally. The use of a standard battery does not excuse the examiner from developing a means of administering the procedure in a manner that allows for reasonably unequivocal interpretation of results in terms of the intent of the items rather than various peripheral considerations. With regard to language specifically, the skilled examiner should have the clinical sophistication to evaluate whether or not the patient understands the instructions, and whether or not other communicative difficulties may confound interpretation of items that are not meant to assess linguistic processing (e.g., word-finding difficulty confounding interpretation of a visual perception test). In any event, neuropsychological assessment of blind, deaf, paralyzed, and aphasic patients is a challenging area requiring ingenuity on the part of the clinician in order to attain a meaningful, interpretable evaluation.

The comments just made are highly relevant to a major point of difference between the two procedures; the relative emphases on language and language-related abilities. The fact that the LNNB is much more language oriented than the HRB could be viewed as reflecting rather basic theoretical differences between Luria and Halstead. In his chapter on the frontal lobes in *The Working Brain* (Luria, 1973) Luria makes it clear that he believes that speech plays a crucial role in the regulation of complex behavior. For example, there is the following sentence.

> This fact clearly shows that *massive lesions of the frontal lobes disturb only the most complex forms of regulation of conscious activity* and, in particular *activity which is controlled by motives formulated with the aid of speech*. (p. 199)

Although Halstead was clearly not uninterested in language, apparently in attempts to make his tests as culture fair and free of the influence of "psychometric intelligence" as possible, nonverbal tests, such as the Category and Tactual Performance tests, were emphasized. In effect, the relative emphasis on language in one procedure and relative lack of emphasis in the other appears to have occurred for unrelated reasons. However, it seems clear that although one in essence "talks the patient" through the LNNB, it is conceivable that many components of the HRB could be administered with minimal verbal instruction. Furthermore, Christensen (1975a) makes the following crucial, but often neglected remark in her preface to *Luria's Neuropsychological Investigation*. "It is to be remembered that the whole investigation is mainly considered to evaluate the functions of the left (dominant) hemisphere." (p. 11). Because the LNNB in almost its entirety consists of the Christensen item set (1975b,c), we obviously have to take that remark quite seriously with regard to our discussion here. It seems clear that the inclusion of so much language related material in the LNNB is not so much because of a specific interest in aphasia, but because of a theoretical orientation that places heavy empahsis on the role of language in the conduct of human behavior. In a sense, this situation is unfortunate for the aphasic patient because interpretation of so much of the LNNB can be contaminated by the presence of the language disturbance. On the other hand, the LNNB might provide a more thorough assessment of the aphasia itself, relative to the HRB, because it is far richer in linguistic material. In those situations in which the aphasia was thoroughly assessed with a comprehensive aphasia examination, the HRB might be the method of choice for assessing the patient's nonverbal abilities.

The other area in which there is a major content discrepancy between the HRB and LNNB is memory. One might put the matter flippantly by indicating that the discrepancy is that whereas the LNNB contains memory testing, the HRB does not. Put more precisely, however, it is the case that the LNNB contains *formal* memory testing procedures, of the type generally associated with human learning experiments or such instruments as the Wechsler Memory Scale (Wechsler & Stone, 1945), but the HRB does not. The absence of formal memory testing as part of the HRB remains puzzling, especially because Halstead was quite interested in memory, and made some major contributions to the area. Halstead has little to say about it in *Brain and Intelligence* (1947), except, perhaps, to suggest that memories, in reorganized and simplified forms, become incorporated into the central integrative field. Informally, Reitan has suggested that memory is a part of all higher cognitive processes, and therefore is involved in many parts of the HRB. One can surely make the case that the WAIS verbal subtests tap the long term storage component of memory, whereas the patient must retain the correct concept in working memory in order to sustain successful performance on the Category test. The difficulty with this position is that there is by now extensive empirical evidence suggesting that patients with pure amnesic syndromes, that is, patients who have poor memories but are otherwise reasonably intact, appear to demonstrate their deficit pattern most clearly and in most detail on measures of the type we have described as formal memory tests (Butters & Cermak, 1980). That is, they have outstanding difficulties with such tasks as learning lists of words, recalling stories, or learning new associations. Furthermore, these difficulties appear to accurately reflect the clinical impression of these patients, particularly with regard to their incapacity to recall events and learn new information.

With regard to the LNNB content scales other than the language and memory scales (Motor, Rhythm, Tactile, Visual, and Intelligence), one can see general family resemblance with corresponding components of the LNNB, although formal correlational studies of the type done by Goldstein and Shelly (1984) for the corresponding language tests and by numerous investigators for LNNB generated IQ estimates vs. the WAIS (e.g., Prifitera & Ryan, 9981) have not been accomplished as yet. Nevertheless, it would appear that certain items of the Rhythm scale are much like the Seashore Rhythm test, whereas several items on the Tactile scale greatly resemble the perceptual disorders subbattery of the HRB. It will be interesting to see how future research will turn out in terms of empirically demonstrating correlations between the two procedures in regard to their content.

A major difference between the nature of the two batteries involves the degree of detail with which specific functions are assessed. As examples, speech discrimination is tested with a 60-item procedure as part of the HRB (the Speech Perception test), whereas the same function is assessed by the LNNB with a relatively small number of items.

Abstract reasoning and related conceptual abilities are assessed by the HRB with a lengthy, challenging procedure: the Category test. Although the Intellectual Processes scale of the LNNB contains a number of items that involve conceptual reasoning, the assessment clearly is not as thorough as the one provided by the Category test. In essence, the HRB consists of tests in the traditional sense in which numerous items are contained in a single procedure that measures some particular function or ability. The LNNB consists of items that are included in scales, but that do not have the psychometric characteristics of conventional test items. Indeed, one of the criticisms of the LNNB has been that there is an insufficient number of items to make a reliable judgment about any particular function. This problem is thought to be particularly acute for the visual-spatial and memory items. Although it may be accurate in certain respects, this criticism becomes less crucial if one adopts the suggestion we have offered that the LNNB is a screening procedure that may be effectively used to identify areas in need of further evaluation with more detailed testing. The converse criticism is often heard concerning the HRB, in that although it assesses certain areas in great detail, notably conceptual abilities and nonverbal problem solving, it really does not come to grips with certain areas at all, notably major aspects of memory, and is somewhat superficial in other areas, such as the assessment of aphasia. With regard to the latter matter however, Reitan has explicitly described his aphasia test as a screening instrument and not as a definitive aphasia examination.

Russell (1980) has proposed neuropsychological deficits may occur on an all-or-none or a dimensional basis. Either the patient has the deficit or does not have it, or the deficit is present to some measurable degree. It can be argued that whereas the HRB incorporates both all-or-none and dimensional principles, the LNNB is essentially limited to the all-or-none concept, because all individual items are scored on no more than a 3-point scale, with many items simply scored on a pass-fail basis. It seems clear that Luria himself did his assessment work largely on an all-or-none basis, indicating in his case reports that the patient did or did not have the sign in question (e.g., "The patient cannot distinguish the necessary acoustic content of a word." [Luria, 1973, p. 140)] vs. "The patient made 27 errors on the Speech Perception test"). The implications of the first statement are that the patient has essentially totally lost an aspect of phonemic hearing, whereas the latter phraseology suggests that the patient has what appears to be a very similar deficit, but only to a certain quantifiable extent. Historically, there has been extensive controversy over whether or not the loss of the "abstract attitude" is a qualitative or quantitative loss (Goldstein & Scheerer, 1941; Goldstein, Neuringer, & Olson, 1968; Reitan, 1958, 1959).

It is not at all clear that the developers of the LNNB specifically advocate an all-or-none or pathognomonic-sign approach. Whereas it is true that many of the individual items are scored on a pass-fail basis, the items are incorporated into major content and factor scales. It also seems necessary to point out that one should not confuse an all-or-none methodology with an insufficient examination. For example, although the inability to make a 2-point discrimination may be a qualitative or all-or-none phenomenon, that does not mean that the presence or absence of the phenomenon may be evaluated on the basis of a single trial evaluation. The examiner is still obligated to run numerous trials with varying distances between the points and stimulation of different parts of the body. In the example from Luria given previously, one would think that when Luria did his examination, he must have used many different words and may have examined the patient on a number of occasions before he made the formulation quoted concerning acoustic analysis. In some cases, the LNNB may contain an insufficient number of items to determine the presence or absence of a specific deficit, but the difficulty in that case is not inappropriate adherence to an all-or-none model when the ability in question may be dimensional in nature, but rather may be related to the adequacy of the evaluation. It would appear that the LNNB does in some ways incorporate both dimensional and all-or-none measurement, but also suffers from brevity of the item list in certain cases. On the other hand, the HRB has the converse difficulty of neglecting certain possibly crucial areas in their entirety, although what it does assess it tends to assess quite thoroughly. One might view the difference in terms of preference for a great deal of information about a relatively small number of areas or relatively less information about a larger number of areas.

One could gain the impression that the LNNB can be interpreted through an examination of passed and failed items in a manner that ultimately leads to identification of the impaired functional system or systems, perhaps leading to the detection of an established syndrome. For example, the patient may have normal word finding, repetition, verbal comprehension, reading and writing abilities, but when asked to produce a narrative about some given topic completely fails to do so, perhaps saying only a few words following an excessively delayed response. Perhaps the patient can read fluently, and has no other language related difficulties, except that when asked about the sounds associated with letters (e.g., What sound is made by the letters g-r-o?) there is complete failure. We have actually seen such cases and others for which that type of analysis is possible. Short of this individual-item approach, the clinician may use the component factor scales to identify these patterns of correlation and dissociation. For example, the patient with a left-posterior lesion may obtain an impaired score on the Verbal

Memory factor scale but not on the Visual and Complex Memory scale. Furthermore, it is also possible to examine the quality of response to any particular item. For example, when a language-impaired patient is asked to produce a narrative, that narrative may be improverished, halting and sparse, but understandable. On the other hand, it may be lengthy and fluent, but make no sense at all. Thus, the speech sample derived from asking the patient to produce a narrative may aid in the differential diagnosis between Broca's and Wernicke's aphasia.

The extent to which this approach to assessment is available to the HRB user is unclear. One can make certain dissociations of the type described above, but they would appear to be more limited. For example, one could determine whether a patient has a naming or perceptual deficit by contrasting recognition of the square, cross, and triangle on visual presentation with recognition by touch. One could also make many pertinent and clinically relevant observations during the patient's performance on the Category or Tactual Performance tests. Even the smoothness of performance on the Finger Tapping test may provide significant information about the patient's motor system that supplements the test scores. Nevertheless, the opportunity to make specific functional correlations and dissociations would appear to be more limited with the HRB than with the LNNB and may reflect a genuine difference between the two procedures. Advocates of the LNNB have expressed this point through indicating that one of its advantages is that it provides a clearer picture of specifically what the patient can and cannot do, relative to other procedures. As we will see in subsequent discussion, there are major difficulties in regard to this level of analysis involving both the HRB and LNNB. Here, we are simply indicating that this apparent difference exists.

In summary, there would appear to be practical and substantive differences between the HRB and LNNB. Practically, the LNNB is briefer, more portable, and less expensive. On the other hand, the HRB is more well established, and provides highly detailed information in a number of areas that are less thoroughly evaluated by the LNNB. For example, the HRB generally includes a complete intellectual evaluation (the WAIS or WISC) whereas the LNNB can only estimate IQ. There are clearly different areas of emphasis. The LNNB appears to stress language and language-related skills, whereas the HRB tends to place more weight on complex problem-solving abilities, and nonverbal abilities in particular. The presence of formal memory testing in the LNNB and its absence in the HRB were noted. The LNNB consists of a large number of discrete, relatively simple tasks, whereas the HRB consists of a smaller number of mostly complex tasks. The possibility was raised that the LNNB may be more amenable to specific qualification of neuropsychological deficit patterns than the HRB, whereas the HRB may be more predictive of the individual's capacity to function in everyday life.

QUESTIONS AND ANSWERS

Perhaps a useful way of summarizing, illustrating, and elaborating on certain aspects of the similarities and differences between the two procedures would be through posing and attempting to answer a number of questions with regard to their capacity to provide an optimal standard neuropsychological assessment.

1. How do the two procedures stand with regard to their general sensitivity to brain dysfunction? Are there specificity differences?

Essentially, all of the literature we have in hand at present indicates that the two procedures achieve almost identical accuracy levels with regard to simple prediction of presence or absence of brain damage. Goldstein and Shelly (1984) have demonstrated that the predictive accuracies of both procedures are superior to what is obtained for the WAIS. Equality with regard to predictive accuracy of this type between the HRB and LNNB is obtained regardless of whether the study in question involves statistical comparisons or clinical judgments. The relative specificities of the two procedures has not been systematically studied, but there is some suggestion that in certain populations, notably chronic schizophrenics, the specificity of the LNNB may be greater than that of the HRB.

2. Which of the two procedures do better at distinguishing between brain-damaged and chronic schizophrenic patients?

The literature gives the nod to the LNNB in this case, as there has been one very positive study (Purisch, Golden & Hammeke, 1978) and one less impressive but nonetheless generally positive study (Shelly & Goldstein, 1983) attesting to the capacity of the LNNB to distinguish between brain-damaged and chronic schizophrenic patients. On the other hand, there have been numerous well known negative studies regarding the capacity of the HRB to make the same discrimination. Perhaps the study by Watson, Thomas, Andersen, and Felling (1968) is the best known of this group of investigations. Nevertheless, the issues invovled in this type of research are quite complex, and we probably have not heard the last word as yet.

3. How do the two procedures stand with regard to degree of association between performance levels and objective evidence of structural brain damage such as that provided by the CT scan?

CT scan studies have been done with both batteries, and in both cases there is some degree of association between extent of deficit and various CT scan atrophy and density indexes. The general problem in this field appears to be that there is no simple association between the degree of neuropsychological impairment and the extent of CT scan abnormality. However, both the HRB and LNNB have achieved similar levels of modest but significant correlations.

4. Is one battery better than the other for lateralization and further regional localization of brain damage?

Despite claims made by advocates of both procedures, based largely, it would appear, on clinical experiences, this question is difficult to answer. As is well known, there is a long history of successful attempts at lateralization with the HRB, going back at least to the early Wechsler-Bellevue studies (e.g., Reitan, 1955b). The LNNB also has two different kinds of lateralization scales, one based entirely on sensorimotor asymmetries and the other on an empirical study of patients with lateralized brain damage (McKay & Golden, 1979a). Whereas the clinical skill of Reitan and his associates with regard to localization of brain damage is quite well known throughout the field, it is somewhat surprising that little has been published in the way of localization studies with the HRB. Perhaps the best known study was reported in Reitan (1964), in which it was demonstrated that patients with discrete lesions in the four quadrants of the brain could be correctly classified with a high degree of accuracy on the basis of clinical judgment, but not through the use of univariate statistical comparisons. Halstead (1947), of course, felt that the tests in his battery were largely associated with frontal lobe function, but subsequent research did not fully support that view. The LNNB has a series of localization scales based on small samples of patients with focal lesions (McKay & Golden, 1979b), but there seems little question, because of the small sample sizes and the variety of lesion types within each locus, that these scales have to be viewed as preliminary in nature. For example, the left-frontal group reported on by McKay and Golden (1979b) contains three tumor cases, five cases with vascular lesions (occlusions), and four trauma cases. Because there are known neuropsychological differences among patients with these different lesion types, even when the locus is the same, further studies must be done to sort out the differential effects, despite overall success in regard to general classificatory accuracy.

The question, as put, cannot really be answered because no one has conducted a comparative study addressed to the issue of whether the HRB or LNNB achieves superior accuracy levels with regard to lateralization and localization of brain lesions. We can only say that, with regard to lateralization, the HRB has received more extensive study and has generally achieved impressive results. There have been remarkably few studies of regional localization with standard neuropsychological batteries, and so we do not know whether one procedure is better than the other for identifying, for example, frontal lobe brain damage. The LNNB does have the advantage of the availability of quantitative localization scales, but the clinical applicability of these scales would appear to be somewhat limited because of their preliminary nature. The published empirical evidence supporting the capacity of the HRB to identify specific regional lesion localization is exceedingly limited. It is quite interesting to note that in Boll's (1981) review of the HRB, the only study cited in the section on specific neurological diagnostic validity is the 1974 Reitan study that we just cited. It is possible that with the development of the CT scan and even more advanced methods, it may be possible to do many more regional localization studies.

With regard to the matter of localization, we feel compelled to note that there is an enormous literature on localization of brain function, some of which is touched on in other chapters of this volume. We would be remiss if we did not point out that certain neuropsychological tests have well-demonstrated capacity to localize, although many of those tests are not incorporated into the standard batteries under discussion here. The discriminative capacity of the Wisconsin Card Sorting Test for frontal lobe damage, the capability of certain memory tests to localize the thalamic and mammilary body lesions associated with Korsakoff's syndrome, and the impressively well established relationships between particular patterns of language disturbance and corresponding lesions in the language zone of the brain all serve as exemplars of the successes neuropsychologial methods have had with regard to regional localization. In this regard, one cannot neglect the many years of impressive work done by Han-Lukas Teuber and his many collaborators in the localization of function area. In essence, many neuropsychological tests have extensive value in localization, but, perhaps too often, these tests or their analogues are not incorporated into the standard assessment batteries.

5. What is the relationship between the test performance on each battery and the chronicity of an illness?

We know of two major studies addressed to this matter, one using the HRB and the other the LNNB. Goldstein and Halperin (1977) found that there was a relationship between length of schizophrenic illness and level of performance on the HRB, with long-term patients performing more poorly than short-term patients. Lewis, Golden, Purisch, and Hammeke (1979) showed no such relationship for the LNNB. We know from two studies done some years ago (Fitzhugh, Fitzhugh, & Reitan, 1961, 1962) that the neuropsychological signs of lateralization that are quite prominent in patients with recently acquired lateralized brain damage, in this case verbal-performance discrepancies on the Wechsler-Bellevue, are found to be much less prominent in patients with chronic, long-standing lateralized lesions. Russell, Neuringer, and Goldstein (1970) reported similar differences utilizing their neuropsychological-key approach. One might conclude that whereas the HRB appears to have some degree of sensitivity to chronicity of illness, similar sensitivity has not yet been demonstrated in the case of the LNNB.

   6. Do the two batteries do equally well at generating specific test profiles for patients with varying neurological disease entities?

A major thrust of neuropsychological research has been that of delineating the neuropsychological concomitants of a number of disease entities. Beginning with the more traditionally studied neurological and psychiatric disorders, this field of inquiry expanded substantially into such varied areas of physical illness as diabetes, hypertension, endocrinological disorders, chronic obstructive pulmonary disease, liver and kidney disease, and numerous other medical disease entities. There has been substantial interest in hypertension and alcoholism, and growing interest in the possible effects of environmental toxins and pollutants such as lead, dioxin, and solvents. In neurology, there has also been an expanding interest in such diagnoses as multiple sclerosis, Huntington's disease, head injury, and epilepsy. With regard to comparing the HRB and LNNB, both procedures have reported success with regard to making appropriate differentiations among these groups, notably alcoholics, epileptics, Huntington's Disease, and MS patients. There is little that would suggest the superiority of one procedure over the other, except perhaps to note that there has been substantially more research in the alcoholism area done with the HRB than with the LNNB, so that the clinician may feel more comfortable making inferences about the matter of alcoholism from HRB data than from LNNB data.

   7. Do the two batteries do equally well at generating a technically sophisticated neuropsychological assessment with regard to specificity of diagnosis, delineation of brain-behavior rela-

tionships, and providing prognostic and rehabilitation planning information?

We will treat this question through the asking of a number of subsidiary questions addressed to commonly occurring assessment challenges. Much of the technical discussion of the two procedures has involved the specifics of each procedure regarding various aspects of the neuropsychological assessment, such as the evaluation of language or of memory. The overall state of the art has produced a number of tentative or more well established methods of resolving these issues, and the question is to what extent the standard batteries reflect these solutions.

A. It is felt by many authorities in the field that aphasia, rather than being a unitary disorder, is actually a variety of syndromes, each of which, in its pure form, involves a unique pattern of language disturbance. Thus, numerous attempts have been made to classify the aphasias, because such classification has implications for the nature of the underlying brain–behavior relationship and the type of treatment that would be most beneficial. To what extent can the two standard batteries contribute to this differential diagnosis of aphasia?

In general, we would be of the opinion that neither battery can provide a differential diagnosis of aphasia with the level of sophistication that can be achieved by the currently used aphasia examinations, such as the Western Aphasia Battery (Kertesz, 1979) and the Boston Diagnostic Aphasia Examination (Goodglass & Kaplan, 1972). The basic difficulty is that the developers of both batteries never mounted the extensive type of study of various groups of aphasic patients that is necessary to build a satisfactory diagnostic system. It should be stipulated that in the case of the HRB, there never was any stated intent to provide differential aphasia diagnoses, but rather to screen for aphasia, that is, to itemize the various aphasic symptoms that are elicited on the brief aphasia screening test included in the HRB. In the case of the LNNB, the intent is not as clear, and there have been suggestions that the procedure has some capacity to provide differential aphasia diagnoses (Golden, Ariel et al., 1982). However, Crosson and Warren (1982) have provided a report that suggests that the current version of the LNNB would have significant difficulties in achieving that end. In the absence of the necessary research, and the necessary test items needed to make crucial diagnostic decisions, it would appear that although both batteries clearly have the capacity to identify the presence or absence of aphasia in general, and can provide catalogs of the language symptoms noted (e.g., word finding difficulty), they cannot incorporate this material into a specific diagnosis of the aphasia subtype (e.g., Broca's aphasia). The clinician would therefore be well advised to utilize one of the extensive aphasia examinations if differential diagnosis is an important consideration. Alternatively, such distinguished clinicians as Harold Goodglass have suggested that highly pertinent diagnostic information can be elicited through educated listening to a speech sample, even in those cases in which a comprehension defect is the most prominent symptom.

B. Extensive research involving the neuropsychological aspects of memory disorder clearly point to the conclusion that amnesia is not a unitary disorder, but like aphasia, can be classified into meaningful subtypes (Butters, 1984). Thus the amnesia of the alcoholic Korsakoff patient is different in important respects from the amnesia of the patient with bilateral hippocampal destruction, which is different in turn from the kinds of amnesia seen in patients with encephalitis or Huntington's disease. Can the standard batteries discriminate among these and other forms of amnesia, in a manner that provides information concerning the specific form of amnesia the patient has?

With regard to the HRB, the answer must clearly be in the negative because of the absence of formal memory testing from this procedure. However, the answer is probably also no for the LNNB as well, because although it does have a Memory scale, it consists of only 13 items, thereby providing a useful screening for memory difficulties, but surely not a comprehensive memory assessment. The amnesic disorders typically have a retrograde and an anterograde component, and it is necessary to administer specialized tests that provide detailed evaluations of recent and remote memory, and that can make the necessary dissociations among primary, short-term, and long-term memory. For example, the Famous Faces Test (Albert, Butters, & Levin, 1979) and the Squire, Slater, and Chace (1975) procedure involving recognition of television programs provide ways of assessing the status of relatively remote and recent memory. Similarly, tests that systematically evaluate the effects of cuing help to identify the encoding deficit thought to characterize the short-term memory of alcoholic Korsakoff patients. In general, it is important to elicit to the greatest extent possible the defect in the information processing chain, perhaps through assessing the traditional areas of encoding, storage, and retrieval. It is not that the LNNB does not have some appropriate items for evaluation of some of these aspects of memory, but as in the case of aphasia, there is no scheme for classification and the crucial dissociations (e.g., sharpness of the anterograde-retrograde gradient) cannot always be made with the items available. Furthermore, there has been no systematic research involving administration of the LNNB to groups of amnesic patients.

C. Neuropsychological investigation of functions of the right-cerebral hemisphere has shown that there are several cognitive functions that seem specifically mediated by the right hemisphere, or at least that the right hemisphere does better than the left hemisphere. Patients with right-hemisphere brain damage typically have visual-spatial deficits, may neglect their left-visual field and may have difficulties with various aspects of nonverbal perception, such as an impairment of the ability to recognize faces. Certain of these deficits are dissociable. For example, a visual-spatial problem may be primarily perceptual or constructional in nature, with different corresponding localizations. To what extent can the two batteries delineate these varying manifestations of right-hemisphere dysfunction?

The HRB systematically evaluates tactile, auditory, and visual neglect, whereas the LNNB does not. However, the LNNB contains both visual-perceptual and constructional items, whereas the HRB only appears to contain constructional items and does not assess perception of complex visual perception with overlapping- or embedded-figures tasks, or with pictures in which figure–ground differentiation is not clear. Neither procedure contains a facial recognition task. The clinician wishing to perform a detailed evaluation of right-hemisphere function may therefore wish to use some combination of the HRB, the LNNB and the rather specific tests prepared by Benton and collaborators (Benton, Hamsher, Varney, & Spreen, 1983) that involve judgment of angles of lines, recognition of faces, and visual form discrimination. In summary, it would appear that the HRB and LNNB evaluate different aspects of right-hemisphere function, but aspects evaluated by one battery are not evaluated by the other and vice versa. Furthermore, more sophisticated tests are generally available, notably the Benton tests, to provide refined analyses of specific right-hemisphere functions.

D. In addition to the major areas of language, memory, and visual-spatial and other nonverbal function disorders, how do the two standard procedures do at providing the necessary dissociations and double dissociations needed to identify specific syndromes?

It seems clear that when one enters the more technical and detailed provinces of neuropsychological assessment, one enters an area in which there are major difficulties with confounding, with mistaking one syndrome or deficit for another. A finger agnosia may really be an anomia, or a sensory deficit. A distorted drawing may be produced because of constructional or motor difficulties, or both. Failure at a task directed toward evaluation of right-parietal function may actually be associated with left-temporal brain damage, because the patient did not comprehend the instructions. Syndromes may be mistakenly identified because of incomplete understanding of their phenomenology. For example, we recently examined a patient who had extensive difficulty reading letters, but not words, and whose handwriting was normal. We immediately considered the possibility of alexia without agraphia (Albert, Goodglass, Helm, Rubens, & Alexander, 1981), but as it turns out, patients with this syndrome typically have the reverse reading disorder from this patient; that is, they can read letters but not words. Furthermore, they typically have color agnosia, which this patient did not have. Thus, the patient had some degree of alexia, and he did not have agraphia, but he clearly did not have Dejerine's syndrome of "alexia without agraphia." It would therefore appear that the definitive identification of these syndromal disorders is highly dependent on the skill and knowledge base of the clinician. It is also unclear as to whether either of the standard batteries can do the job in this aspect of assessment, but it is also unclear whether they should. For example, with regard to the patient just mentioned, neither battery tests specifically for color agnosia. However, color agnosia is an exceedingly rare condition, and one could question the value of testing for it routinely and without some prior indication that it may be present.

CONCLUDING REMARKS

In our series of questions and answers, we intentionally asked questions that could be answered affirmatively by means of a Halstead-Reitan approach to neuropsychological assessment and negatively by a flexible approach of the type advocated by Luria and later by Christensen and others. We also asked questions that would be answered in the opposite way. This strategy was adopted in order to provide the basis for our formulation of the actual similarities and differences between the two standard procedures, the HRB and the LNNB. It is our general view that both procedures are equally effective at answering assessment-related questions of the type that would be formulated within the framework of the empirical, quantitatively oriented neuropsychology advocated and productively practiced for many years by Reitan and his students and collaborators. Both procedures appear to be generally sensitive to brain dysfunction, achieving approximately equal concurrent validity levels. Both procedures do well at distinguishing among patients with right-hemisphere, left-hemisphere, and diffuse brain damage. Unfortunately, however, neither procedure has compiled impressive evidence regarding its capacity to make more precise localizations. However, that is a very arduous task because of the extreme difficulty in obtaining samples of sufficient size of patients with well documented localized lesions in the major brain regions. Perhaps only two people in the history of the field, Teuber (1959) and Newcombe (1969, 1974) truly achieved that goal, and only after many years of research. Both procedures have generated performance profiles that characterize relatively specific neuropsychological characteristics of patients with particular disease entities, such as MS and Huntington's disease. Both seem sensitive to the effects of chronic alcoholism. Both have made a contribution to the differential diagnosis of functional versus organic conditions, but the matter of chronic schizophrenia remains a difficult area in the case of both batteries. Both batteries can identify areas of deficit that may require evaluation by more detailed testing, particularly in the language and memory areas.

It seems quite clear that both standard batteries have been eminently successful with regard to identification of brain dysfunction, somewhat less successful but nevertheless producing important findings in the area of functional versus organic differential diagnosis, and reasonably successful with regard to provision of a comprehensive screening procedure that reveals relative strengths and weaknesses, and that may point to the need for further evaluation. The HRB made major contributions to such areas as cognitive function in alcoholics, performance patterns in various disorders, notably Huntington's disease, MS, various learning disabilities in children and functional hemisphere asymmetries. The LNNB, tracing a similar sequence with more advanced diagnostic and statistical methodologies, appears to be tracing a similarly successful course. In general, the research and practice with these procedures provide the basis for a great deal of what many, if not most, clinical neuropsychologists do, at least as a preliminary evaluation.

The situation with regard to identification of specific syndromes is quite different. The reason for the last set of questions was really to illustrate that the standard battery approach would not appear to be the best way to identify syndromes. Furthermore, the necessary research has not been done with either battery to provide empirical demonstrations of the capacity of these procedures to do syndromal level diagnosis. For example, we do not have the large research program mounted by Goodglass and Kaplan that led to the establishment of the Boston Diagnostic Aphasia Examination (Goodglass & Kaplan, 1972), or the extensive program of Butters, Cermak, and their collaborators (Butters & Cermak, 1980) that is providing increasing understanding of the amnesias, and sophisticated tools for their assessment. At a practical level, it would appear eminently impossible for any single battery to completely assess in full detail all of the enormous variety of syndromes, dissociations, and disconnections that may emerge as a result of local or generalized brain damage. Neuropsychological deficit is enormously complex, and the hope of capturing all of the subtleties and intricacies with a single procedure is surely a vain one. Therefore, it is important to remain aware that the standard comprehensive batteries should not be viewed as substitutes for the standard specialized batteries in such areas as aphasia, memory disorder, motor function, gnostic functions, and hemisphere asymmetries. The comprehensive batteries can be more reasonably viewed as belonging to a more preliminary phrase of assessment during which one is interested in making an initial diagnosis and looking for areas that may require more detailed evaluation. Even the specialized batteries may have to give way to highly individualized assessment, when there is the possibility of a rare or unique disorder that cannot be adequately understood through the use of any of the available instruments.

We do not accept the view that the LNNB provides a means of making Luria's neuropsychological investigative methods available and acceptable to the psychometrically, quantitatively oriented English speaking professional public. Similarly, the HRB approach as it currently exists is quite distant from Halstead's original concepts of biological intelligence and the role of the frontal lobes in regard to that concept. Luria's tests in the hands of Golden and collaborators and Halstead's tests in the hands of Reitan and his associates appear to have been developed into very different entities from what they were in the hands of their originators. The only issue I would take with some of my colleagues is that I do not view these deviations as at all unfortunate. Rather, through the persistence and productivity of these contemporary investigators, we now have two viable, effective, and extensively researched assessment instruments that appear to be extremely helpful in the evaluation and perhaps ultimately the treatment of brain-damaged patients. Furthermore, there seems little point in criticizing these procedures on the basis of unrealistic expectations for them, just as it appears to be unwise to use them assuming that they can in fact do things that they cannot reasonably be expected to do. It is hoped that advocates of these procedures can reduce the amount of unproductive controversy that has revolved around them through refraining from any suggestion that the procedure in question can do more than the available data suggest. With regard to the LNNB, Stambrook (1983) has made the following wise comment.

> The clinical utility of the LNNB does not depend upon either the publisher's and test developers' claims, or on conceptual and methodological critiques, but upon carefully planned and well executed research. (p. 266)

In our view, the same sentiment can be expressed with regard to claims made concerning the superiority of any particular procedure or method of neuropsychological assessment relative to other procedures and methods.

REFERENCES

Albert, M., & Kaplan, E. (1980). Organic implications of neuropsychological deficits in the elderly. In L. Poon, J. Fozard, L. Cermak, D. Arenberg, & L. Thompson (Eds.), New directions in memory and aging. Hillsboro, NJ: Lawrence Erlbaum.
Albert, M. S., Butters, N. & Levin, J. (1979). Temporal gradients in the retrograde amnesia of patients with alcoholic Korsakoff's disease. Archives of Neurology, 36, 211–216.
Albert, M. L., Goodglass, H., Helm, N. A., Rubens, A. B., & Alexander, M. P. (1981). Clinical aspects of dysphasia. New York: Springer-Verlag/Wein.
American Psychological Association. (1984). Joint technical standards for educational and psychological testing. Unpublished document.
Bender, L. (1938). A visual motor gestalt test and its clinical use. American Orthopsychiatric Association, Research Monographs No. 3.
Benton, A. L., Hamsher, K. deS., Varney, N. R., & Spreen, O. (1983). Contributions to neuropsychological assessment. New York: Oxford University Press.
Bergman, H. (1984, May). Brain dysfunctions related to alcoholism. Paper presented at NIAAA Conference on Clinical Implications of Recent Neuropsychological Findings, Boston, MA.
Boll, T. J. (1981). The Halstead-Reitan neuropsychology battery. In S. B. Filskov & T. J. Boll (Eds.), Handbook of clinical neuropsychology. New York: Wiley-Interscience.
Butters, N. (1984). The clinical aspects of memory disorders: Contributions from experimental studies in amnesia. Journal of Clinical Neuropsychology, 6, 17–36.
Butters, N., & Cermak. L. S. (1980). Alcoholic Korsakoff's syndrome: An information processing approach to amnesia. New York: Academic Press.
Chapman, L. J. & Chapman, J. P. (1973). Disordered thought in schizophrenia. New York: Appleton-Century-Crofts.
Chelune, G. J. (1982). A reexamination of the relationship between the Luria-Nebraska and Halstead-Reitan batteries: Overlap with the WAIS. Journal of Consulting and Clinical Psychology, 50, 578–580.
Chimelewski, C., & Golden, C. J. (1980). Alcoholism and brain damage: An investigation using the Luria-Nebraska Neuropsychological Battery. International Journal of Neuroscience, 10, 99–105.
Christensen, A. L. (1975a). Luria's neuropsychological investigation. New York: Spectrum.
Christensen, A. L. (1975b). Luria's neuropsychological investigation: Manual. New York: Spectrum.
Christensen, A. L. (1975c). Luria's neuropsychological investigation: Test cards. New York: Spectrum.
Christensen, A.-L. (1979). A practical application of the Luria methodology. Journal of Clinical Neuropsychology, 1, 241–247.
Christensen, A.-L. (1984). The Luria method of examination of the brain-impaired patient. In P. E. Logue & J. M. Schear (Eds.), Clinical neuropsychology: A multidisciplinary approach. Springfield, IL: Charles C Thomas.
Crosson, B., & Warren, R. L. (1982). Use of the Luria-Nebraska Neuropsychological Battery in aphasia: A conceptual critique. Journal of Consulting and Clinical Psychology, 50, 22–31.
Delis, D. C., & Kaplan, E. (1982). The assessment of aphasia with the Luria-Nebraska neuropsychological battery: A case critique. Journal of Consulting and Clinical Psychology, 50, 32–39.

Drewe, E. A. (1975). An experimental investigation of Luria's theory on the effects of frontal lobe lesions in man. *Neuropsychologia, 13,* 421–429.

Ferris, S. H., & Crook, T. (1983). Cognitive assessment in mild to moderately severe dementia. In T. Crook., S. Ferris, & R. Bartus (Eds.), *Assessment in geriatric psychopharmacology.* New Canaan, CT: Mark Powley.

Fitzhugh, K. B., Fitzhugh, L. C., & Reitan, R. M. (1961). Psychological deficits in relation to acuteness of brain dysfunction. *Journal of Consulting and Clinical Psychology, 25,* 61–66.

Fitzhugh, K. B., Fitzhugh, L. C., & Reitan, R. M. (1962). Wechsler-Bellevue comparisons in groups of 'chronic' and 'current' lateralized and diffuse brain lesions. *Journal of Consulting Psychology, 26,* 306–310.

Goldberg, E., & Tucker, D. (1979). Motor perseveration and long-term memory for visual forms. *Journal of Clinical Neuropsychology, 1,* 273–288.

Golden, C. J. (1977). Validity of the Halstead-Reitan Neuropsychological Battery in a mixed psychiatric and brain damaged population. *Journal of Consulting and Clinical Psychology, 45,* 1043–1051.

Golden, C. J., Hammeke, T. A., & Purisch, A. D. (1978). Diagnostic validity of a standardized neuropsychological battery derived from Luria's neuropsychological tests. *Journal of Consulting and Clinical Psychology, 46,* 1258–1265.

Golden, C. J., Osmon, D. C., Moses, J. A., & Berg, R. A. (1980). *Interpretation of the Halstead-Reitan Neuropsychological Battery: A casebook approach.* New York: Grune & Stratton.

Golden, C. J. (1981). A standardized version of Luria's neuropsychological tests: A quantitative and qualitative approach to neuropsychological evaluation. In S. B. Filskov & T. J. Boll (Eds.), *Handbook of clinical neuropsychology.* New York: Wiley-Interscience.

Golden, C. J., Ariel, R. N., Moses, J. A., Wilkening, G. N., McKay, S. E., & MacInnes, W. D. (1982). Analytic techniques in the interpretation of the Luria-Nebraska Neuropsychological Battery. *Journal of Consulting and Clinical Psychology, 50,* 40–48.

Golden, C. J., Hammeke, T., Purisch, A. D., Berg, R. A., Moses, Jr., J. A., Newlin, D. B., Wilkening, G. N., & Tuente, A. (1982). *Item interpretation of the Luria-Nebraska Neuropsychological Battery.* Lincoln, NE: University of Nebraska Press.

Goldstein, G. (1984, April). Current issues in neuropsychology. In T. Incagnoli (Chair), *Current issues and future directions in clinical neuropsychology.* Symposium conducted at the meeting of the New York State Psychological Association Convention, New York, NY.

Goldstein, G., & Halperin, K. M. (1977). Neuropsychological differences among subtypes of schizophrenia. *Journal of Abnormal Psychology, 86,* 36–40.

Goldstein, K., & Scheerer, M. (1941). Abstract and concrete behavior: An experimental study with special tests. *Psychological Monographs, 53*(2, Whole No. 239).

Goldstein, G., & Shelly, C. (1984). Relationships between language skills as assessed by the Halstead-Reitan battery and the Luria-Nebraska language related factor scales in a nonaphasic patient population. *Journal of Clinical Neuropsychology, 6* (2), 143–156.

Goldstein, G., Neuringer, C., & Olson, J. (1968). Impairment of abstract reasoning in the brain damaged: Qualitative or quantitative? *Cortex, 4,* 372–388.

Goodglass, H., & Kaplan, E. (1972). *The assessment of aphasia and related disorders.* Philadelphia, PA: Lee ' Febiger.

Graham, F. K., & Kendall, B. S. (1960). Memory-for-Designs Test: Revised general manual. *Perceptual and Motor Skills Monograph, II* (Suppl. 2), 147–188.

Halstead, W. C. (1935). The effects of cerebellar lesions upon the habituation of postrotational nystagmus. *Comparative Psychology Monographs, 12,* 130.

Halstead, W. C. (1945). A power factor (P) in general intelligence: The effect of brain injuries. *Journal of Psychology, 20,* 57–64.
Halstead, W. C. (1947). *Brain and intelligence: A quantitative study of the frontal lobes.* Chicago: The University of Chicago Press.
Hamsher, K. deS. (1984). Specialized neuropsychological assessment methods. In G. Goldstein & M. Hersen (Eds.), *Handbook of psychological assessment.* New York: Pergamon Press.
Heaton, R. K., & Pendleton, M. G. (1981). Use of neuropsychological tests to predict adult patients' everyday functioning. *Journal of Consulting and Clinical Psychology, 49,* 807–821.
Jastak, J. F., & Jastak, S. P. (1965). *The Wide Range Achievement Test: Manual of instructions.* Wilmington, DE: Guidance Associates.
Katz, J. J., & Halstead, W. C. (1950). Protein organization and mental function. *Comparative Psychology Monographs, 20,* 1–38.
Kertesz, A. (1979). *Aphasia and associated disorders: Taxonomy, localization and recovery.* New York: Grune & Stratton.
Klove, H., & Matthews, C. G. (1974). Neuropsychological studies of patients with epilepsy. In R. M. Reitan & L. A. Davison (Eds.), *Clinical Neuropsychology: Current status and applications.* Washington, DC: V. H. Winston & Sons.
Lewis, G., Golden, C. J., Purisch, A. D., & Hammeke, T. A. (1979). The effects of chronicity of disorder and length of hospitalization on the standardized version of Luria's neuropsychological battery in a schizophrenic population. *Clinical Neuropsychology, 4,* 13–18.
Lezak, M. (1983). *Neuropsychological assessment.* New York: Oxford University Press.
Luria, A. R. (1970). *Traumatic aphasia.* The Hague: Mouton.
Luria, A. R. (1973). *The working brain.* New York: Basic Books.
Luria, A. R. (1980). *Higher cortical functions in man* (2nd ed.). New York: Basic Books.
Luria, A. R., & Hutton, J. T. (1977). A modern assessment of the basic forms of aphasia. *Brain and Language, 4,* 129–151.
Majovski, L. V., Tanguay, P., Russell, A., Sigman, M., Crumley, K., & Goldenberg, I. (1979). Clinical neuropsychological evaluation instrument: A clinical research tool for assessment of higher cortical deficits in adolescents. *Clinical Neuropsychology, 1,* 3–8.
McKay, S., & Golden, C. J. (1979). Empirical derivation of neuropsychological scales for the lateralization of brain damage using the Luria-Nebraska Neuropsychological Test Battery. *Clinical Neuropsychology, 1,* 1–5.
McKay, S., & Golden, C. J. (1979). Empirical derivation of experimental scales for localizing brain lesions using the Luria-Nebraska Neuropsychological Battery. *Clinical Neuropsychology, 1,* 19–23a.
McKay, S. E., & Golden, C. J. (1981). The assessment of specific neuropsychological skills using scales derived from factor analysis of the Luria-Nebraska Neuropsychological Battery. *International Journal of Neuroscience, 14,* 189–204.
Moses, J. A., Jr., Golden, C. J., Ariel, R., & Gustavson, J. L. (1983). *Interpretation of the Luria-Nebraska Neuropsychological Battery (Vol. I and II).* New York: Grune & Stratton.
Newcombe, F. (1969). *Missile wounds of the brain: A study of psychological deficits.* Oxford: Clarendon Press.
Newcombe, F. (1974). Selective deficits after focal cerebral injury. In S. J. Dimond & J. G. Beaumont (Eds.), *Hemisphere function in the human brain.* London: Elck Science.
Prifitera, A., & Ryan, J. J. (1981). Validity of Luria-Nebraska intellectual process scale as a

measure of adult intelligence. *Journal of Consulting and Clinical Psychology, 49,* 755–766.

Purisch, A. D., Golden, C. J., & Hammeke, T. A. (1978). Discrimination of schizophrenic and brain-injured patients by a standardized version of Luria's neuropsychological tests. *Journal of Consulting and Clinical Psychology, 46,* 1266–1273.

Reitan, R. M. (1955a). An investigation of the validity of Halstead's measures of biological intelligence. *Archives of Neurology and Psychiatry, 73,* 28–35.

Reitan, R. M. (1955b). Certain differential effects of left and right cerebral lesions in human adults. *Journal of Comparative and Physiological Psychology, 48,* 474–477.

Reitan, R. M. (1958). Qualitative versus quantitative mental changes following brain damage. *The Journal of Psychology, 46,* 339–346.

Reitan, R. M. (1959). Impairment of abstraction ability in brain damage: Quantitative versus qualitative changes. *Journal of Psychology, 48,* 97–102.

Reitan, R. M. (1964). Psychological deficits resulting from cerebral lesions in man. In J. M. Warren & K. Akert (Eds.), *The frontal granular cortex and behavior.* New York: McGraw-Hill.

Reitan, R. M. (1966). A research program on the psychological effects of brain lesions in human beings. In N. R. Ellis (Ed.), *International review of research in mental retardation.* New York: Academic Press.

Reitan, R. M. (1976). Neuropsychology: The vulgarization Luria always wanted. *Contemporary Psychology, 21,* 737–738.

Russell, E. W. (1980). An all-or-none effect of cerebral damage. *Journal of Clinical Psychology, 36,* 858–864.

Russell, E. W. (1981). The pathology and clinical examination of memory. In S. B. Filskov & T. J. Boll (Eds.), *Handbook of clinical neuropsychology.* New York: Wiley-Interscience.

Russell, E. W., Neuringer, C., & Goldstein, G. (1970). *Assessment of brain damage: A neuropsychological key approach.* New York: John Wiley & Sons.

Shelly, C., & Goldstein, G. (1983). Discrimination of chronic schizophrenia and brain damage with the Luria-Nebraska battery: A partially successful replication. *Clinical Neuropsychology, 5,* 82–85.

Smith, A. (1975). Neuropsychological testing in neurological disorders. In W. J. Friedlander (Ed.), *Advances in neurology: Volume 7. Current review of higher nervous system dysfunction.* New York: Raven Press.

Spiers, P. A. (1981). Have they come to praise Luria or to bury him? The Luria-Nebraska battery controversy. *Journal of Consulting and Clinical Psychology, 49,* 331–341.

Squire, L. R., Slater, P. C. & Chace, P. M. (1975). Retrograde amnesia: Temporal gradient in very long term memory following electroconvulsive therapy. *Science, 187,* 77–79.

Stambrook, M. (1983). The Luria-Nebraska Neuropsychological Battery: A promise that may be partly fulfilled. *Journal of Clinical Neuropsychology, 5,* 247–269.

Teuber, H.-L. (1959). Some alterations in behavior after cerebral lesions in man. In A. D. Bass (Ed.), *Evolution of nervous control from primitive organisms to man.* Washington, DC: American Association for the Advancement of Science.

Walsh, K. W. (1978). *Neuropsychology: A clinical approach.* Edinburgh: Churchill Livingston.

Watson, C. G., Thomas, R. W., Andersen, D., & Felling, J. (1968). Differentiation of organics from schizophrenics at two chronicity levels by use of the Reitan-Halstead organic test battery. *Journal of Consulting and Clinical Psychology, 32,* 679–684.

Wechsler, D., & Stone, C. P. (1945). *Wechsler memory scale manual.* New York: The Psychological Corporation.

Wheeler, L. (1964). Complex behavioral indices weighted by linear discriminant functions for the prediction of cerebral damage. *Perceptual and Motor Skills, 19,* 907–923.

Wheeler, L., & Reitan, R. M. (1963). Discriminant functions applied to the problem of predicting cerebral damage from behavioral test: A cross-validation study. *Perceptual and Motor Skills, 16,* 681–701.

Wheeler, L., Burke, C. J., & Reitan, R. M. (1963). Application of discriminant functions to the problem of predicting brain damage using behavioral variables. *Perceptual and Motor Skills, 16,* 417–440.

9

Comparison of Halstead-Reitan and Luria-Nebraska Neuropsychological Batteries
Research Findings

ROBERT L. KANE

The field of neuropsychology is one in which the data base is growing at an explosive rate. The tools used to attempt to measure and understand complex human brain-behavior relationships remain varied. Presently, two test batteries have gained wide acceptance by clinicians favoring a standard-battery approach. The two batteries, the Halstead-Reitan Battery (HRB) and the Luria-Nebraska Neuropsychological Battery (LNNB) will be examined in this chapter in a comparison of research data from studies using both batteries with the same group of patients.

For clinicians favoring a standard battery approach, the predominant test battery has been the HRB. It was largely through the employment of this battery that neuropsychology was first recognized as a legitimate subspeciality of psychology. Parsons (1984) has noted that although the HRB began as an extension of theory (Halstead's theory of biological intelligence), the empirical validation of the test has had far more effect on the field than the theory itself. Nevertheless, the lack of a

---

ROBERT L. KANE • Veterans Administration Medical Center, East Orange and Department of Neurosciences, University of Medicine and Dentistry, Newark, NJ 07103.

concomittant development of the underlying theory has placed a ceiling on the utility of the HRB. For example, limitations are noted in understanding the specifics of why a patient's performance may be impaired on a given test and in planning strategies for cognitive retraining. Despite these limitations, the HRB remains a functional tool for clinicians providing evaluations of suspected brain-injured patients.

Over 20 years after the introduction of the HRB, Golden and his coworkers (Golden, Hammeke, & Purisch, 1978; Purisch, Golden, & Hammeke, 1978) introduced a new battery of tests. This battery was a standardized version of examination techniques first described by A. R. Luria (1966) and further delineated by Christensen (1975). There were two potential advantages to this newer test battery. One, the items making up the LNNB are designed to assess cognitive impairment according to Luria's theory of higher cortical functions. Luria stressed the necessity for individualization of the testing procedures to ascertain the reason a patient exhibits an impaired performance. The standardization of his techniques held the promise of providing a more refined way of examination that was acceptable to most American-trained psychologists, in contrast to their earlier rejection of his qualitative approach to assessment. The second advantage of the LNNB was one of practicality. The LNNB requires half the administration time as the HRB. Present-day issues of health care cost containment make shortened administration time a significant asset.

Despite the potential advantages, questions have been raised regarding the construction of the LNNB and its thoroughness as a neuropsychological battery (Stambrook, 1983). Whether the LNNB will live up to the promise of being a comprehensive standardized battery with a valid theoretical base remains to be established. What does seem clear is that a great deal of data has already been amassed with regard to the sensitivity of the LNNB to various forms of cerebral dysfunction (cf. Golden, this volume).

Early in the development of the LNNB, it became clear to the present author and his collaborators that for the putative advantages of the LNNB to be of consequence, it must show a sensitivity to the presence and extent of cerebral impairment comparable to the HRB. Hence, a series of investigations were undertaken to compare the two test batteries with respect to their relative diagnostic accuracies and the degree to which they assess level of impairment. It was understood from the beginning that comparability in assessing the presence and level of impairment does not insure comparability for all purposes for which neuropsychological assessments are utilized. Nevertheless, these studies were an important first step. Were the LNNB to perform less well than the HRB in assessing the presence and extent of brain impairment, its general usefulness as a neuropsychological battery would be called into question.

This chapter will focus on studies directly comparing the two neuropsychological batteries. The initial studies to be reviewed will be those that involved using raters. In these studies, clinicians were asked to classify patients as brain damaged or non-brain-damaged, using only results from the HRB and the LNNB. The next section will present findings from studies comparing the batteries statistically. Such studies involved correlating patients' performances on the various scales and subtests of the LNNB and the HRB, correlating measures of the degree of impairment derived from the two batteries and comparing the percentages of patients correctly classified as either brain damaged or non-brain-damaged using discriminant function techniques.

The third section of this chapter will be devoted to data that directly bare on the relationship of both the HRB and the LNNB to psychometric intelligence as measured by the Wechsler Adult Intelligence Scale (WAIS). Last to be discussed is the currently available data regarding the relationship between the Memory scale of the LNNB and the Wechsler Memory Scale (WMS). The WMS is often used in conjunction with the HRB to aid in the assessment of memory, because the HRB is often felt to be lacking in the area of memory assessment.

STUDIES USING RATERS

The first study to compare directly results obtained when both the HRB and the LNNB were given to the same group of patients (Kane, Sweet, Golden, Parsons, & Moses, 1981) involved having raters classify patients as brain damaged or non-brain damaged based solely on findings from the two neuropsychological batteries. This particular study employed two raters, one of whom classified patients on the basis of profiles obtained from the LNNB and one of whom classified patients based on data from the HRB. This particular design had the disadvantage of confounding raters with test batteries. However, at the time of this initial study, the only available expert rater for the LNNB was the third author. The subjects used in the study were referrals seen at the Oklahoma City or Palo Alto V. A. Medical Centers. The neurological group ($N=23$) included 12 patients who had incurred direct trauma to the head, 4 patients with dementia, 3 with vascular lesions, 1 with a neuroplasm, and 3 with unspecified brain disease. The controls ($N=22$) used in this study were a mixed group of psychiatric patients. In the strict sense, severely impaired psychiatric patients may not be "control" patients, as some form of organic impairment has been postulated to be part of these disorders (Lishman, 1978).

On the other hand, when the question is asked, "Is the patient organic?," the most frequent differential is between patients with psychiatric disturbance versus those with a neurological diagnosis. For this initial comparison study, the psychiatric group consisted of 12 patients who carried diagnoses of personality disorders, 6 diagnosed as schizophrenic, 3 with affective disorders, and 1 patient with the diagnosis of posttraumatic stress disorder. Patients were assigned to the brain-damaged or psychiatric group on the basis of independent medical criteria. The raters were given only the subjects' test scores, along with age, education level, and handedness of each subject. The raters were asked to classify the patient as either brain-damaged or psychiatric. In this first study, there was a nonsignificant tendency for the HRB rater to correctly identify a higher percentage of psychiatric patients (86% vs. 77% in 23 patients). The LNNB rater correctly identified a higher percentage of the brain-damaged patients (87% vs. 70% in 22 patients). This latter difference did reach statistical significance ($\chi^2 = 4.0$, $p < .05$). The overall hit rates of both batteries (78% and 82% respectively) were not significantly different. In addition, the two raters agreed in 37 of 45 cases, producing an agreement of 82%. Despite the caveats of confounding raters with batteries and the tendency of psychiatric patients to perform poorly on cognitive tests, these initial results were encouraging and supported the comparability of the two test batteries.

Kane, Parsons, Goldstein, and Moses (study in progress) recently undertook to replicate and expand on this initial study. They used data derived from the same 92 patients (46 brain damaged and 46 controls) used in the Kane, Parsons, and Goldstein (1984) study that is reviewed later in this chapter. Although the subjects will be described in more detail later in this paper, it should be noted that the control subjects in this study comprised a mixed group of medical and nonschizophrenic psychiatric patients.

By the time of this follow-up study, it was possible to enlist neuropsychologists with experience using both the LNNB and the HRB, permitting all raters to rate all subjects on both batteries. As in the first study, they were given only the age, education, handedness, and sex (all male veterans) of the subjects, in addition to the test data for each subject. The raters were provided with 184 protocols (92 from the HRB and 92 from the LNNB) and were asked to make blind judgments on the presence or absence of brain damage, the degree of damage if present, the location of the damage if present, and their confidence in making the ratings. At the present time, data from this study have not been

Table 1. Hit Rates for Each Rater by Test Battery and Patient Group

	Rater					
	No. 1		No. 2		No. 3	
	OBS	Control	OBS	Control	OBS	Control
LNNB	93%	87%	85%	85%	87%	89%
HRB	93%	78%	87%	78%	80%	76%

completely analyzed. However, the hit rates for each rater for both the HRB and the LNNB have been computed and will be reported here.

Table 1 presents the percentage of correct classifications by test battery and rater for both the organic and control patients. It will be noted that all three raters achieved high rates of correct classification for both the organic and controls, regardless of the test battery they employed. The hit rates are especially impressive considering the limited information given each rater in addition to the basic test results. Averaging accuracy ratings across raters, the percentage of correct classification using the LNNB was 87.7% for the control and brain-damaged groups combined. For the HRB, the overall correct classification rate was 82.3%. Assessing the average performance of the raters by patient group as well as by test battery, the following percentages emerge. Using the LNNB, the raters correctly identified 88.3% of the organic patients and 87% of the controls. Using the HRB, the raters correctly identified 86.7% of the organic patients and 77.3% of the controls.

The hit rates across both test battery and patient group were in the high 80s except for the HRB ratings for the control patients. In this case, the rate of correct classification was in the upper 70% range, indicating there were a greater number of patients misdiagnosed as brain-damaged when the HRB was employed in contrast to the LNNB. This may have been the result of the HRB not providing a standard way of controlling for the effects of age and education, unlike the LNNB. In addition, the complexity of the tasks on the HRB may have presented difficulties for some medical patients.

STATISTICAL RELATIONSHIPS

Early in the development of the LNNB, Golden and others interested in the development of this battery (Golden et al., 1981) compared results obtained from 108 patients who had been given both the HRB and the LNNB. The breakdown of the sample included 30 patients with a schizophrenic diagnosis, 48 with documented organic brain pathology, and 30 normal control subjects. Golden et al. (1981) compared the two batteries by first using data from each in a discriminative analysis to assess their effectiveness in classifying each group of patients. Secondly, Golden et al. (1981) compared scores derived from each battery using multiple correlation. This type of comparison allowed the researchers to determine the degree to which scores from the two batteries conformed to each other. It also allowed them to assess the degree to which a patient's performance on one battery could be used to predict his performance on the other.

The first discriminant analysis reported by Golden et al. (1981) compared the ability of the two batteries to accurately classify the brain-damaged (n=48) and control patients (n=60). The overall hit rate for the LNNB for the neurological group was 87% (42/48), and 88% (53/60) for the control group. Of 108 patients, the LNNB correctly classified 95, yielding a combined hit rate of 88%. For these same patients, the HRB correctly classified 90% (43/48) of the neurological group and 84% (50/60) of the control patients. The overall hit rate using the HRB was 86% (93/108). None of the above differences between the two batteries were statistically significant. It should also be noted that there was 89% agreement between the two batteries for the 108 cases; that is, data derived from both batteries led to the same conclusion in 96 of the 108 cases. Similar hit rates were also obtained for the schizophrenic group—the Luria correctly identifying 80% (24/30) and the Halstead 77% (23/30).

It is interesting to note that in the first comparison study (Kane et al., 1981), where the two raters were asked to classify patients as either brain damaged or psychiatric, the rater using the LNNB obtained an overall hit rate of 82%, whereas the rater using the HRB obtained an overall hit rate of 78%. These hit rates were comparable to those obtained in the Golden et al. (1981), study for the schizophrenic group. In the follow-up study by Kane et al. (in progress), the overall hit rate for the three raters using the LNNB was 87.7%. When using the HRB, the combined rate was 82%. These ratings were comparable to the overall hit rates in the Golden et al. (1981) study for the brain-damaged and control groups employing discriminant function analysis. The agreement between these studies using raters and discriminant analysis is impressive in light of the limited information given the raters and the ability of discriminant analysis to capitalize on both relevant and chance variation to reach the correct conclusion.

Returning to the Golden et al. (1981) study, the second mode of comparison involved correlating 14 scores from the HRB with the scores from 14 scales of the LNNB. In addition, a multiple regression analysis was employed such that various combinations of scores from each of the two batteries could be used to predict each of the individual scores obtained from the other.

For these computations, T scores from 14 Luria summary scales were assessed in relationship to 14 scores taken from the HRB. Tests from the HRB used in this comparison were the Category test; the Tactual Performance test, including scores for total time, memory, and location; the Seashore Rhythm test; Speech Sounds Perception; Finger Tapping for both the dominant and nondominant hand; the Trail Making tests, parts A and B; the Sensory-Perceptual Examination; the Aphasia Screening Test; and the Verbal and Performance IQ scores from the WAIS. Raw scores were used for the HRB, with the exception of the total time measure for Tactual Performance test. As some patients finished a different number of blocks and others failed to work the entire 10 minutes on a given trial, the total time for the three trials was divided by the total number of blocks placed. This computation yielded a score that reflected a time-per-block measure computed in seconds.

The results of the multiple correlation analysis between the two neuropsychological test batteries are reproduced in Tables 2 and 3. As can be seen from the second table, the Rs for the HRB ranged from .71 to .96. The majority of scores (9 of 14) exceeded .85. The Rs for the LNNB ranged from .77 to .94. Here 8 of the 14 exceeded a value of .85. All Rs were significant at the .05 level or better. Not surprisingly, the lowest correlation was between the LNNB scales and finger tapping with the dominant hand, the R reaching a value of only .71. This lower correlation for finger tapping seemed to result from the fact that pure motor speed items on the LNNB are not found on a separate scale of motor speed. Rather, the Motor Scale on the LNNB is comprised of items relating to the complex functional system of motoric behavior (Luria, 1966). Hence, some items measure motor speed whereas others relate to the kinesthetic feedback component of the motoric act. Still others relate to the orientation of motor movements in space. In addition, there are items relating to the role of language in regulating motoric behavior and in assessing the patient's ability to engage in bucco-facial movements. To compensate for this problem, the authors recomputed the R using items 1–4 from the Motor scale. These items relate directly to motoric speed. The result of this procedure changed the R for the dominant hand tapping to .85.

An interesting finding that emerged from this study was that the WAIS IQ scores were the strongest predictors from the HRB of a patient's performance on the LNNB. This finding resulted in the authors extending the initial study to compute the correlations between WAIS

Table 2. *Means and Standard Deviations for Halstead-Reitan Variables and Three Highest and Overall Multiple Correlations with Luria-Nebraska Scores*

Variable	M[b]	SD	Correlation with Luria-Nebraska[a]							
			1		2		3			
			Scale	r	Scale	r	Scale	r	Overall	
Category	70.06	35.92	Memory	.67	Visual	.65	Intelligence	.63	.89	
TPT–Total time per block[c]	38.16	49.81	Motor	.75	Tactile	.74	Visual	.74	.91	
TPT–Memory	6.91	3.32	Receptive	−.35	Intelligence	−.30	Rhythm	−.28	.77	
TPT–Location	3.00	2.64	Intelligence	−.55	Memory	−.49	Arithmetic	−.42	.77	
Rhythm (correct)	23.73	4.97	Memory	−.59	Receptive	−.57	Right	−.50	.76	
Speech sounds perception	8.69	6.89	Arithmetic	.62	Memory	.61	Pathognomonic	.60	.90	
Finger tapping (dominant)	41.89	11.64	Left	−.36	Receptive	−.31	Tactile	−.26	.71	
Finger tapping (non-dominant)	38.36	21.30	Right	−.32	Left	−.32	Tactile	−.30	.74	
Trail making (A)	50.41	34.16	Right	.73	Motor	.70	Tactile	.63	.89	
Trail making (B)	136.45	96.53	Right	.60	Motor	.56	Memory	.54	.92	
Sensory-Perceptual[d]	6.13	8.91	Writing	.50	Expressive	.39	Motor	.34	.86	
Aphasia	8.39	12.04	Expressive	.76	Reading	.72	Pathognomonic	.72	.95	
Verbal IQ	99.42	17.80	Intelligence	−.84	Writing	−.82	Memory	−.79	.93	
Performance IQ	91.21	25.70	Visual	−.89	Writing	−.83	Pathognomonic	−.82	.96	

Note. TPT = Tactual Performance Test.
[a] $df = 106$ for individual correlations; $df = 93$, 104 for overall multiple correlations. All correlations significant at better than .05. All multiple correlations significant at .001.
[b] $n = 108$.
[c] In sec.
[d] $n = 102$; $df = 88$, 14 for multiple correlation.

From "Relationship of the Halstead-Reitan Neuropsychological Battery to the Luria-Nebraska Neuropsychological Battery," by Golden et al. (1981) in *Journal of Consulting and Clinical Psychology, 49*, 410–417. Copyright 1981 by the American Psychological Association. Reprinted by permission of the publisher and the author.

Table 3. Means and Standard Deviations for Luria-Nebraska Variables and the Three Highest and Overall Correlations With Halstead-Reitan Scores

Variable	M[a]	SD	Correlation with Halstead-Reitan[b]							
			1		2		3		Overall
			Scale	r	Scale	r	Scale	r	
Motor	52.34	15.02	PIQ	−.79	TPT–total	.75	Trails A	.69	.90
Rhythm	53.71	14.68	PIQ	−.68	VIQ	−.61	Aphasia	.61	.77
Tactile	50.05	12.03	TPT–total	.74	PIQ	−.68	Trails A	.63	.81
Visual	56.33	12.08	PIQ	−.89	TPT–total	.74	VIQ	−.71	.94
Receptive	51.15	13.67	VIQ	−.73	Aphasia	.69	PIQ	−.61	.83
Expressive	52.42	12.98	VIQ	−.76	Aphasia	.76	PIQ	−.65	.87
Writing	59.46	9.49	PIQ	−.83	VIQ	−.82	Aphasia	.52	.93
Reading	51.92	18.56	VIQ	−.73	Aphasia	.72	PIQ	−.69	.84
Arithmetic	61.17	18.56	VIQ	−.74	TPT–total	.73	Aphasia	.64	.89
Memory	54.52	15.18	VIQ	−.79	PIQ	−.79	Category	.67	.89
Intelligence	57.65	13.62	VIQ	−.84	PIQ	−.78	TPT–total	.70	.92
Pathognomonic	59.95	14.41	PIQ	−.82	VIQ	−.76	Aphasia	.72	.89
Right	48.00	19.20	Trails A	.73	Trails B	.60	PIQ	−.60	.84
Left	45.25	14.20	PIQ	−.67	VIQ	−.64	Aphasia	.61	.78

Note. PIQ = Performance IQ; VIQ = Verbal IQ; Trails A = Trail Making (A); Trails B = Trail Making (B); TPT = Tactual Performance Test.
[a] $N = 108$.
[b] $df = 106$ for individual correlations; $df = 14, 93$ for multiple correlations. All correlations significant at .001.
From "Relationship of the Halstead-Reitan Neuropsychological Battery to the Luria-Nebraska Neuropsychological Battery," by Golden et al. (1981) in Journal of Consulting and Clinical Psychology, 49, 410–417. Copyright 1981 by the American Psychological Association. Reprinted by permission of the publisher and the author.

IQ scores and the subjects' performance on the other tests that make up the extended HRB. This latter comparison yielded correlations between the Verbal IQ and the HRB ranging from .37 to .76 in absolute value. Correlations between the Performance IQ and other HRB measures ranged from .31 to .71 in absolute value. Hence, a large portion of the common variance between the two neuropsychological test batteries was also shared by the WAIS.

When we compare the results of the studies reviewed thus far, certain patterns begin to emerge. First, when both the LNNB and the HRB were given to the same group of patients, both batteries correctly identified the brain-damaged and non-brain-damaged patients with about the same level of accuracy. When using discriminant function techniques, neither battery was able to achieve an advantage over the other. However, when clinicians were asked to blindly rate patients' performances, they tended to be more accurate with the LNNB for some comparisons. Such was the case in the first study using raters (Kane, et al., 1981) for the organic group, and in the second (in progress) for the controls.

Second, when patients' performances on one battery were correlated with their performances on the other, there appeared to be a significant degree of common variation. Therefore, it would be expected that both batteries are sensitive to similar aspects of higher cortical functioning. Predictions which could be made with one battery could most likely be made with the other, though this latter point remains to be validated.

Third, patients' performances on both the LNNB and on various measures comprising the extended HRB, exclusive of the Wechsler scales, also correlated strongly with their performance on the Wechsler scales.

This latter point caused Chelune (1982) to question the comparability of the two test batteries. Rather, he argued, if one removes the variance that these two neuropsychological test batteries share with the Wechsler scales, there remains only a small amount of variation (16%) common to both batteries. Chelune (1982) concluded that the overlap in variance between the HRB and the LNNB is essentially a function of each battery's relationship to psychometric intelligence. The conclusion that the HRB and the LNNB were, for all practical purposes, identical, was called into question, as was the contribution of the two neuropsychological batteries independent of the WAIS.

Golden, Gustavson, and Ariel (1982) later responded to Chelune's comments. They pointed out that removing the common variance shared by the two batteries with the WAIS, in effect, unfairly removed effects of organic pathology from results obtained with both the HRB and the LNNB. Golden et al. argued that the percentage of common variance shared by the HRB, the LNNB, and the WAIS is, in fact, a reflection of common sensitivity all these measures have to the effects of organic brain pathology. In his response to Golden et al., Chelune (1983) conceded this point. However, he felt the question of the independent contribution of the two batteries, other than that of psychometric intelligence, remained at issue. He also stressed that lack of equivalence could be a virtue giving neuropsychologists distinct batteries with distinct properties rather than two similar and overlapping instruments. This issue of the contribution of both the HRB and the LNNB independent of psychometric intelligence will be discussed later in this chapter.

If we put this question aside temporarily, the results of the earlier studies comparing the HRB and the LNNB are seen as forming the beginnings of a pattern already detailed above. For it was clear that acceptable hit rates could be obtained using either battery under optimal circumstances of discriminant analysis or under the more difficult procedure of rating protocols blindly. In addition, it was becoming clear that neither battery could gain a consistent superiority over the other in separating organic from control or psychiatric patients.

Neuropsychological test batteries are employed not only to aid in identifying the presence of brain damage, but also to quantify the extent of the impairment in a given patient. Measuring the degree of impairment resulting from organic pathology has implications for planning future interventions with patients. The intervention may be active attempts at rehabilitation or helping patients and their families cope with the consequences of brain impairment.

Assessing the extent of the behavioral deficit is also relevant for monitoring the changes that may occur in patients either as the effects of brain pathology worsen or resolve. A logical question to follow, then, was to what degree do the HRB and the LNNB provide similar assessments of the level of pathology present in a given patient?

The first study to appear addressing the issue of comparability with regard to level of impairment was that of Shelly and Goldstein (1982). In this study, the authors compared various indexes of impairment derived from the two batteries. They ran a combined factor analysis using scores derived from the two batteries. They also subjected measures from the batteries sensitive to laterality to a combined factor analysis.

Shelly and Goldstein compared the percentage of test scores falling in the impaired range and the Average Impairment Rating, the AIR (Russell, Neuringer, & Goldstein, 1970), from the HRB with several indicators of severity of impairment from the LNNB. LNNB measures included the percentage of scale scores falling above a T score of 60, the percentage of scale scores falling above a T score of 70, the average T score computed for all the LNNB scales, and the T score for the Pathonomonic scale. The results of their correlations between these measures are reproduced here in Table 4. It is clear from visual inspection that the correlations between the various estimates of the extent of impairment derived from the two neuropsychological batteries is high. Especially noteworthy is that the Russell et al. (1970) method for computing the AIR and the average T score from the LNNB summary scales correlated .82. Of the measures employed in the study, these two probably yield the best single estimates of level of impairment.

The most elaborate study to date comparing the HRB and the LNNB in both diagnostic accuracy and in concordance for measuring level of impairment is that done by Kane, Parsons, and Goldstein (1985). In this study, 46 patients with brain damage and 46 controls were administered both test batteries. The organic group was composed of 27 patients who had suffered head trauma, 12 cases with vascular

Table 4. Intercorrelations of Halstead-Reitan and Luria-Nebraska Indices of Impairment

	% Impaired	AIR	% Impaired (T>60)	% Impaired (T>70)	Average T score	Pathonomonic scale
Halstead-Reitan Indexes:						
% Impaired	1.00	.89	.69	.60	.71	.63
Average impairment rating (AIR)		1.00	.80	.74	.82	.75
Luria-Nebraska indices:						
% Impaired (T>60)			1.00	.90	.96	.84
% Impaired (T>70)				1.00	.93	.81
Average T score					1.00	.87
Pathonomonic T score						1.00

Note. From "Psychometric Relations between the Luria-Nebraska and Halstead-Reitan Neuropsychological Test Batteries in a Neuropsychiatric Setting" by C. Shelley and G. Goldstein, 1982, Clinical Neuropsychology, 4, 128–133. Copyright 1982 by Owens Press. Reprinted by permission.

lesions, 6 dementia patients, and 1 patient with a neoplasm. The control group was made up of 26 patients with medically diagnosed conditions not involving higher CNS disturbance, 13 psychiatric patients, most diagnosed as personality or affective disorders (excluding patients who were schizophrenic), and 7 functional neurological patients. This latter group of patients was taken from a neurology unit where they were being worked up for a variety of complaints that proved to be factitious or somatoform in nature. The average age and education for the brain-damaged group was 39.7 and 11.6 years, respectively. The control group averaged 38.9 years in age and 12.3 years in education. None of these group differences were significant.

Results obtained from the two test batteries were compared in several ways. Scores obtained from the HRB were converted to T scores using the normative work of Vega and Parsons (1967). This yielded T scores for the following 12 HRB measures: the Category test; Tactual Performance test scores for total time, memory and location; Rhythm test; Speech Perception test; Trail Making parts A and B; Finger Oscillation; and Hand Dynamomoter. For these latter two measures, T scores were employed for both the dominant and nondominant hands.

The raw scores for the LNNB are converted to T scores as part of the standard scoring of the battery. The T score conversions developed in the original standardization of the LNNB were employed in this study.

The first step in the Kane et al. (1985) study was to reaffirm the sensitivity of the first 12 LNNB scales (exclusive of the left and right hemisphere scales) and the 12 above mentioned measures from the HRB to the presence of brain damage. This was done by subjecting these scores to separate multivariate analyses of variance (MANOVAS) using group membership (organic versus control) as the dependent variable and the 12 scores from each battery as independent variables. The results of the MANOVAS reconfirmed the measures employed as sensitive to the effects of brain impairment. All 12 of the LNNB scales yielded significant differences in group performance with probability values at or less than $p=.0001$. Nine of the 12 measures from the HRB showed group differences equal to or less than $p=.0001$. Of the remaining HRB measures, tapping with the nondominant hand distinguished the groups at $p=.0015$. The probability that groups differed on the dynamometer with the dominant and nondominant hands was $p=.0405$ and $p=.0685$, respectively.

However, the main emphasis of the Kane et al. (1985) study was not to reaffirm that both batteries were sensitive to the effects of brain lesions, but to provide a more detailed comparison of their relative diagnostic accuracy. The study also investigated the agreement between the two batteries in assessing the extent of impairment present in a given patient. A variety of comparisons were employed. First, the T scores for the 12 LNNB and the 12 HRB measures were subjected to separate discriminant analyses in order to compare their relative diagnostic accuracies with this group of mixed organic and control patients. The result of the discriminant analysis again supported the equivalency of the two batteries. The LNNB produced an overall correct classification rate of 96.7% (100% of the controls and 93.48% of the brain damaged). The HRB produced an overall classification rate of 93.8% (95.4% of controls and 91.67% of the brain damaged).

The second method for assessing the diagnostic power of the two neuropsychological batteries was to compute the median T score for both groups on each of the two batteries. The number of brain-damaged and control patients scoring above and below the combined group medians was calculated. For the LNNB, T scores were constructed in such a manner that the higher the T score, the more impaired the performance. For the initial standardization group, the mean T score was set at 50 with a standard deviation of ±10. In the Kane et al. (1985) study, the median T score for all subjects was 52.65 with 87% of the brain-damaged patients scoring above this score and 87% of the controls scoring below. There was a clear tendency for the organic group to score above the mean of the original normative sample. They also scored above the combined group median on this independent patient sample.

The T scores for the HRB (Vega & Parsons, 1967) were constructed in an inverse manner from those of the LNNB. That is, on the HRB, the higher the T score, the better the performance. Similar to the LNNB, the mean T score for the control patients used in the normative sample was 50 with a standard deviation of ±10. In the Kane et al. (1985) study, the combined group median for the 12 T scores derived from the HRB was 50.05, nearly identical to the mean of the normative sample of Vega and Parsons. In the Kane et al., study, 84.8% of the brain-damaged patients scored below (more impaired) this score and 84.8% of the non-brain-damaged control patients scored above it (less impaired). Using the combined group medians as cutting points for both batteries produced hit rates of 87% for the LNNB and 84.8% for the HRB. The percentages of correct classification for both batteries were equivalent.

The use of a common metric for both batteries also allowed for a direct comparison of the HRB and the LNNB in assessing the level of impairment of cognitive functions. However, it will be recalled that the T scores from the LNNB were constructed such that the higher the score, the more impaired was the performance, whereas the reverse was true for the HRB. To facilitate comparing results for the two batteries,

Kane and his collaborators inverted the T scores obtained from the LNNB. This was done by subtracting the obtained T score from 100. The result of this transformation was to make consistent the directional interpretation of scores obtained from the two batteries with lower scores, indicating a more impaired performance.

In addition to using T scores, Kane et al. used impairment ratings to compare the similarity with which the batteries measure level of impairment. For the HRB, the Average Impairment Rating (AIR) of Russell et al. (1970) was employed. This method is widely used with the HRB and has been recently revalidated by Goldstein and Shelly (1982). The Russell et al. method for computing the AIR involves assigning a rating of 0 to 5 for each of 12 measures derived from the HRB, with 0 representing a better than average performance, 1 representing an average performance, and 2 through 5 indicating increasing levels of impairment. The 12 measures from the HRB used in computing the AIR include: the Category test; time, memory, and location scores from the Tactual Performance test; Speech Sounds Perception test; Seashore Rhythm test; Finger Tapping speed of the slower hand; part B of the Trail Making test; a score computed from the Sensory-Perceptual examination; a spatial-relations score based on the patient's performance copying a Greek cross and on the Block Design subtest of the WAIS; a score based on the Digit Symbol subtest of the WAIS; and an aphasia score computed from the Screening Test for Aphasia. These individual scores are then summed and divided by 12 to obtain the AIR.

Initially, no such system had been worked out for the LNNB. For the purpose of their comparison study, Kane and his colleagues developed a technique for rating impairment on the LNNB that paralleled the AIR. The system developed for the LNNB involved assigning a rating of 0 to 5 to each of the first 12 scales of the LNNB (exclusive of the left- and right-hemisphere scales), based on the relation of the obtained T scores to the mean and standard deviation of the initial standardization group. A score of 1 was assigned to a T score falling between 40 and 60. That is, all scores falling within plus or minus one standard deviation of the standardization group mean were considered within the range of a normal or average performance. Scores falling below a T score of 40 were assigned a rating of 0, as they were considered better than average. T scores above 60 were assigned ratings of 2 through 5, depending on their elevations. Impairment ratings were increased by a unit of one for each one-half standard deviation above a T score of 60. Therefore, a T score between 61 and 65 received a rating of 2, a T score between 66 and 70 received a rating of 3, and so forth. Any score greater than a T of 76 was assigned a rating of 5 to be consistent with the AIR for the HRB where the maximum impairment rating assigned to any score is 5. To compute the overall impairment rating for the LNNB, the individual ratings for the 12 scales were summed and divided by 12 to obtain an average impairment rating (L-AIR) similar to that computed for the HRB. Following the above procedures, it was possible to correlate average impairment scores from both the HRB and the LNNB.

Table 5 presents the relationships between the average T scores computed from the two batteries and their respective average impairment ratings. The average T scores from the two batteries correlated at .78. The positive correlation resulted from using inverted T scores from the LNNB, keeping consistent the relationship between direction of the T scores and the presence of impairment. Interestingly, the average impairment ratings from the two neuropsychological batteries also correlated .78. Both measures showed the same magnitude of relationship between the two batteries and suggested a similarity in the degree to which the HRB and the LNNB measure cognitive performance.

This degree of concordance with which both batteries assess level of impairment can be seen by inspecting the mean T score and impairment rating obtained in the Kane et al. (1985) study. Table 6 presents the means and standard deviations for the T scores and the average impairment ratings for the brain-damaged and control groups. It will be noted that for the control group, the mean T scores differed by only .09 of a point. As mentioned previously, the average T scores for the control patients on both batteries were close to those of the original stan-

Table 5. *Correlation Matrix for Halstead-Reitan and Luria Nebraska Summary Measures*[a]

	LNNB		HRB	
	Mean T score (inverted)	L-AIR	Mean T score	AIR
LNNB				
Mean T score (inverted)	1.00	−.98	.78	−.83
L-AIR		1.00	−.75	.78
HRB				
Mean T score			1.00	−.93
AIR				1.00

Note. From "Statistical Relationships and Discriminative Accuracy of the Halstead-Reitan, Luria-Nebraska, and Wechsler I.Q. Scores in the Identification of Brain Damage" by R. L. Kane, O. A. Parsons, and G. Goldstein, 1985, *Journal of Clinical and Experimental Neuropsychology, 7,* 211–223. Copyright 1985 by Swets and Zeitlinger. Adapted by permission.
[a]The negative signs on correlations involving the AIR and the L-AIR scores are due to the fact that higher scores on these measures reflect greater impairment while higher T scores indicate better performance.

Table 6. Summary Scores for the Halstead-Reitan and Luria-Nebraska Batteries

	Control	Brain damaged
Luria-Nebraska mean T score[a]	53.78 ± 6.7	35.12 ± 10.55
Halstead-Reitan mean T score[a]	53.87 ± 5.80	44.04 ± 6.25
Luria-Nebraska mean impairment rating	.97 ± .55	2.57 ± 1.15
Halstead-Reitan mean impairment rating	1.30 ± .59	2.58 ± .76

Note. From "Statistical Relationships and Discriminative Accuracy of the Halstead-Reitan, Luria-Nebraska, and Wechsler I.Q. Scores in the Identification of Brain Damage" by R. L. Kane, O. A. Parsons, and G. Goldstein, 1985, *Journal of Clinical and Experimental Neuropsychology, 7*, 211–223. Copyright 1985 by Swets and Zeitlinger. Adapted by permission.
[a]The higher the mean T score, the better the performance.

dardization groups on whose performance the T scores were based. For the brain-damaged group, the scores were not as close, differing by 8.92 points. However, both means reflect impairment of cognitive functions as would be expected for the brain-damaged group. In addition, there is a difference in range between T scores developed for the HRB and those developed for the LNNB. When the LNNB scales were standardized (Golden et al., 1978), T scores for the various scales ranged from lows in the 30s to highs of 100 (although these have recently been extended beyond 100). When the T scores for the HRB were developed by Vega and Parsons (1967), they were truncated, ranging between 27 and 73. Apparently, this difference was of little consequence for the controls, who scored close to the mean of the standardization groups. For the brain-damaged group, whose scores deviated farther from the mean of the normative sample, this difference in range resulted in the LNNB appearing to measure the level of impairment as more severe. It would appear, then, that the tendency for the LNNB to rate the brain-damaged group as more severely impaired with respect to T scores is artifactual. This interpretation is supported by looking at the other measure used in the study to assess level of impairment: the average impairment ratings for the two batteries. As noted in Table 6, the AIR computed for the HRB and the L-AIR computed for the LNNB differed by only .01 points for the brain-damaged group.

The ratings for both batteries for the brain-damaged group were in the right direction to be consistent with the presence of impairment and were significantly greater (more impaired) than were the impairment ratings for the control patients. The difference for the control patients was just 0.37. Based on the correlations between the average T scores and the impairment ratings, and on the similarity of the scores obtained, it seemed clear that both batteries were estimating the level of impairment in a highly similar fashion.

Is it then fair to say that the HRB and the LNNB are identical? A further analysis of the data would suggest that such is not the case. Although the two batteries were remarkably similar in their assessments of brain damage, a further inspection of the data reveals less than perfect agreement on an individual case basis. When the T scores from the HRB and the LNNB were subjected to a discriminant function to compare hit rates obtained from each battery, a total of eight patients (two controls and six brain damaged) were misclassified. There was no case in which the same patient was misclassified by both batteries. When the summary scores derived from the two batteries were likewise placed in a discriminant function, that is, the average HRB T score together with the AIR, and the average LNNB T score along with the L-AIR, a total of 20 patients were misclassified. Of these, only four were misclassified by both sets of scores. What is clear is that despite their high intercorrelations and the strong agreement between the batteries regarding the presence and extent of cerebral dysfunction, there are distinctive features that each battery possesses having relevance for the individual case. These distinctive features have not been delineated precisely at this time. However, Shelly and Goldstein (1982) factor analyzed both the LNNB and the HRB and found that the Luria tended to load more strongly on a language-related factor and the Halstead more robustly on a nonverbal cognitive factor. It should be noted that in the Kane et al. (1985) study, the VIQ score for the WAIS correlated more highly with the average Luria T score and the L-AIR. The PIQ score showed a higher correlation with the average T score and the AIR from the Halstead battery.

THE HRB, LNNB, AND PSYCHOMETRIC INTELLIGENCE

As noted earlier in this chapter, the Golden et al. (1981) study, which demonstrated strong statistical relationships between the various subscales of the HRB and LNNB, was challenged by Chelune (1982), who contended that much of the shared variance between the two test batteries could be accounted for by their relationship to psychometric intelligence. Chelune used data presented by Golden and his associates and calculated that the shared variance between the two neuropsychological test batteries was only 16% when WAIS intelligence scores were statistically partialed out. Golden, Gustavson, and Ariel (1982) argued, and Chelune (1983) agreed, that cognitive functions assessed by the WAIS are important in the appraisal of brain damage. The issue of to what extent the different measures contribute similar or independent information to the overall assessment remained. It was for this reason that Kane et al. (1985) compared the discriminant validity of summary scores derived from the LNNB and the HRB with those derived from the WAIS.

When the summary measures derived from the HRB (average T score and AIR) were placed in a discriminant function, 85.9% of all patients were classified correctly. When summary measures from the LNNB (average T score and L-AIR) were used, an identical overall rate of correct classification was obtained. When the Verbal and Performance IQ scores of the WAIS were employed in a discriminant analysis with this same group of patients, an overall correct classification rate of 84.8% was achieved: a percentage equivalent to that obtained with either the LNNB or the HRB summary measures.

Identical hit rates do not necessarily imply exact equivalence between the measures employed. Individual case analyses can reveal differences in cognitive measures despite similar hit rates and high degrees of intercorrelation. This same point which was made for the LNNB and the HRB can also be made if summary measures derived from the WAIS are included.

A total of 12 control subjects were misclassified by summary measures computed from the HRB, LNNB, or WAIS for both the organic and control groups. Of these, 2 were missed by the HRB alone, 3 by the LNNB alone, and 3 by the WAIS alone. With respect to the organic group, a total of 17 cases were misassigned by 1 or more of the summary measures employed in the study. Of these, 4 cases were missed only by the HRB, 4 only by the LNNB, and 3 only by the WAIS. We again have the situation where obtaining identical or nearly identical hit rates with different measures does not necessarily imply a functional identity between test batteries. Although different measures of cognitive functions may show similar abilities to screen for the presence of brain damage, the independent contribution provided by the measures in question to the assessment of brain dysfunction needs further exploration.

THE LNNB AND THE ASSESSMENT OF MEMORY

Although this chapter is devoted to comparing the HRB and the LNNB with respect to research findings, an equivalent comparison in the specific area of memory assessment is not possible, as the HRB does not address this issue in any comprehensive way. Typically, clinicians supplement the HRB with specific tests of mnestic functioning. The Wechsler Memory Scale (WMS) is frequently used for this purpose.

The LNNB includes a scale for assessing memory functions as part of the test battery. The Memory scale of the LNNB employs techniques based on Luria's (1976) investigation into the memory process and, as such, relies heavily on interference and distractor techniques. Items related to learning verbally meaningful material and rote learning a list of unrelated words are also included.

Recently, data has become available on the relationship of the Memory Scale of the LNNB to the WMS. Although preliminary, these findings are worthy of mention. An initial study was done in this area by Ryan and Prifitera (1982). The authors gave both the WMS and the Memory scale from the LNNB to a group of 32 psychiatric patients. Ryan and Prifitera reported that the Luria Memory scale T score correlated with the Memory Quotient (MQ) from the WMS at $-.65$, indicating that the two measures shared 42% common variance. They also used cutting scores from the two batteries to compute the percentage of agreement in classifying patients as having normal or impaired memories. On the LNNB memory scale, T scores greater than 60 were considered impaired. On the WMS, a MQ 12 or more points below the Full Scale WAIS IQ score was considered as indicating memory impairment. Using these decision rules, the two memory measures agreed 71% of the time in classifying subjects as having normal memories and 73% of the time in indicating impaired mnestic functioning.

Whereas the study by Ryan and Prifitera (1982) addressed the concurrent validity of the LNNB Memory scale and its relationship to the MQ derived from the WMS, a study by Larrabee, Kane, Schuck, and Francis (1985) investigated the construct validity of the LNNB Memory scale. First of all, Larrabee, Kane, and Schuck (1983) began by assessing the construct validity of the WMS using a factor analytic approach. They factor analyzed scores from 256 subjects who had been given both the WAIS and the WMS. Of these subjects, 145 had clear evidence of neurological dysfunction, 74 were a mixed group of nonneurological psychiatric and medical patients, and 37 patients could not be reliably classified into either a neurological or a nonneurological classification. The results of this first study supported the construct validity of both the Logical Memory and Associate Learning subscales of the WMS. It failed to do so for the Visual Reproduction subscale. When factor analyzed in conjunction with the WAIS, the Visual Reproduction subscale loaded most strongly on a factor that related to perceptual organization rather than learning and memory. In a follow-up study, Larrabee et al.

(1985) again employed a factor analytic technique using selected scales from the WAIS to provide marker variables for both verbal comprehension and perceptual organization. They also used subscales from the WMS, incorporating both immediate and delayed recall conditions, and other presumptive measures of memory, such as the Benton Visual Retention Test (BVRT) and two subscales derived from the LNNB Memory scale.

McKay and Golden (1981) published factor scales to be used with the LNNB. These factor scales are essentially subscales within the 12 major scales of the LNNB, focusing on more specific abilities within each scale. The two factors reported by McKay and Golden (1981) for the Memory scale were Me1 (Verbal Memory) and Me2 (Visual and Complex Memory). Scores derived from these two factors were utilized in the Larrabee et al. (1985) construct validity study. In brief, the follow-up study by Larrabee et al. (1985) confirmed the results of the initial study with respect to the WMS subscales. An important exception was that, with a delayed recall administration, the Visual Reproduction subscale loaded most heavily on a learning and memory factor while still retaining a substantial secondary loading on a nonverbal cognitive factor. Both Me1 and Me2 factor scores from the LNNB showed substantial loadings on a factor that related to learning and memory. In addition, Me1 showed a secondary loading on a factor relating to attention and concentration, whereas Me2 showed a secondary loading on a visual-spatial intelligence factor.

In summary, in the comparison of different measures of neuropsychological functioning, the major findings from the above memory studies are that the overall T score from the LNNB Memory scale has at least a moderate degree of validity when judged against the MQ from the WMS; the Logical Memory and Associate Learning subscales from the WMS, along with Me1 and Me2, provide items sensitive to mnestic functioning; and, the Visual Reproduction subscale of the WMS provides a measure of nonverbal memory if given in a delayed recall condition, as recommended by Russell (1975). This latter finding appears consistent with previous research, which suggests that the discrepancy between scores obtained in the Logical Memory and Visual Reproduction subtests of the WMS by patients with lateralized lesions is best seen with a delayed recall presentation (Delaney, Rosen, Mattson, & Novelly, 1980; Russell, 1975).

The issue of whether the WMS or Memory scale from the LNNB is more sensitive to the various forms of memory impairment seen with different lesions or with different neurological disorders has not been tested. Indeed, it appears from a review of the memory literature (Butters, 1984; Butters, this volume) that various tests of anterograde and retrograde memory may be needed to measure the types of mnestic deficits associated with different neurological conditions. Nevertheless, if the Logical Memory, Associate Learning, and Visual Reproduction (delayed recall) subscales of the WMS are used as marker variables, then it becomes clear that the types of items found on the Memory scale of the LNNB are related to learning and memory.

CONCLUDING COMMENTS

The goal of this chapter was to review the research to date comparing the HRB and the LNNB. In the area of memory, the Memory scale of the LNNB was compared to the WMS, as the HRB does not measure mnestic functioning. When one examines the research done with the HRB and the LNNB, there appears to be strong support for a relative comparability between the two batteries, though differences do emerge.

To address the similarities first, it seems clear that both batteries are effective in screening for the presence of organic brain dysfunction and that both provide similar estimates of the level of impairment. The shared variance between the HRB and the LNNB is high, and although in part this common variance is also shared by the WAIS, it would appear that the cognitive, motoric, and sensory effects of brain damage affect the ability of subjects to perform on tasks making up all three of these measures.

Differences in the two neuropsychological batteries are also apparent. The combination of clinical experience, together with the factor analytic study of Shelly and Goldstein (1982), would suggest that the HRB may be more sensitive to the issues of complex problem solving and nonverbal abilities. The LNNB places greater emphasis on verbally mediated skills. This particular hypothesis requires further validation before it can be accepted. However, it is clear that on an individual-case basis, different conclusions are sometimes reached, depending on which test battery is employed.

Despite these differences, the overall hit rates expected from the selection of either battery are quite comparable. Issues of overall sensitivity to either the presence or severity of brain damage would not lead to selection of one battery over the other. It does need to be stated that if the purpose of the examination is to screen for the presence and degree of impairment, the relative comparability of the two batteries tends to favor the use of the LNNB. It would appear, at least for the basic questions of the presence and extent of dysfunction, the LNNB can perform at the same level as the HRB and do so in a more efficient manner because of its shortened administration time. It also must be stated that neuropsychological measures are put to a variety of uses over and above those used thus far in comparing the two test batteries.

Several areas have not been addressed in the literature. These include the comparability of the two batteries for localizing lesions, for comparing the specific types or patterns of deficits seen with different neurological disorders, for generating information useful in rehabilitation, for studying the progression of various neurological disorders, for example, Alzheimer's type dementia, or for making specific predictions about how a patient will function in the day-to-day environment outside the laboratory. Studies done to date are clearly first steps in understanding the relationship of the batteries.

Findings from the research reviewed are significant as the use of standardized batteries has advantages in neuropsychological practice and research. Nevertheless, it is possible in neuropsychology, as often happens in clinical psychology, to become too test oriented, placing a large emphasis on the examination instruments and too little emphasis on the logic of the examination itself. To borrow an analogy from medicine, the neuropsychological examination can be viewed as a review of systems for higher cortical functions. Just as in the medical review of systems where one looks at the functioning of various organ systems in an organized fashion, the neuropsychological examination reviews various components of cognitive, perceptual, mnestic, linguistic, and sensorimotor functions. The exact methodology for doing so can change, depending on scientific and theoretical knowledge of the functional organization of the brain. Examination methods may also depend on the nature of the referral question and on the particular population in question. The HRB was based on a theory of biological intelligence (Halstead, 1947). The LNNB uses items that come from Luria's investigation into the nature and organization of higher cortical functions (Christensen, 1975; Luria, 1966). Despite the theoretical roots of both batteries, their basic strength to date has come from their empirical validation as instruments sensitive to the effects of brain lesions. Although such an approach is necessary at the beginning states of test validation, it is important that work not stop there. Ultimately, the selection of a test battery will depend on what research in the cognitive and neurosciences is able to tell us about the neural organization of behavior and about the types of tasks that are most representative of underlying abilities. Increasingly, neuropsychological test batteries are being called on to delineate subtypes of what were once considered general disorders (e.g., Butters, 1984), and to yield information to be applied to the rehabilitation process.

Both the HRB and the LNNB have demonstrated a clear sensitivity to the effects of various brain lesions, and as such, have and continue to play an important role in the practice of neuropsychology. It remains for future research to demonstrate under which conditions each battery can be expected to make its maximum contribution.

REFERENCES

Butters, N. (1984). The clinical aspects of memory disorders: Contributions from experimental studies of amnesia and dementia. *Journal of Clinical Neuropsychology, 6,* 17–36.

Chelune, G. J. (1982). A re-examination of the relationship between the Luria-Nebraska and Halstead-Reitan Batteries: Overlap with the WAIS. *Journal of Consulting and Clinical Psychology, 50,* 578–580.

Chelune, G. J. (1983). Effects of partialing out postmorbid WAIS scores in a heterogeneous sample: Comment on Golden, et al. *Journal of Consulting and Clinical Psychology, 51,* 932–933.

Christensen, A. L. (1975). *Luria's Neuropsychological Investigation.* New York: Spectrum.

Delaney, R. C., Rosen, A. J., Mattson, R. H., & Novelly, R. A. (1980). Memory function in focal epilepsy: A comparison of nonsurgical unilateral temporal lobe and frontal lobe samples. *Cortex, 16,* 103–117.

Golden, C. J., Hammeke, T. A., & Purisch, A. D. (1978). Diagnostic validity of a standardized neuropsychological battery derived from Luria's neuropsychological tests. *Journal of Consulting and Clinical Psychology, 46,* 1258–1265.

Golden, C. J., Kane, R., Sweet, J., Moses, J. A., Cardellino, J. P., Templeton, R., Vincente, P., & Graber, B. (1981). Relationship of the Halstead-Reitan Neuropsychological Battery to the Luria-Nebraska Neuropsychological Battery. *Journal of Consulting and Clinical Psychology, 49,* 410–417.

Golden, C. J., Gustavson, J. L., and Ariel, R. (1982). Correlations between the Luria-Nebraska and the Halstead-Reitan neuropsychological batteries: Effects of partialing out education and postmorbid intelligence. *Journal of Consulting and Clinical Psychology, 50,* 770–771.

Goldstein, G. & Shelly, C. (1982). A further attempt to cross-validate the Russell, Neuringer, and Goldstein Neuropsychological Keys. *Journal of Consulting and Clinical Psychology, 50,* 721–726.

Halstead, W. C. (1947). *Brain and intelligence: A quantitative study of the frontal lobes.* Chicago: University of Chicago Press.

Kane, R. L.. Sweet, J. J., Golden, C. J., Parsons, O. A., & Moses, J. A. (1981). Comparative diagnostic accuracy of the Halstead-Reitan and Standardized Luria-Nebraska Neuropsychological Batteries in a mixed psychiatric and brain-damaged population. *Journal of Consulting and Clinical Psychology, 49,* 484–485.

Kane, R. L., Parsons, O. A., & Goldstein, G. (1985). Statistical relationships and discriminative accuracy of the Halstead-Reitan, Luria-Nebraska, and Wechsler I. Q. scores in the identification of brain damage. *Journal of Clinical and Experimental Neuropsychology, 7,* 211–223.

Larrabee, G. J., Kane, R. L., Schuck, J. R., & Francis, D. J. (1985). The construct validity of various memory testing procedures. Journal of Clinical and Experimental Neuropsychology, 7, 239–250.

Larrabee, G. J., Kane, R. L., & Schuck, J. R. (1983). Factor analysis of the WAIS and

Wechsler Memory Scale: An analysis of the construct validity of the Wechsler Memory Scale. *Journal of Clinical Neuropsychology, 5,* 159–168.

Lishman, W. A. (1978). *Organic psychiatry: The psychological consequences of cerebral disorders.* Oxford: Blackwell Scientific.

Luria, A. R. (1976). *The Neuropsychology of Memory.* Washington, DC: V. H. Winston & Sons.

Luria, A. R. (1966). *Higher Cortical Functions in Man.* New York: Basic Books.

McKay, S. E., & Golden, C. J. (1981). The assessment of specific neuropsychological skills using scales derived from factor analysis of the Luria-Nebraska Neuropsychological Battery. *International Journal of Neuroscience, 14,* 189–204.

Parsons, O. A. (1984). Recent developments in clinical neuropsychology. In G. Goldstein (Ed.), *Advances in clinical neuropsychology: Vol. 1.* New York: Plenum Press.

Purisch, A. D., Golden, C. J., & Hammeke, T. A. (1978). Discrimination of schizophrenic and brain-injured patients by a standardized version of Luria's neuropsychological tests. *Journal of Consulting and Clinical Psychology, 46,* 1266–1273.

Russell, E. W. (1975). A multiple scoring method for the assessment of complex memory functions. *Journal of Consulting and Clinical Psychology, 43,* 800–809.

Russell, E. W., Neuringer, C., & Goldstein, G. (1970). *Assessment of brain damage: A neuropsychological key approach.* New York: Wiley-Interscience.

Ryan, J. J., & Prifitera, A. (1982). A concurrent validity of the Luria-Nebraska Memory Scale. *Journal of Clinical Psychology, 38,* 378–379.

Shelly, C., & Goldstein, G. (1982). Psychometric relations between the Luria-Nebraska and Halstead-Reitan Neuropsychological Test Batteries in a neuropsychiatric setting. *Clinical Neuropsychology, 4,* 128–133.

Stambrook, M. (1983). The Luria-Nebraska Neuropsychological Battery: A promise that may be partially fulfilled. *Journal of Clinical Neuropsychology, 5,* 247–269.

Vega, A. & Parsons, O. A. (1967). Cross-validation of the Halstead-Reitan tests for brain damage. *Journal of Counsulting Psychology, 31,* 619–625.

10

A Comparison of the Halstead-Reitan, Luria-Nebraska, and Flexible Batteries Through Case Presentations

GERALD GOLDSTEIN and THERESA INCAGNOLI

INTRODUCTION

The purpose of this chapter is that of illustrating, through the presentation of three cases, the similarities and differences among various approaches to neuropsychological assessment discussed elsewhere in this volume. The first two patients were given the Halstead-Reitan Battery (HRB) and Luria-Nebraska Neurological Battery (LNNB); one is a patient with a neurological illness, whereas the other is a psychiatric patient. The third case is also a neurological patient; he received the LNNB and a specialized series of tests selected on the basis of the preliminary information provided by the LNNB results. Thus, we have a comparison of two standard batteries with a neurological and a psychiatric patient, and a combined standard battery and flexible approach in a patient with an unusual neuropsychological syndrome. In each case, efforts were made to correlate the findings for the patients with pertinent literature regarding the disorder each of the patients repre-

---

GERALD GOLDSTEIN • Veterans Administration Medical Center, Highland Drive, Pittsburgh, PA 15206, and Department of Psychiatry and Psychology, University of Pittsburgh, Pittsburgh, PA 15213.   THERESA INCAGNOLI • Veterans Administration Medical Center, Northport, NY 11768, and School of Medicine, State University of New York, Stony Brook, NY 11790.

sents. The first two cases were selected, in part, because there is an available literature for both standard batteries covering the disease entities they represent, Huntington's disease and schizophrenia. With regard to the third presentation, there is a case literature available, and the patient was compared to similar cases.

A CASE OF HUNTINGTON'S DISEASE

This case was chosen for several reasons. First, there is little question about the diagnosis. The patient has a striking family history of Huntington's disease. Two younger sisters had already died of it when we evaluated the patient and his father was currently hospitalized for Huntington's disease in a neighboring community hospital. Furthermore, the patient had the full blown clinical phenomenology of the disorder, including the choreiform movements, ataxia, dysarthria, intellectual impairment and the history of violent episodes sometimes seen in Huntington's disease patients. The other reason for presenting him is that there have been systematic studies of Huntington's disease with both the HRB and LNNB (Boll, Heaton, & Reitan, 1974; Moses, Golden, Berger, & Wisniewski, 1981). Therefore, we would have some expectations about what would be found on both procedures. Although the pattern of neuropsychological deficit varies in this condition with the stage of the illness, the disorder at some point involves generalized intellectual impairment, memory deficit and psychomotor dysfunction, perhaps associated with the chorea, or movement disorder, itself.

The patient was 28 years old at the time of testing, and had completed 11 years of education. Before being hospitalized, he was employed as a canner and had previously been on active duty in the air force. His initial symptomatology appeared to involve outbursts of violent behavior, and there is a record of psychiatric hospitalization five years prior to the time we evaluated him. At the time of admission to our facility, the disorder had not been diagnosed, although the family history was known. It soon became apparent, however, that the patient had Huntington's disease, and had the motor-system symptoms in a particularly severe form, involving speech articulation as well as ataxia and adventitious movements of his limbs. The date of onset of the symptoms could not readily be determined, because the psychiatric symptomatology appeared to blend into the movement disorder and dementia. In any event, it was clear that the onset was relatively early for this disease, and its frank expression was preceded by an apparently lengthy prodromal period. During hospitalization he was treated with Prolixin Decanoate, which he was using at the time of the neuropsychological assessment. The medication appeared to relieve the symptoms to some extent, but he remained significantly ataxic, sometimes requiring the use of a wheelchair.

NEUROPSYCHOLOGICAL ASSESSMENT

Halstead-Reitan Battery

The HRB results, administered and scored using the Russell, Neuringer, and Goldstein (1970) method are presented in Table 1. It may be noted that the patient obtained a Wechsler Adult Intelligence Scale

Table 1. *Halstead-Reitan Battery Data*

Lateral dominance: patient is right handed, eyed, and footed

Perceptual disorders examination	Right	Left	Wide Range Achievement Test	
Tactile suppressions	8%	17%	Reading	3.9 Grade
Auditory suppressions	0%	0%	Spelling	2.6 Grade
Visual suppressions	0%	0%	Arithmetic	2.2 Grade
Finger agnosia errors	20%	5%		
Finger tip writing errors	10%	30%	11 years of education	
Dichotic listening errors	54%	52%		

Halstead and related tests			WAIS subtest	Score
Test	Score	Rating	Information	5
			Comprehension	5
Halstead category	94 errors	3	Arithmetic	3
Formboard—total time	30 min.	5	Similarities	2
Formboard—memory	3 blocks	3	Digit span	4
Formboard—location	3 blocks	2	Vocabulary	5
Speech perception	26 errors	4	Digit symbol	4
Rhythm	20 correct	3	Picture completion	9
Tapping—right	20 taps	4	Block design	6
Tapping—left	23 taps	—	Picture arrangement	6
Trails A	89 sec.	—	Object assembly	7
Trails B	300 sec.	5		
Digit symbol	4		Verbal IQ	64
Aphasia screening	21 errors	3	Performance IQ	77
Spatial relations	4 errors	2	Full scale IQ	68
Perceptual disorders	—	2		
Percentage of ratings in the impaired range	100%			
Average impairment rating	3.33			

Note. Rating of 0 = superior, 1 = average, 2 = mildly impaired, 3 = moderately impaired, 4 = severely impaired, 5 = very severely impaired.

(WAIS) Verbal IQ of 64, a Performance IQ of 77 and a Full Scale IQ of 68. The Average Impairment Rating was 3.33, with all of the tests in the Halstead series performed in the impaired range. The computer generated reports suggested severe impairment, with greater involvement of the left hemisphere. It may be noted that this predicted lateralization was never confirmed, but was predicted on the basis of a lower Verbal than Performance IQ and slightly reduced tapping speed with the right hand relative to the left hand (20 taps for the right hand vs. 23 for the left). As a point of reference, our patient may be compared with the mean scores for the Huntington's group studied by Boll et al. (1974). These comparative data are presented in Table 2. It is apparent, first of all that our patient had a substantially lower IQ than Boll et al.'s aver-

Table 2. Comparisons Between Boll et al. (1974) Mean Scores on Halstead-Reitan Tests with Scores Obtained by Patient

Tests	Boll et al. Mean	Patient's score
Tapping—dominant	26.27	20
Tapping—nondominant	23.18	23
Formboard—dominant*	6.28	2.5
Formboard—nondominant*	8.41	2.5
Formboard—both hands*	4.6	2.5
Formboard—total*	19.31	7.5
Digit symbol	5.27	4
Trails A—seconds	61.81	89
Trails B—seconds	203.54	300
Picture arrangement	5.36	6
Block design	6.9	6
Object assembly	6.09	7
Information	9.81	5
Comprehension	7.63	5
Similarities	8.27	2
Vocabularly	9.63	5
Speech perception—errors	14.36	26
Seashore rhythm—correct	20.72	20
Formboard memory—blocks	4.36	3
Formboard location—blocks	1.45	3
Picture completion	7.72	9
Arithmetic	7.0	3
Category	96.07	94
Verbal IQ	95.54	64
Performance IQ	92.9	77
Full scale IQ	92.81	68

*Minutes per block.

age patient. However, roughly comparable scores were obtained for the Tapping, Picture Arrangement, Block Design, Object Assembly, Rhythm, and the Category tests. Our patient did somewhat better on the Tactual Performance test, but substantially worse on tests with significant language content—Trails, the WAIS verbal subtests and Speech Perception. Although our patient had a severe articulation defect, that probably does not account entirely for his poor verbal scores. However, the patient was a "D" student in high school, and quit at the beginning of his senior year.

In addition to the language difficulties, which well may have antedated the onset of the Huntington's disease, the patient also had significant difficulties with conceptual reasoning and attention, even when assessed by tasks that do not involve smooth speech articulation or other aspects of motor function. Obviously, motor skills are impaired as well with speed, dexterity and, to some extent, strength involved. Perceptual abilities are relatively well preserved particularly in the tactile modality. Although he made some errors, he did reasonably well at identifying his fingers by touch, and at recognizing numbers written on his fingertips.

Luria-Nebraska Battery

The profile and quantitative results for the LNNB are presented in Table 3. It is immediately apparent that the wide range of abnormal scores noted on the Halstead-Reitan are also seen here. All of the scores for the content or ability scales are above the patient's critical level of 58.62. Again for reference purposes, our patient's scores were compared with the mean scores obtained by Moses et al. (1981). This comparison is presented in Table 4. Because these investigators divided their cases into early, middle, and late stages, we have presented both their middle- and late-stage data. Our patient was clearly not in the early stage of the illness. Looking at the non-language-related scales first, it would appear that our patient lies between the early and late stage patients of Moses and collaborators. On the Motor, Tactile, and Visual scales, he looks more like the middle-stage cases, whereas on Rhythm, Memory, Intellectual Processes and the Pathognomonic scale, he is more like the late-stage cases. He is more like the late-stage cases on all of the language related ability scales, but as indicated above, that may be because of a preexisting language processing disorder, possibly in combination with his articulation defect. Thus, although the patient has perceptual and motor deficit levels characteristic of the middle stage of Huntington's disease (3–5½ years post onset), his cognitive abilities seem more characteristic of individuals in the late stage of the

Table 3. Luria-Nebraska Scores

Major scales		Localization scales	
Scale	T	Scale	T
Motor	75	Left hemisphere	83
Rhythm	81	Right hemisphere	81
Tactile	63	Left frontal	86
Visual	65	Left sensorimotor	69
Receptive speech	73	Left parietal-occipital	86
Expressive speech	89	Left temporal	73
Writing	89	Right frontal	79
Reading	81	Right sensorimotor	76
Arithmetic	97	Right parietal-occipital	88
Memory	82	Right temporal	77
Intellectual processes	90		
Pathognomonic	95	Factor scales	
Right hemisphere	67		
Left hemisphere	56	Kinesthetic-based movements	75
		Drawing speed	71
Critical level	58.62	Fine motor speed	62
		Spatial-based movement	46
Estimated verbal IQ	= 66	Oral motor skills	80
Estimated performance IQ	= 73	Rhythm & pitch perception	72
Estimated full scale IQ	= 69	Simple tactile sensation	74
		Stereognosis	56
		Visual acuity & naming	67
		Phonemic discrimination	54
		Using relational concepts	69
		Concept recognition	89
		Verbal-spatial relationships	48
		Word comprehension	47
		Logical grammatical relations	46
		Simple phonetic reading	91
		Word repetition	61
		Reading polysyllabic words	72
		Reading complex material	79
		Reading simple material	112
		Spelling	82
		Motor writing skills	126
		Arithmetic calculations	90
		Number reading	74
		Verbal memory	86
		Visual & complex memory	62
		General verbal intelligence	71
		Complex verbal arithmetic	72
		Simple verbal arithmetic	94

Table 4. Comparisons Between Moses et al. (1981) Mean Scores on Luria-Nebraska with Scores Obtained by Patient

Scale	Moses et al. middle-stage mean T score	Moses et al. late-stage mean T score	Patient T score
Motor	84	103	75
Rhythm	73	82	81
Tactile	68	106	63
Visual	70	91	65
Receptive speech	70	100	73
Expressive speech	73	93	89
Writing	69	84	89
Reading	60	80	81
Arithmetic	77	115	97
Memory	68	85	82
Intellectual processes	72	101	90
Pathognomonic	75	102	95
Right hemisphere	72	109	67
Left hemisphere	69	105	56

disorder. It may be noted that the cases used in the Boll et al. (1974) study tended to be in the late stage, with a mean length of illness of 5.24 years (SD=3.41). Outstandingly high LNNB factor-scale scores for the nonlanguage related factor scales were obtained for the Oral Motor Skills (80) and Verbal Memory (86) scales.

Discussion

It should be stipulated at the beginning that these protocols were not submitted for blind clinical judgement, and so one cannot say whether one battery or the other is better for specific diagnosis of Huntington's disease. However, one can also comment on similarities and differences by means of comparing the two procedures with regard to the major aspects of neuropsychological assessment: intellectual and conceptual abilities, language, perceptual and motor skills, memory, attention, and visual-spatial abilities. With regard to general intellectual assessment, the WAIS IQ estimates predicted by the LNNB (VIQ=66, PIQ=73, FSIQ=69) are quite close to the values obtained from the WAIS itself (VIQ=64, PIQ=77, FSIQ=68). In both cases, the Verbal IQ was found to be lower than the Performance IQ, which is of some clinical significance in this case. Both procedures also suggest marked impairment of higher conceptual abilities. Both batteries also indicate substantial impairment of language abilities. The HRB does it through the WAIS verbal subtests, the Wide Range Achievement Test (WRAT), the Speech Perception and the Aphasia Screening Test. On the LNNB comparable deficits are noted on the Receptive Speech, Expressive Speech, Writing, Reading and Arithmetic scales. It is our inference that a major component of the language difficulty antedated acquisition of the symptoms of Huntington's disease. Such an inference could be made in the case of the LNNB on the basis of a study by McCue, Shelly, Goldstein, and Katz-Garris (1984) in which it was found that learning disabled young adults did more poorly on the complex language-related factor scales (e.g., Reading Complex Material) than on the simpler language related scales (e.g., Phonemic Discrimination). This patient obtained a score of 54 on the Phonemic Discrimination scale and 79 on the Reading Complex Material factor scale. In this comparison and others, he generally followed the pattern noted in the study of McCue and collaborators. From the standpoint of the HRB, the low grade levels obtained on the WRAT are also indicative of developmental language disability, particularly because Huntington's disease does not generally impair the ability to read, spell, or calculate.

With regard to memory, the comparison is difficult to make because of the absence of formal memory testing on the HRB. However, if the patient were administered the Wechsler Memory Scale (WMS) or similar test, he probably would have demonstrated a memory deficit, if only on the basis of his significant attentional disturbance. Both procedures suggest an attentional deficit. It can be inferred from the impaired level of performance on the Rhythm scale in the case of the LNNB and the Digit Span and Rhythm test from the HRB. Similarly, the performance tests of the WAIS and the Visual scale of the LNNB both suggest visual-spatial difficulties.

In summary, both batteries would provide the impression of a significantly impaired patient with a severe movement disorder, but relatively well preserved tactile function. Both batteries identify a language deficit, and it could be inferred from both procedures that at least some component of this deficit was developmental in nature and not an adult-acquired aphasia. It might be suggested that it is not surprising to see such substantial agreement in a patient with such global, severe impairment. Although the point may be well taken to some extent, there are other considerations. First, the LNNB has received some criticism in clinical circles for being "too easy" and not sufficiently sensitive to the magnitude of deficit found to be associated with various brain disorders. In this case, however, one would have to agree that the extent and magnitude of this patient's deficits are equally reflected in both procedures. Second, both procedures seem equally sensitive to the presence of a significant movement disorder. It is reflected in the tapping and pegboard tasks associated with the HRB, as well as in the difficulties the patient had with simple movements, coordination, oral movements and rapid copying of figures; tasks that constitute sections of the LNNB Motor scale. The relative preservation of tactile abilities are reflected in the perceptual disorders examination of the HRB and the Tactile scale of the Luria-Nebraska. As indicated previously, it could be inferred from both procedures that the patient had a developmental language disorder. There are, therefore, dissociations that can be made by both procedures, and we do not simply have a case of a globally impaired patient with no discernible pattern of strengths and weaknesses. It would appear that both procedures have the capability of making the pertinent dissociations in this case, with equal levels of sophistication. Furthermore, both procedures can be viewed as adequately reflecting the widespread and devastating behavioral manifestations of Huntington's disease.

A CASE OF SCHIZOPHRENIA

The role of neuropsychological assessment of schizophrenic patients remains a controversial area, with some authorities feeling that because the standard neuropsychological test batteries were not designed to diagnose schizophrenia, they should not be used for that purpose. Others feel that neuropsychological assessment can produce important diagnostic material for schizophrenic and for brain-damaged patients, but unfortunately, our tests do not appear to be sensitive to differences between brain-damaged and schizophrenic patients. Still others feel that the methods they employ are useful, and are sensitive to that difference. This case was chosen as an illustration of what a schizophrenic patient may look like on both standard batteries. He was selected on the basis of being a fairly typical, young chronic schizophrenic without complicating factors, such as a history of alcoholism or head injury. The diagnosis is also quite clear in this case, in that the patient has a history of classic signs of a schizophrenic disorder, including delusional thinking, flatness of affect, and tangential speech. There is no well documented history of hallucinations, but one episode involving visual and auditory hallucinations is described in his medical records. When tested, he was receiving 2mgm. daily of Cogentin and 5 mgm. of Haldol, three times a day.

The patient was 31 years old at the time of testing, and had a history of intermittent hospitalization of at least 10 years prior to that time. His medical history was essentially negative, except for some use of LSD and street drugs. There was one episode of Stelazine overdose that occurred about ten years prior to testing. When not hospitalized, the patient lived a nomadic life, traveling from city to city, earning a living by panhandling. He completed two years of college, and so it can be inferred that he had his initial psychotic episode during college, but that is not clear from the history. The circumstances of his first hospitalization involved an alleged physical attack on his father, and there were other episodes of alleged or documented assaultiveness throughout his history. Indeed, his most recent hospitalization was precipitated by an assault. At the time of testing, the patient was being maintained on medication in a day treatment center, which he attended three times a week.

NEUROPSYCHOLOGICAL ASSESSMENT

Halstead-Reitan Battery

The HRB, administered in the Russell, Neuringer, and Goldstein (1970) version yielded the results in Table 5. The patient obtained a WAIS Verbal IQ of 94, a Performance IQ of 88 and a Full Scale IQ of 91. The Average Impairment Rating was 2.0, with 58.33% of the tests performed in the impaired range. This level of performance is thought to reflect a mild degree of impairment. There were no suggestions of any test pattern that could reflect a lateralized or localized brain lesion. As can be seen in Table 5, he did exceptionally poorly on the Category and Rhythm tests, with normal scores on the Memory component of the Tactual Performance test, Finger Tapping, Aphasia Screening and Digit Symbol. Digit Symbol was rated as normal because it did not deviate greatly from the other WAIS performance tests. On the WAIS, tests involving overlearned verbal knowledge were performed at average levels, but tests involving attention, concentration, and problem-solving abilities were not done well. One might get the impression of a previously reasonably bright individual who is not doing well at present; an inference that would be compatible with his level of education.

The test results can be viewed as reflecting a relatively clear dissociation between complex and simple functions. The patient maintains normal performance levels on tests evaluating elementary perceptual and motor abilities such as the Finger Tapping test and the Perceptual Disorders examination. As indicated, he also does well at tasks that only require recourse to information learned in the past. However, on tests requiring attention, sustained concentration, conceptual reasoning, and problem solving, he does uniformly poorly. Even relatively straightforward attentional tasks, such as dichotic listening to digits, are done poorly. In general, his test findings are quite con-

Table 5. Halstead-Reitan Battery Data

Lateral dominance: patient is right handed, eyed, and footed

Perceptual disorders examination	Right	Left
Tactile suppressions	8%	0%
Auditory suppressions	0%	25%
Visual suppressions	8%	8%
Finger agnosia errors	10%	10%
Finger tip writing errors	5%	20%
Dichotic listening errors	44%	46%

Wide Range Achievement Test	
Reading	8.1 Grade
Spelling	7.4 Grade
Arithmetic	6.1 Grade

14 Years of Education

Halstead and related tests

Test	Score	Rating
Halstead category	117 errors	4
Formboard—total time	18.5 min	2
Formboard—memory	8 blocks	1
Formboard—location	4 blocks	2
Speech perception	17 errors	3
Rhythm test	15 errors	4
Tapping—right	52 taps	1
Tapping—left	45 taps	—
Trails A	30 sec.	—
Trails B	123 sec.	2
Digit symbol	7	1
Aphasia screening	5 errors	1
Spatial relations	2 errors	1
Perceptual disorders	—	1
Percentage of ratings in the impaired range	58.33%	
Average impairment rating	2.00	

WAIS subtest

Subtest	Score
Information	11
Comprehension	9
Arithmetic	7
Similarities	11
Digit span	7
Vocabulary	10
Digit symbol	7
Picture completion	9
Block design	7
Picture arrangement	7
Object assembly	10
Verbal IQ	94
Performance IQ	88
Full scale IQ	91

Note. Rating of 0 = superior, 1 = average, 2 = mildly impaired, 3 = moderately impaired, 4 = severely impaired, and 5 = very severely impaired.

sistent with the literature that clearly suggests that schizophrenics have difficulties with such complex procedures as the Category and Tactual Performance tests, and also tend to have significant attentional dysfunction.

This patient, in a way, reflects the difficulty clinicians have in using the HRB with chronic schizophrenic patients. They perform in a range consistent with the presence of brain dysfunction, but no such dysfunction can be documented on the basis of the medical history or standard neurodiagnostic precedures. Perhaps the CT scan will eventually clarify this issue to some extent, but the current data indicate that most schizophrenics do not have abnormal CT scans. Perhaps it would be more reasonable to view test results of this type as reflecting the well known cognitive and attentional deficits found in chronic schizophrenics.

Luria-Nebraska Battery

The patient was administered the LNNB about three weeks after he received the HRB. The results are presented in Table 6. It will be noted that the profile is abnormal, with several scores falling above the patient's critical level of 55. The predicted WAIS IQs (VIQ=97, PIQ=93, FIQ=95) were quite close to the actual values (VIQ=94, PIQ=88, FIQ=91). Scores on the Motor and Expressive Speech scales were the only ability scales below critical level, but the right- and left-hemisphere-scale scores also fell within the normal range. The worst score (69) was on the Rhythm scale, whereas the best score was obtained on the Motor scale (50). There are two psychometric criteria that the patient meets for presence of brain dysfunction; the score on the Pathognomonic scale is above critical level, and there are more than two T scores on the ability scales above critical level. On the localization scales there is an elevated score on the Right Parietal-Occipital scale, with the next highest elevation appearing on the Left-Occipital scale. Relatively high scores are noted on the Simple Tactile Sensation and Oral Motor Skills factor scales. Other factor-scale scores that are also elevated include the Concept Recognition, Verbal Memory and Simple Verbal Arithmetic scales.

The most outstanding deficit found for this procedure appeared on the Rhythm scale, which is generally thought of as a measure of auditory attention. However, difficulties with memory and various academic skills were also found. We would be inclined to interpret the high score on the Oral Motor Skills factor scale as a medication effect, in that the patient had been taking neuroleptics for many years. The elevated score on the Tactile scale is somewhat puzzling. The patient had extensive difficulty in identifying objects by touch, accompanied by a somewhat lesser amount of difficulty with tactual point localization and discrimination between sharp and dull tactile stimuli. As will be noted shortly, however, his score was not really deviant from what was found in two samples of chronic schizophrenic patients.

There have been two studies reporting data on LNNB profiles of chronic schizophrenics, one done by Purisch, Golden, and Hammeke,

Table 6. Luria-Nebraska Data

Major scales		Localization scales	
Scale	T	Scale	T
Motor	50	Left hemisphere	55
Rhythm	67	Right hemisphere	55
Tactile	62	Left frontal	56
Visual	60	Left sensorimotor	45
Receptive speech	60	Left parietal-occipital	62
Expressive speech	54	Left temporal	58
Writing	63	Right frontal	60
Reading	58	Right sensorimotor	49
Arithmetic	63	Right parietal-occipital	73
Memory	55	Right temporal	56
Intellectual processes	61		
Pathognomonic	63	Factor scales	
Right hemisphere	48		
Left hemisphere	48	Kinesthetic-based movements	52
		Drawing speed	62
Critical level	55	Fine motor speed	45
		Spatial-based movement	46
Estimated verbal IQ	= 97	Oral motor skills	80
Estimated performance IQ	= 93	Rhythm & pitch perception	69
Estimated full scale IQ	= 95	Simple tactile sensation	97
		Stereognosis	41
		Visual acuity & naming	62
		Visual-spatial organization	66
		Phonemic discrimination	40
		Using relational concepts	64
		Concept recognition	74
		Verbal-spatial relationships	48
		Word comprehension	63
		Logical grammatical relations	66
		Simple phonetic reading	45
		Word repetition	43
		Reading polysyllabic words	51
		Reading complex material	45
		Reading simple material	45
		Spelling	62
		Motor writing skills	46
		Arithmetic calculations	55
		Number reading	89
		Verbal memory	76
		Visual & complex memory	57
		General verbal intelligence	48
		Complex verbal arithmetic	58
		Simple verbal arithmetic	72

(1978) and the other done by Shelly and Goldstein (1983). The mean profiles reported in these studies were found to be quite similar to each other, and are reported in tabular form in Table 7, along with the profile obtained by our patient. Several aspects of this comparison are noteworthy. First, all of the profiles are quite similar to each other with regard to both level and pattern. Second, the Rhythm scale was the most elevated of the scales in two of the three cases and is highly elevated in all instances. There are always several scales above 60, which is generally used as the cutoff score for impairment when the critical level method is not used. In all cases, the scores for the Rhythm, Tactile, Reading and Writing, Arithmetic and Memory scales were 60 or above. In general, our patient performed in a manner that was quite comparable to the average level and pattern of performance found in the two studies. His intellectual functioning was somewhat better than what was found for the average patient in these two studies, but he had 14 years of education. The mean educational levels in both of the studies was approximately 11½ years. He is also about nine years younger than the average patient in the two studies.

It seems clear that the LNNB points out the attentional deficit commonly seen in chronic schizophrenic patients, but also reveals not remarkably severe but documentable deficits in numerous other areas, including academic skills, tactile perception, language comprehension, and memory. The present patient did not demonstrate the degree of

Table 7. Comparisons between Schizophrenic Patient and Shelly and Goldstein (1983) and Purisch, Golden, and Hammeke (1978) Studies

Major scale	Shelly & Goldstein T	Patient T	Purisch et al. T
Motor	57	50	67
Rhythm	66	69	73
Tactile	60	60	62
Visual	58	58	60
Receptive speech	61	59	69
Expressive speech	58	54	64
Reading and writing	60	60	60
Arithmetic	66	63	66
Memory	62	62	70
Intellectual processes	68	58	68
Pathognomonic	61	59	59
Left hemisphere	54	49	63
Right hemisphere	54	49	61

intellectual impairment found in the two studies, but was clearly comparable with regard to the findings for the great majority of scales. In general, his profile is almost identical to the mean profile reported in the Shelly and Goldstein (1983) study, but reflects slightly less impairment than was the case for the average patient in the Purisch, Golden, and Hammeke (1978) study. It may be noted that the Shelly and Goldstein cases were, on the average, younger and better educated than the Purisch, Golden, and Hammeke cases. Therefore, the mean profile they obtained is probably a more appropriate basis of comparison for the present patient than would be the case for the Purisch, Golden, and Hammeke study.

DISCUSSION

Both procedures reflect the common dilemma found when attempting to do neuropsychological assessments with schizophrenic patients. They perform in a range consistent with the presence of brain damage, but no brain damage is found using conventional neurodiagnostic methods and historical data. This patient, a young man with some higher education and a negative neurological history, clearly demonstrates cognitive deficits, on both the HRB and LNNB. Assuming, then, that schizophrenics will show deficits on neuropsychological tests, one might go on to try to find a deficit pattern characteristic of schizophrenics. Chelune, Heaton, Lehman, and Robinson, (1979) were eminently unsuccessful at achieving that goal with the HRB. In the case of the LNNB, Purisch, Golden, and Hammeke (1978) found that scores on the Rhythm, Receptive Speech, Memory, and Intellectual Processes scales did not discriminate at statistically significant levels between their brain-damaged and schizophrenic patients, but there were statistically significant differences for the other scales, which were viewed as assessing more basic abilities, such as motor function. Thus, impaired performance on the former scales in combination with normal performance on the latter ones might constitute a characteristic schizophrenic profile. However, Shelly and Goldstein (1983) were unable to replicate that pattern, commenting that its presence or absence may depend largely on the nature of the sample of brain-damaged patients used in the comparison.

It would be our view that neuropsychological assessments of schizophrenia can serve two useful purposes, one of which is not discrimination of schizophrenia from brain damage. The first important role it can fulfill is descriptive in nature, providing a reasonably comprehensive picture of the patient's cognitive and perceptual assets and deficits. There is surely enough variability among schizophrenic patients with regard to level and type of impairment to make this endeavor worthwhile. Such descriptive material can be useful for formulating a prognosis, rehabilitation planning, monitoring treatment, and various research applications. The second role is that of providing correlative data with other findings. Although we have particular reference to the CT scan, other laboratory and clinical findings may also be of significance. Blood and spinal fluid levels of various neurochemicals may be significantly correlated with neuropsychological test results, as may be clinical manifestations of the disorder, particularly the negative symptoms. It is now thought that there are two major types of schizophrenia. One type has primarily negative symptoms, atrophy is often noted on the CT scan, prognosis is poor, and there is little if any responsivity to neuroleptics. The other type is just the opposite, with a predominance of positive symptoms, a normal CT scan, good responsiveness to neuroleptics, and relatively good prognosis. If neuropsychological test results fit into this picture, then neuropsychological assessment can potentially be of significant value with regard to diagnosis, prognosis, and treatment planning.

A CASE OF MNESTIC DYSFUNCTION IN THALAMIC INFARCTION

This case illustrates the use of the LNNB as a screening instrument. Golden and Maruish (this volume) note that the examiner might want to further investigate deficits suggested by the LNNB or evaluate skills (e.g., remote memory) not measured by the battery. Goldstein (this volume) has recommended the use of the LNNB as a screening device in which one looks for areas that require more detailed evaluation. Furthermore, this case is noteworthy because it highlights that when the CT scan is utilized as a criterion for lesion localization in thalamic infarction, unilateral infarction can be associated with both verbal and nonverbal mnestic dysfunction.

Medical History

The patient is a 60-year-old right-handed man who completed 11 years of formal schooling and a GED and who was employed as a sexton. In November, 1983, he developed left-sided weakness and slurred speech. Neurological examination was entirely normal except for a left-upper and lower extremity paralysis with a positive Babinski sign and clonus on the left. Past medical history was otherwise unremarkable except for a history of a childhood seizure disorder. The seizures were not well documented, with the last one reported to have occurred in

1945. There was no history of drug or alcohol abuse. At the time of the above reported incident, the patient was not on any anticonvulsive medication. Medication at the time of this evaluation consisted of dilantin 100 mg. p.o. t.i.d., aldomet 250 mg. b.i.d., dialose p.o. q.i.d. and multi-vitamins.

The patient underwent four CT evaluations. Initial evaluation in November, 1983, with and without contrast, noted a hemorrhagic infarct in the region of the right thalamus with a question of some bleeding into the right lateral ventricle posteriorly. There was no shift of midline structures and minimal compression was noted. The last CT, performed April, 1984 (the time at which this neuropsychological evaluation was conducted), again noted infarction of the right thalamus with no additional infarction noted. No hemorrhagic component was demonstrated. Independent reading by three radiologists noted the infarction to be confined to the right thalamic region unilaterally. Because an arteriovenous malformation could unify the history of seizures and cerebral hemorrhage, an arteriogram was performed in December, 1983. However, there was no evidence of a mass lesion, tumor, aneurysm, or AV malformation.

NEUROPSYCHOLOGICAL ASSESSMENT

It should be noted that only selective test data consistent with the purposes for which this case was chosen will be presented. Examination began with standard administration and scoring of the LNNB.

The profile and quantitative results for the LNNB are presented in Figure 1. Whereas it is true that three scales (Rhythm, Receptive Speech, and Intellectual Processes) are just above the critical level (63.79), of greater concern is the striking elevation of the Memory scale and less so of the Expressive Speech scale. The deficits in expressive speech appear to be related to hearing loss; for example, he repeated "table" as "cable", "ball" as "bar" and "hat" as "cat". Audiometric evaluation of February, 1984, noted sensory-neuro loss for the right (mild to profound) and left (moderate to profound) ears. When the patient was reevaluated on this scale after he had secured a new hearing aid, repetition greatly improved and overall performance on this scale was below the critical level.

The striking elevation on the Memory scale was characterized by scores of 2 (on a scale of 0, 1, 2) on 11/13 items. When required to learn a series of 7 unrelated words over 5 trials, he made several errors. The observation that he utilized the actual result on each trial as the prediction for his response on the subsequent trial is of interest because such pathological features in level of aspiration have been reported in pa-

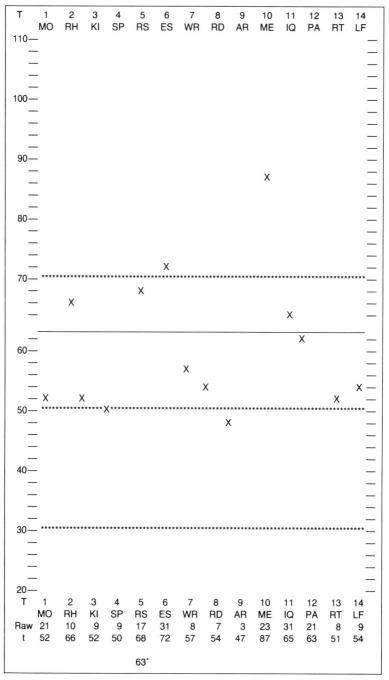

Figure 1. Luria profile.

tients with frontal syndromes (Luria, 1980). Immediate sensory trace recall was impaired for visual and tactile visual memory. Verbal memory was severely impaired under interference conditions as was logical memory for a paragraph and pictorial-verbal association.

To more fully evaluate such mnestic dysfunction, the patient was

Table 8. Comparison of Test Results in Two Patients with Right Thalamic Lesions

	Score	
Test	Speedie & Heilman	Patient
Intelligence		
WAIS-R VIQ	92	101
PIQ	83	88
FSIQ	88	95
Frontal lobe		
Wisconsin card sorting	2 categories	3 categories
Verbal fluency	60–64th percentile	89th percentile
Design fluency free condition	0 novel drawings	2
4 line condition	4 novel drawings	1
Stroop test	—	20, 23, 31
Verbal memory		
WMS logical memory		
Immediate	8	5
Delayed	8	3
WMS paired associate	6, 7, 7	4, 5, 5
Rey AVLT		
Trials 1–5	4, 5, 7, 8, 9	1, 3, 4, 6, 7
Postinterference	8	0
Recognition	15	6
Visual memory		
WMS visual reproduction		
Immediate	5	6
Delayed	2	0
Benton Visual Retention Test—correct	3	5
Rey-Osterreith		
Copy	29th percentile	20th percentile
Immediate	9.5 percentile	<10th percentile
Delayed	9.5 percentile	<10th percentile
7/24 Test		
Trials 1–5	—	4, 1, 2, 4, 3
Postinterference	—	2
Pictorial verbal learning test		
Trials 1–5	—	6, 4, 7, 7, 11
Postinterference	—	6

Note. — = data unavailable.

administered a specialized battery of neuropsychological tests. For comparative purposes his performance is contrasted with that of a right-handed male reported to have suffered an infarction of the right-dorsomedial nucleus of the thalamus (Speedie & Heilman, 1983).

Intelligence

The patient's IQ values were in the average range with a Verbal IQ of 101, a Performance IQ of 88 and a Full Scale IQ of 95. The WAIS IQ estimates predicted by the LNNB (VIQ=98, PIQ=87.5 and FSIQ=95) approximate these values. The performance IQ was lower than the verbal IQ on both measures and on the WAIS-R such a 13-point verbal/performance disparity is significant beyond the .05 level of confidence for a person of his age (Wechsler, 1981). This pattern replicates that reported for Speedie and Heilman's (1983) patient where the 9-point differential between verbal and performance IQ approached statistical significance at the .05 level.

Although performance on the Block Design and Object Assembly subtests was generally correct, the patient was penalized for slow speed.Such motoric retardation cannot be accounted for on the basis of depression, because neither formal psychometric evaluation nor clinical interview of the patient yielded any indication of depression.

Frontal-Lobe Functions

Because of the reciprocal connections between the dorsomedial nucleus of the thalamus and the frontal lobes, tests sensitive to frontal lobe functions were administered. A dissociation for verbal/nonverbal material was noted where the patient demonstrated intact performance on verbal fluency (89th percentile, Controlled Oral Word Association Test of the Multilingual Aphasia Examination (Benton & Hamsher, 1978) and markedly impaired performance on design fluency (Jones-Gotman & Milner, 1977). Here, performance was characterized by few novel drawings under both the free[1] and 4-line[2] conditions, and perseveration was pervasive. On the Wisconsin Card Sorting Test (Grant & Berg, 1948), severe perseveration was again noted (62 perseverative responses). The patient was only able to establish three categories, a performance level consistent with Milner's (1963) dorsolateral frontal

---

[1] In the free recall condition subjects are instructed to invent drawings that represent neither actual objects nor derivations from such objects. Drawing of an abstract pattern which could be named, such as a geometric form, is also prohibited.

[2] In the fixed condition subjects are instructed to invent drawings consisting of four lines.

patient's preoperative attainment. Severe impairment was also noted on the Stroop test under all administration conditions. In summary, the dissociation for verbal/nonverbal material and the perseveration that our patient demonstrated replicates that reported for Speedie and Heilman's (1983) patient with unilateral right thalamic infarction.

Verbal Memory

On the Wechsler Memory Scale our patient attained a Memory Quotient of 89 and demonstrated impaired immediate recall of semantic material when compared with Hulicka's (1966) norms. The decay (60% recall) in delayed logical memory that occurs when Russell's (1975) method was employed stands in contrast to the lack of such loss in the comparison patient. Impaired performance was also demonstrated on paired associate learning (> -1 SD according to Hulicka's norms) where the patient demonstrated little incremental learning of easy items over repeated trials and no learning of "hard" associations.

Our patient also performed poorly on the Auditory Verbal Learning Test (AVLT) (Rey, 1964; Taylor, 1959) when his performance was compared to that of manual laborers (Rey, 1964). He was particularly sensitive to proactive interference where his performance dropped to zero after the interference condition and is at variance with the comparison patient.

In summary, our patient performed more poorly than the comparison patient on all verbal measures. The decay in verbal material after interference conditions is noteworthy and particularly striking when viewed against the performance of Speedie and Heilman's patient.

Visual Memory

Our patient demonstrated fairly intact immediate figural memory; on the Wechsler Memory Scale immediate recall of figural material was just above the mean for the age group 60–69 (Hulicka, 1966) and on the Benton Visual Retention Test (Benton, 1974) he reproduced 5 geometric designs correctly when six were expected. This is in contrast to our comparison patient, who demonstrated defective immediate figural recall. Interestingly, under the delayed recall condition our patient was unable to reproduce any design correctly.

Overall performance on the Rey Osterreith Complex Figure (1944) was comparable for the two patients under the copy and immediate and delayed recall conditions (see Table 8). However, our patient was able to maintain the external configuration, experiencing difficulty with internal details, whereas the converse was true for Speedie and Heilman's (1983) patient.

Performance on the 7/24 Spatial Recall Test[3] (Rao, Hammeke, McQuillen, Khatri, & Lloyd, in press) was characterized by a negative learning curve for nonverbal stimuli and extremely impaired performance on the postinterference condition. On the Pictorial Verbal Learning Test (Lezak & O'Brien, 1981), the visual analogue of the Auditory Verbal Learning Test, our patient was somewhat impaired, although less so than in the verbal sphere.

In summary, although our patient demonstrated intact immediate figural recall, delayed recall of nonverbal material was severely impaired. He performed better on a pictorial-verbal learning test on recall and after an interference condition than on either a purely verbal (Rey AVLT) or nonverbal test (7/24 Spatial Recall Test). Furthermore, a dissociation for verbal/nonverbal stimuli was noted for frontal lobe functions where design fluency was extremely impaired, whereas verbal fluency was intact. This lateralizing pattern was replicated in intellectual functioning where Verbal IQ was greater than Performance IQ.

DISCUSSION

Ample evidence for memory deficits that are material specific accoording to the hemisphere involved have been provided by stimulation studies and documented cases reported in the literature. Ojemann and Fedio (1968) demonstrated specific short-term verbal recall deficits after stimulation to the posterior superior portion of the left lateral thalamus (anterolateral pulvinar). No such deficits were found with stimulation of corresponding right thalamic regions. In additional studies (Ojemann, Blick, & Ward, 1971) of stimulation to left ventrolateral thalamic sites anterior to those in the pulvinar, the relationship between the dominant lateral thalamus and short-term verbal memory deficits was again reported with no such deficits occurring with right ventrolateral thalamic stimulation.

Unilateral thalamic lesions provide additional evidence for material-specific memory deficits. Besides the comparison patient presented here with anterograde memory deficits for visual-spatial material after right thalamic infarction (Speedie & Heilman, 1983), other cases of unilateral damage are noted. The famous case N.A., who demonstrated

[3]In this revision 7 poker chips are randomly placed on a 6 × 4 checkerboard. After a 10 second exposure, the subject is asked to reproduce the original 7 chip pattern with 9 chips and an empty board. Learning trials are repeated 4 times with the same pattern (design A). One trial with a new pattern (design B) then ensues, followed by a free recall of design A.

marked amnesia for verbal material (Teuber, Milner, & Vaughn, 1968; Squire & Slater, 1978), was noted on CT scan to have a lesion localized to the left dorsomedial nucleus of the thalamus (Squire & Moore, 1979). More recently, Crosson, Parker, Warren, LaBreche, and Tully (1983) reported a case of left thalamic tumor with moderately impaired short-term verbal recall and intact short-term nonverbal recall.

When these findings are considered, this case becomes noteworthy because it illustrates several phenomena that would not be expected on the basis of CT scan readings of a unilateral right thalamic infarction. These include immediate intact figural memory with corresponding deficits for verbal material. Sensitivity to interference and loss of recall under delayed conditions, which would be expected for nonverbal material on the basis of his lesion, was also present for verbal material. On the basis of a unilateral right thalamic lesion, it could be expected that one would experience difficulty with the external gestalt of a figure as compared to internal details. However, for our patient the converse was true. On the other hand, performance on measures of intelligence and frontal-lobe functions were lateralized as would be expected.

In conclusion, this case demonstrates the utilization of the LNNB as a screening instrument to highlight deficit areas in need of further investigation. Furthermore, this case highlights that when the CT scan is utilized as a criterion for lesion localization in thalamic infarction, unilateral damage can be associated with verbal and nonverbal mnestic dysfunction.

SUMMARY

These case presentations appear to make several points. In the case of the first two patients, it was noted that both standard batteries provided results that were quite consistent with their own literature regarding the disorder in question. However, it was necessary in both instances to take extenuating circumstances into account to explain discrepancies; notably, age, education, and assumed premorbid-level conditions. Both standard batteries, in their own ways, appeared to provide comprehensive assessments, and in many respects would agree with regard to the nature of the functional deficits found, although the tests used to assess them were different from each other. We do not know what would have happened had we submitted these cases to blind interpretation, but it seems clear that both batteries are describing more or less the same individual in both cases. The third patient provides a clear example of how a standard battery can be used as a screening instrument, signalling the clinician to do a more thorough assessment in some specific area; memory in this case.

With regard to the third case, it seems fair to say that neither the Halstead-Reitan nor the Luria-Nebraska Batteries could have described the patient's amnesia in the detail provided by the specialized memory tests. However, if the patient were only given memory tests, many significant points about his functioning would have been missed. The case therefore provides an excellent example of the advantages of a combined comprehensive and specialized approach. The first two cases are of particular interest because of their concurrence with the quantitative research data summarized in Dr. Kane's chapter. He reports that the two standard batteries have extremely high concordance rates for a number of variables used in clinical decision making.

ACKNOWLEDGMENT

Dr. Incagnoli wishes to thank Dr. Gerald Goldstein for his careful review of this case.

REFERENCES

Benton, A. L. (1974). *The Visual Retention Test* (4th ed.). New York: The Psychological Corporation.

Benton, A. L., & Hamsher, K. (1978). *Multilingual Aphasia Examination*. Iowa City: Benton Laboratory of Neuropsychology, Division of Behavioral Neurology, Department of Neurology, University Hospitals.

Boll, T. J., Heaton, R., & Reitan, R. M. (1974). Neuropsychological and emotional correlates of Huntington's chorea. *The Journal of Nervous and Mental Disease, 158,* 61–69.

Chelune, G. J., Heaton, R. K., Lehman, R. A., & Robinson, A. (1979). Level versus pattern of neuropsychological performance among schizophrenics and diffusely brain damaged patients. *Journal of Consulting and Clinical Psychology, 47,* 155–163.

Crosson, B., Parker, J. C., Warren, R. L., La Breche, T., & Tully, R. (1983, August). *Dominant thalamic lesion with and without aphasia*. Paper presented at the meeting of the American Psychological Association, Anaheim, CA.

Grant, D. A., & Berg, E. A. (1948). A behavioral analysis of degree of reinforcement and ease of shifting to new responses in a Weigl-type card-sorting problem. *Journal of Experimental Psychology, 38,* 404–411.

Hulicka, I. (1966). Age differences in Wechsler Memory Scale scores. *The Journal of Genetic Psychology, 109,* 135–145.

Jones-Gotman, M., & Milner, B. (1977). Design fluency: The invention of nonsense drawings after focal cortical lesions. *Neuropsychologia, 15,* 653–674.

Lezak, M. (1983). Neuropsychological assessment, 2nd ed. New York: Oxford University Press.

Lezak, M. D., & O'Brien, K. (1981, February). *A comparison of traumatically brain injured and control subjects on verbal and pictorial presentations of a learning task*.

Paper presented at the meeting of the International Neuropsychological Society, Atlanta, GA.
Luria, A. R. (1980). *Higher cortical functions in man* (2nd ed.). New York: Basic Books.
McCue, M., Shelly, C., Goldstein, G. & Katz-Garris, L. (1984). Neuropsychological aspects of learning disability in young adults. *Clinical Neuropsychology, 6*, 229–233.
Milner, B. (1963). Effects of different brain lesions on card sorting. *Archives of Neurology, 9*, 90–100.
Moses, J. A., Golden, C. J., Berger, P. A., & Wisniewski, A. M. (1981). Neuropsychological deficits in early, middle, and late stage Huntington's disease as measured by the Luria-Nebraska Neuropsychological Battery. *International Journal of Neuroscience, 14*, 95–100.
Ojemann, G. A., & Fedio, P. (1968). Effect of stimulation of the human thalamus and parietal and temporal white matter on short-term memory. *Journal of Neurosurgery, 29*, 51–59.
Ojemann, G. A., Blick, K. I., & Ward, Jr., A. A. (1971). Improvement and disturbance of short-term verbal memory with human ventrolateral thalamic stimulation. *Brain, 94*, 225–240.
Osterreith, P. A. (1944). Le test de copie d'une figure complexe. *Archives de Psychologie, 30*, 206–356.
Purisch, A. D., Golden, C. J., & Hammeke, T. A. (1978). Discrimination of schizophrenic and brain-injured patients by a standardized version of Luria's neuropsychological tests. *Journal of Consulting and Clinical Psychology, 46*, 1266–1273.
Rao, S. M., Hammeke, T. A., McQuillen, M. P., Khatri, B. O., & Lloyd, D. (in press). Memory disturbance in chronic, progressive multiple sclerosis. *Archives of Neurology*.
Rey, A. L. (1964). *L'examen clinique en psychologie*. Paris: Presses Universitaires de France.
Russell, E. W. (1975). A multiple scoring method for the assessment of complex memory functions. *Journal of Consulting and Clinical Psychology, 43*(6), 800–809.
Russell, E. W., Neuringer, C., & Goldstein, G. (1970). *Assessment of brain damage: A neuropsychological key approach*. New York: Wiley-Interscience.
Shelly, C., & Goldstein, G. (1983). Discrimination of chronic schizophrenia and brain damage with the Luria-Nebraska battery: A partially successful replication. *Clinical Neuropsychology, 5*, 82–85.
Speedie, L. J., & Heilman, K. (1983). Anterograde memory deficits for visuospatial material after infarction of the right thalamus. *Archives of Neurology, 40*, 183–186.
Squire, L. R., & Moore, R. Y. (1979). Dorsal thalamic lesion in a noted case of human memory dysfunction. *Annals of Neurology, 6*, 503–506.
Squire, L. R., & Slater, P. C. (1978). Antereograde and retrograde memory impairment in chronic amnesia. *Neuropsychologia, 16*, 313–322.
Taylor, E. M. (1959). *Psychological appraisal of children with cerebral deficits*. Cambridge: Harvard University Press.
Teuber, H.-L., Milner, B., & Vaughan, H. G. (1968). Persistent antereograde amnesia after stab wound of the basal brain. *Neuropsychologia, 6*, 267–282.
Wechsler, D. (1981). *Wechsler Adult Intelligence Scale-Revised*. New York: The Psychological Corporation.

11
Assessment of Aphasia

ANDREW KERTESZ

THE DEVELOPMENT OF APHASIA TESTING

The systematic assessment of aphasic disability is as old as aphasiology itself. Broca (1861) described a rather detailed testing procedure that must have been routinely employed by clinicians at that time in examining aphasics. In fact, much of the interview with Lalonde, his second patient, is similar to what some clinicians do even today. He began by asking his patient conversational questions, commented on the patient's speech output and comprehension, and went on to describe his gestures, tested his tongue movements, his writing, and arithmetic. Hughlings Jackson added sign making, writing, comprehension, repetition, reading, and tongue movements, as well as a description of spontaneous speech, as regular features of the aphasia examination. Pierre Marie (1906) emphasized the importance of deficits in comprehension. Moutier (1908), Pierre Marie's pupil, described a rather complete set of systematic tests for aphasics in his large monograph entitled "L'aphasie de Broca."

The first systematic aphasia examination in English was detailed in two large volumes by Henry Head (1926). He examined head-injured soldiers by tailoring the tests to their deficits and covering the possible modalities affected. Few anecdotal controls were used, the administration was not standardized, and he emphasized flexibility. However, many of his ideas in testing have been subsequently adopted.

ANDREW KERTESZ • Department of Clinical Neurological Sciences, St. Joseph's Hospital Research Institute, London, Ontario N6A 4V2 Canada.

Naming and recognition of common objects used the same six objects in assessing word recognition, nonverbal matching, naming, reading, and writing, much like in the Porch Index of Communicative Abilities (PICA) (see following). He also studied sentence formation by using picture description. Writing, copying, and reading of sentences were tested as well. His comprehension tests consisted of commands to put a number of coins in various bowls. This task can be considered a precursor of the Token Test (DeRenzi & Vignolo, 1962). The Hand-Eye-Ear Test is a similar sequential task with elements of right and left orientation and praxis. The patient was asked to carry out and imitate a series of movements consisting of touching an eye or an ear with the ipsilateral or contralateral hand in various serial combinations.

Head complemented these tests with understanding a paragraph from a newspaper, arithmetic tests, setting a clock, drawing objects from a model and memory, sketching a ground plan of a familiar room, finding the way along some familiar route, completing puzzles, and playing games, such as dominoes and cards.

A standardized test battery was constructed, from various existing intelligence tests, by Weisenberg and McBride (1935). They recognized that the time for the number of tests assembled was too long and constructed shorter batteries, one for use in severe disorders, with an estimated time of two to three hours for administration, and one for milder disorders. However, definitions of severity were not given and standardization, in the modern sense, was incomplete. Nonetheless, they did obtain normative data.

Their test is a precursor of many of the current tests of word recognition and sentence comprehension. Commands of increasing complexity were used in the test of following directions, which also served as a test for apraxia. For example, the patient was requested to put the back of his hand on top of his head (much like more recently used experiments of meaningless movements). They also applied Pierre Marie's "Three-Paper Test," the complex commands from the Stanford-Binet Comprehension test, and Abelson's geometrical figures, with commands such as "point inside the circle and the triangle but not in the square" (very much like the Token Test).

They also included metalinguistic tasks of oral opposites, oral analogies, sentence completion tests, and oral absurdities, as well as the same in print. An extensive array of nonlanguage tests from various performance scales, as well as drawing from a model and from memory were also used. Recognizing nonspeech sounds was specifically tested by having the patient reproduce noises, which were produced behind a screen, by using one of several objects placed in front of him (e.g., dropping a coin, snapping a rubber band, running a finger along a comb).

PRINCIPLES OF APHASIA TESTING

In the 1960s aphasia testing became well established because of a need for diagnostic assessment and planning for therapy and to compare patients for research. Arthur Benton (1967) reviewed the existing aphasia tests and expressed the need for further development. He felt that the existing tests had not been generally adopted, they had not been published in a usable form, and they did not present exact criteria for scoring or offer detailed guidelines for interpretation of performance. He did not think that any single test showed convincing evidence of a clinical utility that was superior to any other. However, since then several comprehensive aphasia examinations that fulfill some of the requirements have been published.

Aphasia testing is difficult to standardize because of the considerable difference between the goals of clinicians and researchers. As our knowledge of the aphasic phenomena increases, testing must change also. There is considerable pressure to incorporate new findings and the tests that are modern today may soon become obsolete. Researchers devise special tests to probe a certain phenomena and, if these become important theoretically or clinically, they will be incorporated into general tests. Nevertheless, a few standardized yet comprehensive aphasia batteries remain in use because of the great clinical need and the need for stability in comparisons of research data between investigators. Very few of the aphasia tests will fulfill the need for both clinical and research goals.

From this author's clinical experience with aphasics and standards derived from intelligence testing and construction of other psychological tests, my summary of what an ideal aphasia test should do runs as follows:

1. Explore all potentially disturbed language modalities.
2. The subtests should discriminate between clinically relevant aphasia types.
3. The test items should include a range of difficulty in order to examine a representative range of severity of deficit (construct validity).
4. There should be enough items to eliminate most of the day-to-day and test-to-test variability (internal consistency).
5. The administration and scoring should be standardized and replicable (intertest and interrater reliability).
6. The effect of intelligence, education, and memory, should be minimized to achieve as purely as possible, a test of language (content validity).

7. It should discriminate between normals, aphasics, and non-aphasic brain-damaged individuals (criterion validity).
8. The length should be practical to accomplish administering of all of the subtests in one sitting.
9. The test items should measure the same factor (internal consistency).
10. The test should measure what is generally acknowledged in the field as a language deficit or aphasia (content validity and intertest reliability).

Most, but not all, aphasia tests rate speech output or expressive function, comprehension, naming, repetition, reading, and writing. Some add arithmetic, praxis, and drawing as regular complementary subtests, whereas others have a more extensive supplementary section that, at times, tends to move the language test towards a more general test of neuropsychological performance. A great number of special items can be added in endless combinations and all the major sections can be further subdivided utilizing the various input and output modalities. It is desirable to sample the input modality used in receiving the instructions, the input modality of the actual stimulus, and the output modality of the response. Sometimes sampling these modalities in a mechanical and rigid fashion obscures the goals of testing. Rigidity in administration and early extensive sampling of modalities can decrease the clinical utility of a test. It is very difficult to achieve a compromise between sampling issues and testing time that will satisfy most people. Tests of excessive length will rarely be used in their entirety, and, although they aim at comprehensiveness, this goal will be infrequently achieved.

Contents of a Comprehensive Examination

What one should measure in an aphasia test is one of the least agreed on and one of the most crucial issues in this field. In general, the following details the major areas of concern. Spontaneous speech, expository speech, speech output, and fluency can be measured by rating various aspects of speech that are obtained by a standard interview, a picture description, or by asking the patient to describe a familiar activity, such as shaving or making spaghetti (as in an Italian test). Fluency, rate of speech, phrase length, grammaticality, articulation, melodic line, and paraphasias are the most often rated factors. Even though the rating of speech production is more difficult than other speech parameters, Goodglass, Quadfasel, and Timberlake (1964) emphasized its importance, in addition to the usual input–output dichotomy, in differentiating aphasic patients. The importance of the fluency–nonfluency dimension has been generally accepted, although there are various methods of measuring it (Benson, 1967; Goodglass & Kaplan, 1972; Howes & Geschwind, 1964; Kerschensteiner, Poeck, & Brunner, 1972; Wagenaar, Snow, & Prins, 1975). Some scales are very extensive but may be too lengthy for practical use, such as the one by Wagenaar, Snow, and Prins (1975).

Another problem with multiple ratings scales is that they use abstract and fractionated categories, such as word finding and number of utterances per minute, that may oppose one another in the same patient. Such ambiguities reflect the complexities of speech production but create difficulties in classification. Other tests use a rating scale that incorporates various aspects of speech production into a list of definitions to be matched to certain clinical behaviors. Speech production rating is essential and tests that do not have it, or have not placed emphasis on it, cannot easily be used for classification and this lessens their clinical applicability.

The information content of spontaneous speech is often incorporated into other rating scales of speech production. At other times it is measured as a scale of severity estimation or specifically defined in a body of elicited (spontaneous) speech. Because it is difficult, if not impossible, to obtain entirely spontaneous speech, elicited speech must be standardized so information content can be compared. Picture description constrains thematic output and vocabulary, thereby providing a standard stimulus for comparability of linguistic features.

Comprehension testing is the cornerstone of aphasia batteries. It is difficult to separate it entirely from speech production or apraxia, as patients need some way of signaling the answer, such as pointing to a multiple choice of items or cards or even shaking their heads. Some comprehension tests that rely only on visual material may occasionally have false-positive results if visual recognition is impaired. Even though auditory comprehension may be good, the patient may not visually recognize differences between the target item and the foils. Hemianopic patients may fail on a horizontal array of stimuli because they cannot see half of the items. A similar source of errors occurs in right-hemisphere-damaged patients with visual-spatial neglect. This is a common reason why some of these patients will be misclassified as "aphasic" by a rigid application of a test. Verbal comprehension overlaps with verbal intelligence and some tests using complex ideation are, in fact, using a problem-solving test that goes beyond verbal comprehension. Some of the longer items also test auditory short-term memory. Therefore, it is wise to use short items, under 8 to 10 lexical units, in order to measure comprehension free from the influence of memory factors.

The difficulty level of comprehension items has to be wide enough to sample an entire aphasic population. Most aphasia tests fail to examine low-level comprehension and, therefore, cannot distinguish between severe global patients, who have a poor prognosis, and severe Broca's aphasics, who have recovered useful comprehension. For example, the Token Test is difficult for severely affected aphasics who will score poorly even though they may have a significant amount of residual comprehension. Similarly, the complex manipulation of syntax in some auditory tests of comprehension is too difficult, except for moderately or mildly affected aphasics. Along with these constraints, the on-and-off nature of auditory verbal comprehension and the variability of a patient's responsiveness in some types of aphasia demand extensive testing or retesting.

Because yes-no questions require only nodding as one-word replies, they are powerful in discriminating between those who are globally affected and those with residual comprehension and a better prognosis. These questions, however, must be short and personally relevant to the patient to be the most effective. Results may also be influenced by chance factors, however. Moderately or mildly affected patients require items that vary syntactic complexity, vocabulary, and sequences of items in one sentence (auditory comprehension span). Most notably, the comprehension of grammar has especially received wide attention from linguistic studies and the number of specific tests that explore one or more aspects of linguistic features are too numerous to be reviewed here. Many are too difficult for a general aphasic population. Others use only a few items with a pointing task and cannot adequately fulfill these testing requirements.

Phonemic discrimination is less crucial for comprehension than semantic and syntactic aspects (Blumstein, Baker, & Goodglass, 1977). It appears that even Wernicke's aphasics with severe comprehension difficulty are able to distinguish phonemes. Yet phonemic discrimination is extensively tested in some batteries. Schuell and Jenkins (1961) emphasized the role of word frequency in comprehension, and Goodglass, Klein, Carey, and Jones (1966) noted the differences between semantic categories, although this has been subsequently questioned by Poeck, Hartje, Kerschensteiner, and Orgass (1973).

Syntactic variations, especially relational and locational prepositions, are particularly difficult even for Broca's aphasics, as pointed out by Ombredane (1951), Luria (1966) and Goodglass, Gleason, and Hyde (1970). Most modern aphasia tests contain items of comprehension that assess these features. In order to demonstrate mild comprehension difficulty shown even by Broca's aphasics, syntactically complex sentences and sequential material are required. They are related to the severity of the disturbance and are particularly useful in detecting aphasia. However, these tasks may not differentiate between various aphasic types and may even decrease the specificity of differential diagnosis if other features are not considered. Also, they are often abnormal in right-hemisphere-damaged, or even in normal patients.

Changing the conditions of presentation and context, such as speaking slower and providing redundant but semantically related information, can facilitate comprehension (Gardner, Albert, & Weintraub, 1975). Low-level, residual comprehension can be in evidence in severely affected aphasics when gestures of refusal or quizzical expressions are taken into consideration as appropriate responses, rather than restricting them to correct answers (Boller & Green, 1972). Some of these "appropriate" responses are subject to the examiner's interpretation and need to be further standardized.

Repetition is often neglected in aphasia tests because it is not considered a natural language function. However, it makes a significant theoretical and practical contribution to the analysis of aphasic disorders and is therefore desirable. Repetition is important in making the differentiation between conduction and transcortical aphasia. Without an established repetition subtest these entities cannot be diagnosed. Repetition testing often includes items that test oral agility or articulatory phonemic competence. It is usual to include words, numbers, and sentences of increasing complexity and articulatory difficulty. Goodglass and Kaplan (1972) added high and low probability sentences to their repetition tasks, noting that aphasics did poorly in the latter group. Some tests include the repetition of nonsense words or grammatically incorrect sentences to observe phonological or syntactic abilities. Tongue twisters, individual phonemes, and sentences consisting of only closed class or grammatical words have also been used.

Naming is universally tested at various levels of complexity. A single bedside visual presentation of objects is termed "confrontation naming" (the patient is confronted with the object). The difficulty of test items can be controlled for stimulus frequency, imageability, level of abstraction, manipulability, and word length. Some of these may have more influence than others, but extensive studies on the semantic structure of language and lexicon have yielded relatively little clinical correlation as yet. Various semantic categories of objects are generally lost evenly in aphasia, although there are occasional reports to the contrary. These are termed modality specific anomias (Dennis, 1976; Goodglass, Klein, Carey, and Jones, 1966). Using different modalities for presentation, the examiner may uncover a dissociated naming ability between tactile and visual stimulation, leading to a diagnosis of visual agnosia. Some tests incorporate a tactile naming task, whereas others use it as a control task if visual naming falls. Color anomia, often associated with pure alexia without agraphia, may be better classified with visual agnosias (Geschwind & Fusillo, 1966). Response modality may also be differentially affected. At times, written naming may be superior to oral naming (Bub & Kertesz, 1982; Hier & Mohr, 1977), but often this is not tested regularly.

There are various methods for the administration of the naming tasks. Some tests allow the patient to look at the stimulus without time restrictions. Others specify the exposure duration and measure response time. These time restrictions create severe difficulties for aphasics. For the amount of information gained at upper level of functioning, such limitations will lower the score of the more severely affected patients and may even mask residual ability. Nevertheless, a degree of practical limitation should be set on the time each particular item takes. The supplying of cues is not allowed by some tests and encouraged by others. If residual functioning is important, cueing may help bring it out, but very few tests standardize the procedure or incorporate it in the scoring. Phonemic cueing usually provides the initial sound of a word, whereas semantic cueing uses the first half of composite words or associates to elicit the target word. Naming is fairly consistent for each patient at each given time, and the number of items need not be as large as is necessary for the comprehension task.

Word finding in spontaneous speech is different from confrontation naming, but it is related to the concept of semantic and lexical access. It can be further elaborated by testing sentence completion, responsive speech (the context facilitates word retrieval), and word fluency, a task that is used in many intelligence tests and "frontal-lobe" batteries. In this task the patient is asked to produce as many words as possible belonging to a semantic category, such as animals, or words beginning with a certain letter, within one minute. Although this is affected in almost every instance of aphasia and persists even after other functions have recovered, it is not specific for aphasia and appears to be indicative of brain damage in general. It may be the earliest aphasia sign in Alzheimer's disease (Appell, Kertesz, & Fisman, 1982). Standardized versions of word fluency tasks were produced by Spreen and Benton (1977) and Wertz (1979). Automatized, overlearned serials, such as the days of the week or months or nursery rhymes, represent word retrieval at a different level. At times, these tasks are easier for some nonfluent aphasics, although their diagnostic value is doubtful.

Reading and writing has been a traditional part of aphasia examinations, but lately they have received special attention from cognitive psychologists with the subsequent infusion of different terminology from information processing models. The previous divisions of pure alexia, alexia without agraphia, verbal, literal, and sentence alexia are supplanted by deep and surface dyslexia, phonological alexia, and word form alexia (letter-by-letter reading). Most aphasia tests include the reading of words and sentences. Some also include the identification of letters, numbers, the matching of words to a choice of pictures and the reverse, and the matching of auditory input to a choice of written words, which usually include visual, semantic, and phonological foils. In addition to this exploration of modalities in reading, the analysis of the lexical or whole-word reading versus the phonological or grapheme–phoneme conversion route may be examined. This difference is tested by using nonwords (read only by grapheme–phoneme conversion), matching visually different homophones, or contrasting orthographically regular and irregular words. This area is in continuous development, but extensive supplementary alexia and agraphia tests are being used in many laboratories, including our own.

Reading and writing are greatly influenced by educational and occupational factors and are, therefore, more difficult to standardize for a population. There are many dissociations from aphasia and the two can be present independently. Although some consider the disturbances of written language just another modality of aphasia, there is plenty of evidence for independent, although related, mechanisms. Reading can also be dissociated from writing in various combinations of linguistic dimensions, as recent single-case studies have indicated. In pure alexia, there are often associated signs of other visual involvement, such as a color naming defect or visual agnosia. There are also cases of transcortical alexia (reading without reading comprehension) usually in association with transcortical sensory aphasia or recovering Wernicke's aphasia. Therefore, it is important to test reading comprehension as well as reading aloud. Alexia and agraphia can be residual disturbances even after the aphasia resolves and this is also a major source of dissociation. Reading and writing may be impaired from right-hemisphere lesions in which the left side of a word or sentence is neglected or when perseverative loops appear in writing. Motoric aspects of writing represent a substantial portion of the failure to write, even more than the motor impairment in speech. Similarly, visual confusions in reading at a perceptual level can be impaired at earlier stages of processing.

Finally, there is the issue as to whether reading and writing deficits should be considered to be part of a total aphasic impairment in scoring and recovery studies. There is, of course, the argument that illiterates develop aphasic syndromes that are definable without reading or writing. Some aphasia tests incorporate the findings of written tests into their system of evaluation and diagnosis, whereas others keep it separate. There are arguments for both, depending on the intent and the theoretical model for testing. Recently there has been a trend towards separate consideration, although one should not lose sight of the possible interactions between oral and written language that are particularly evident during testing.

Supplementary tests include the testing of functions that are associated with language mechanisms or may be impaired with them. There is less agreement in the use of these tests in aphasia examination. These may include tests of calculation, gesture, pantomime, skilled movements, practised and symbolic movements, and praxis, which is usually tested by intransitive limb gestures, transitive movements that involve objects, bucco-facial movements, and complex sequential actions independent from language. Visual-spatial matching and constructional tasks, such as drawing and block design, may also be used. The inclusion of a wide variety of the tests referred to as "supplemental" could, in some cases, produce a battery resembling a more comprehensive neuropsychological battery. However, such an approach has many drawbacks (e.g., excessive length, norms that are not comparable between tests, etc.). Therefore, supplementary tests should be chosen carefully and for specific reasons.

COMPREHENSIVE APHASIA EXAMINATIONS

Eisenson's Examination for Aphasia

Eisenson's Examination for Aphasia (1954) was constructed as a practical, clinical instrument with informal instructions to be administered in 30 to 90 minutes. A simple scoring system estimated the various levels of ability for each language function. Receptive disturbances were assessed by items that included recognition of common objects, colors, forms, pictures, numbers, letters, and printed words. Patients were required to name these items as well. Sentence comprehension was tested by a series of questions with a choice of four responses. Oral and reading comprehension of paragraphs were also done. Expressive disturbances were tested with automatic speech, writing, spelling, naming, word finding, oral reading, and carrying out actions. Although the test has not been standardized and the scoring system is on an informal 5-point scale, it had been widely used by clinicians as a guide for treatment goals. More recent comprehensive and standardized tests have considerably decreased its use.

The test is published by the Psychological Corporation, and consists of a manual and 25 record forms.

The Language Modalities Test for Aphasia (LMTA)

The LMTA by Wepman and Jones (1961) uses a film strip to present visual stimuli, as well as auditory stimuli presented by the examiner. Responses are both oral and graphic for both kinds of stimuli. A matching response is also used. There is a screening section of 11 items. If necessary, this is followed by another 46 items. The stimuli include objects, words, numbers, and sentences of increasing length. There is a test of comprehension of auditory and written material (reading). The test also employs imitation or repetition, form recognition, arithmetic, spelling, and articulation. A corpus of speech is elicited by asking a story to be made up from four pictures. This obtains enough spontaneous speech to allow an exploration of syntax and vocabulary. The scale is scored as follows: (a) correct response, (b) phonemic or graphic errors, (c) syntactic errors, (d) semantic errors, (e) jargon or unintelligible response, (f) no response. There are scoring examples included in the manual.

The test was standardized using 168 aphasics. Forms I and II show high correlations in the number of correct responses (0.80) and, similarly, equal proportions of correct responses between the two forms in all stimulus and response types. Factor analysis combines the various modalities, such as naming and reading, into single factors. The authors classify aphasics on the basis of oral responses as follows: (a) syntactic aphasia (errors are made with grammatical words, such as prepositions, verb endings and case markers), (b) semantic aphasia (word finding or semantic difficulties), (c) pragmatic aphasia (patients with poor comprehension and inappropriate substitution of words with meaningless sentences), (d) jargon aphasia (unintelligible neologistic speech), (e) global aphasia (very little speech output, except for automatic phrases or meaningless combinations of sounds).

The LMTA is published by Education Industry Service (1961). It consists of an instruction manual, administration manual, film strip, subject response booklet, examiner's record book, and a scoring summary. The administration of the test is estimated at between two to five hours and is therefore given in several installments. Although this test has presaged the current interest in psycholinguistics and grammatical competence, it has not gained popularity. This is probably due to the cumbersome presentation on film strips and the lack of adequate standardization. Currently, it is not widely used.

The Minnesota Test for Differential Diagnosis of Aphasia (MTDDA)

The MTDDA by Schuell (1965) consists of 69 tests with 606 items. This well known aphasia examination is a major, standardized, detailed test battery, but because of its size it is rarely, if ever, used in its complete form. It has the following sections (with the number of subtests in parenthesis): auditory disturbances (10), visual and reading disturbances (11), speech and language disturbances (20), visual-motor and writing disturbances (14), numerical relations and arithmetic processes (8), and tests for body image (6). Scores are based on the number of errors. One hundred and fifty-seven aphasics and 50 normal patients were used for a factor analysis, which was interpreted to show five major factors: Language behavior; Visual Discrimination, Recognition, and Recall; Visual-spatial Behavior; Gross Movements of the Speech Musculature; and Recognition of Stimulus Equivalents.

Factor analysis detected a main language factor that was taken as proof for the unitary nature of language. However, five aphasic syndromes were distinguished: simple aphasia, aphasia with visual involvement, aphasia with sensory motor deficits, aphasia with scattered findings, and irreversible aphasia (Schuell, 1974). Subsequently, two minor syndromes were added (mild aphasia with persistent dysfluency or dysarthria and aphasia with intermittent auditory imperception). Despite its name, the MTDDA does not define the criteria for the differential diagnosis of these syndromes nor does it make an attempt to distinguish aphasia from other disorders that may have some form of language disturbances.

In its administration, an assumption is made as to the patient's ability. Testing is then started at an estimated difficulty level, yet the test is long, requiring two to five hours in more than one session. There is also a clinical rating scale of 0–6 and a diagnostic scale of 0–4 that summarizes performance in various language functions. It has been pointed out that the population used for standardization was rather chronic and that the pooling of a large number of patients tested in various states after brain damage decreased the MTDDA's ability for interpretation of test scores for prognosis (Brookshire, 1979).

Several short versions have been proposed. Thompson and Enderby (1979) found that the majority of the items were too easy for most aphasics and lacked discriminating power. They obtained a significant correlation between a reduced number of subtests and the standard version.

Unfortunately, the classification provided by Schuell is difficult to apply and does not fulfill the needs of most clinicians. Although the test remains in fairly wide use, there has been little recent research work published with it and, also, its use for prognosis is limited. As a consequence of the frequent revision of this battery, the standardization sample for any single form is relatively small.

The revised edition is published by the University of Minnesota (1972). It includes a manual, two packages of stimulus cards, and a package of 25 test booklets.

NEUROSENSORY CENTER COMPREHENSIVE EXAMINATION FOR APHASIA (NCCEA)

The NCCEA by Spreen and Benton (1968) consists of 20 language tests and 4 control tests of visual and tactile functions. The subtests are: 1. Visual naming of common objects, 2. Description of use of the same objects, 3. & 4. Tactile naming, with right and left hand, 5. Sentence repetition of tape recorded sentences, 6. Digit repetition, 7. Digit reversal, 8. Word fluency, using three one-minute trials for all words named, beginning with a specific letter, 9. Sentence construction from five sets of three words, 10. Object identification by name (auditory recognition task), where the patient points to objects named by the examiner, 11. Identification by sentence, using a shortened version (36 items only) of the Token Test (Spellacy and Spreen, 1969), 12. Oral reading of names of objects presented before, 13. Oral reading of the twelve command sentences in test 11, 14. Silent reading of names, which involves matching the printed name of an object to the object selected from an array, 15. Reading sentences for meaning—the patient is instructed to execute the twelve written commands used in test 11, 16. Visual-graphic naming requests the patient to write the names of 10 objects visually presented, 17. Writing names, which scores test 16 for correctness of spelling. If the naming portion is not performed, then the patient is dictated a name and asked to write it, 18. Writing of sentences on dictation, 19. Copying sentences, 20. Articulation, (which is also a test of repetition) of thirty meaningful and eight nonsense words, presented from a tape recording.

Test interpretation requires considerable experience. An age and education correction is available for the scores. The scores may then be plotted on one of three profile sheets for percentile ranks in either an adult aphasic or nonaphasic brain-damaged, or normal population. There is considerable overlap between the nonaphasic brain-damaged and aphasic population because some of the test items are relatively difficult for nonaphasics as well. The test is to be used in its entirety. Several subtests provide a second set of items if errors occur in the first set. This second set of items provides more quantitative information.

The range of item difficulty is limited in some of the subtests. The test was correlated with Sarno's (1969) Functional Communication Profile (FCP) by Kenin and Swisher (1972) and the Western Aphasia Battery (WAB) (Kertesz, 1979), which measure similar changes in language function. Kenin and Swisher (1972) commented that the auditory comprehension scores were adequately sampling behavior. Crockett (1977) found high correlations between some subtests of the NCCEA and a 17-item Verbal Rating Scale (a measure of speech output and also some comprehension).

The administration can be somewhat difficult because of the awkwardness of 4 trays of objects and the use of a tape recorder to present auditory tasks. There are some nonlanguage items included, such as reversal of digits. The NCCEA does not provide for the sampling and scoring of spontaneous speech. Repetition sampling is also limited. On the average, the test takes about an hour and a half to administer.

The NCCEA is published by the University of Victoria, Neuropsychology Laboratory (1969), and consists of a manual, a package of answer booklets, packages of profiles for the various populations—(A, B, and C), a tape containing auditory materials, a set of plastic tokens, reading cards, and photographs of four trays with objects.

The Porch Index of Communicative Abilities (PICA)

The PICA by Porch (1971) uses 10 common objects as stimuli in the 18 subtests. These are subdivided into response modalities, 4 verbal, 8 gestural, and 6 graphic: 1. description of object use (elicited speech), 2. demonstration (gestural) of object use, 3. actual object use, 4. naming the object, 5. matching printed sentences that describe the use of the object (reading and comprehension), 6. choosing one of the sentences in 5. on auditory stimulation, 7. a task of reading aloud that is similar to 5., 8. match a picture to an object, 9. sentence completion, 10. auditory word recognition, 11. object to object matching, 12. repetition after a tape-recorded voice. Many of the subtests, although they are considered gestural because of the response modality, are really more traditional tests of auditory comprehension or reading. The only true gestural tests are 2., which is a test of praxis on command, and 3., which is a test of praxis with object use. There is no test of praxis on imitation. There are six other tests labelled "graphic," which include: 13. written description of the function of each object, 14. writing the name of the object, 15. writing on dictation, 16. writing on auditory spelling, 17. copying written words, and 18. copying geometric forms.

The scoring system is complex and tries to account for all possible varieties of responses. It is described in great detail for each subtest in

16 categories. The scoring level is differentiated along the dimensions of accuracy, responsiveness, completeness, promptness, and efficiency. For instance, a score of 16 is a complex response that is accurate, responsive, immediate, elaborative, etc. The scoring system requires extensive training to master, and a 40-hour workshop is recommended by the author. Porch (1967) reported the validity of the ranking of the 16 categories by showing a high agreement between three speech pathologists. There was good interrater and test–retest reliability. Percentile scores are used to provide recovery curves by using an overall test percentile and the mean percentile of the nine highest and the nine lowest scored subtests. A recovery ceiling is assumed when the three percentiles coincide (Porch, Collens, Wertz, & Friden, 1980). Response levels for modalities and subtests can be plotted and a ranked response summary is achieved by arranging the subtests in order of difficulty. When the patient's overall response level exceeds 10, adequate communication is usually achieved. Modalities that fall in the lower percentiles are expected to improve to a greater degree.

The PICA does not formally distinguish between aphasia types other than (a) severe, verbal formulation difficulty, (b) severe dysarthria, (c) inadequate verbal monitoring. In addition, bilateral damage has a special pattern. The test does not include the assessment of conversational speech or picture description.

The standardization of the PICA in 280 left hemisphere damaged patients is included in the manual. The mean overall PICA score was 10.02, close to the 50th percentile score of 10.64. There were some questions raised about the ordinality of the complex PICA scoring system (McNeil, Prescott, & Chang, 1975). For example, repetitions and self-corrections were judged to be more functional than incomplete or incomplete delayed responses. The mean scores of the PICA do not represent any specific category of behavior, but act only as a severity measure. Nevertheless, the scoring system has face validity in describing aphasic behavior. Holland (1977) correlated the PICA with the Functional Communication Profile (FCP) (0.86), the Boston Diagnostic Aphasia Examination (BDAE) (0.88) and the Communicative Abilities in Daily Living (CADL) (0.93) suggesting a high concurrent validity.

The PICA has also been criticized for not assessing spontaneous speech and auditory comprehension in sufficient detail (Boone, 1972). Another disadvantage is the extensive amount of training required to achieve reliability in using the multi-dimensional scoring system. Although the PICA samples similar modalities to other aphasia tests, the test does not distinguish between various cognitive processes or linguistic and behavioral distinctions. The effectiveness of communication is not fully tested and there is no opportunity for any supportive feedback. The scoring system was also criticized because it does not indicate the actual response, but only one aspect of it and only in the output modality (Martin, 1977).

The PICA requires approximately an hour to administer and another 30 minutes to score. The test administration, including even the seating of the patient and the speed and rhythm of the testing, is rigidly prescribed and special training is required to learn its administration, scoring, and interpretation.

The PICA is published by Consulting Psychologists Press Inc. 1967 (Vol. 1). Volume 2 of the manual, published in 1971, contains further data on standardization and comparison to other tests. The test comes in a kit with two sets of 10 test objects, a test format booklet, the two volumes of the manuals, sets of cards for Reading subtests, Visual Matching subtests, and 50 packages of score sheets (rank response summaries, modality response summaries, aphasia recovery profile sheets and predictive data summary sheets).

APPRAISAL OF LANGUAGE DISTURBANCE (ALD)

The ALD by Emerick (1971) is a systematic stimulus and response modality oriented test with each combination of stimulus and response including heterogenous language functions. Oral to Oral subtests assess automatic speech, repetition, word finding, sentence completion, finding antonyms, and definitions. Oral to Visual group include subtests for pointing to objects, pictures, and words, reading, and auditory comprehension. The Oral to Gesture subtests include tests of praxis, such as coughing, whistling, humming, shaking the head and demonstrating actions, as well as pointing to body parts.

The Oral to Graphic group of subtests include writing to dictation, as well as on request. It incorporates most of the items from the Oral to Oral subtest in writing. The Gestural to Visual subtest is a pantomime comprehension test. Subjects respond from a multiple choice of objects, pictures, or words. The Visual to Gesture subtest is a test of object use. The Visual to Oral subtest includes both reading and naming tasks. The Visual to Graphic subtest includes copying, naming of objects in writing, and a written picture description. The second section is designed to explore central processing (e.g., matching silhouettes to line drawings and pictures, matching pictures to written words). Sorting and arranging colors and shapes, object assembly, a special test of demanding or asking, and the Peabody Picture Vocabulary Test are also included. The third section is designed to examine peripheral mechanisms of communication, such as tongue, lip, and jaw movements and phonation, as well as tactile recognition, and arithmetic. In respect to modality orientation, the test is similar to the LMTA of Wepman.

The test–retest reliability was examined on 76 patients with a correlation of 0.74, but recovering patients were also included in this sample. When only a chronic population was examined, the correlation was higher (0.8). Interrater reliability between two trained clinicians was 0.86.

Although the ALD is rather comprehensive, some of the subtests, such as object assembly and the various category sorting tasks, appear to be supplementary, belonging to standard intelligence tests. There are no instructions for classification and the test manual is brief. The test takes from one to two hours. There is no data concerning its value in prognosis or in the description of aphasic syndromes. There has been no reported use of the test for research.

The ALD is published by Northern Michigan University Press (1971). It consists of a manual with test booklets, stimulus cards, color shapes, and a hand puzzle.

THE BOSTON DIAGNOSTIC APHASIA EXAMINATION (BDAE)

The BDAE by Goodglass and Kaplan (1972; Revised Edition 1983) is a comprehensive aphasia test designed for (a) the diagnosis of presence and type of aphasic syndromes leading to differences concerning cerebral localization; (b) measurements of levels of performance over a wide range for both initial evaluation and detection of change over time; (c) comprehensive assessment of the assets and liabilities of the patients in all areas as a guide to therapy.

The first section of the test examines expository speech elicited by a picture description and interview with six rating scales of 7 points. These scales are melodic line, phrase length, articulatory agility, grammatical form, paraphasia in running speech and word finding. The severity is also estimated on a subjective rating scale that assesses overall communication from zero (no communication) to five (no deficit).

Auditory comprehension is examined at multiple levels. One such level is word discrimination, a task of involving the recognition of words spoken by the examiner in which the patient points to objects, geometric figures, letters, actions, numbers, and colors. Also, there is a similar task involving pointing to body parts and fingers. There is a test of commands of increasing complexity. The section on complex ideational material uses a yes-no task, and for a correct score requires both questions to be answered, following material of paragraph length.

Oral expression includes oral agility, such as alternating movements of the tongue and lips and the rapid repetition of words. Automatized sequences are days, months, numbers, the alphabet, recitation of nursery rhymes, singing, and tapping rhythms. Repetition of words includes letters, numbers, and a tongue twister. Repetition of phrases and sentences incorporates both high probability and low probability items. Word reading is scored with a bonus for speed of response. Responsive naming incorporates an estimation of the delay of response. Visual confrontation naming is tested in the same category as the visual discrimination task in the comprehension section. Body part naming is also included. Fluency in controlled association is similar to the word fluency task of previous intelligence tests, and asks for the patient to name as many animals as possible in one minute. Oral sentence reading completes this section with an all or nothing scoring system.

Reading comprehension is tested with either a phonetic association task or the recognition of a printed word on a multiple choice between five auditory stimulation items, including connotative foils. Symbol and word discrimination is a multiple choice matching of letters and short words across different styles of writing. This is also a visual recognition task. Word picture matching uses assorted objects and colors. Reading sentences and paragraphs requires the patient to complete a sentence from the end of a paragraph by pointing to one of a multiple choice of four words or phrases.

The writing tasks begin with the patient's name and address and the copying of a sentence. The mechanics of writing is scored differently on the new edition. Serial writing of the alphabet and numbers and writing from dictation of letters, numbers, and high frequency short words are employed to measure recall of written symbols. Written-word finding consists of writing on dictation and has the option of the patient using either oral spelling or anagram letters. Written-confrontation naming and written formulation follow, which include picture description in writing and sentences written to dictation.

Supplementary language tests are included in the manual and incorporate the psycholinguistic exploration of auditory comprehension and expression. There are tests of preposition of location using both pointing tasks and before and after yes-no tasks, which are similar to the Token Test. There is also a passive subject/object discrimination task requiring one word answers, but it may be adapted to a yes-no form. Comprehension of possessive relationships is similarly tested. Expressive agrammatism is tested by repetition and sentence manipulation tasks. The repetition task explores the contrast between indicative and interrogative, and uses conditional construction. Manipulation of verb tense is presented in a sentence completion paradigm using adjustment of the verb tense and requires a considerable amount of comprehension. There is a task requiring the patient to ask questions beginning with do, can, or may, or WH questions opening with interrogative pronouns, such as who, where, how, what, and why. There is a repetition task for conduction aphasia including tongue twisters, grammatical words, and the repetition of numbers and number–word combinations. There are also tests for callosal disconnection syndromes, such as the dissociation of modalities in naming, tested by naming by touch in either hand with vision excluded, and writing with the minor hand.

Supplementary nonlanguage tasks include drawing, stick memory, finger comprehension, finger naming, visual finger matching, matching two-finger positions, right and left orientation, arithmetic, clock setting, and three-dimensional block designs. There is also a supplementary apraxia test that includes five items of bucco-facial, intransitive limb, transitive limb, and whole body movements that are tested on oral commands, and, if failed, by movements on imitation and, if that fails, movements with a real object, if possible. There are also three items of serial actions with real objects.

The scoring system is complex and includes some rating scales. There are subjective but detailed instructions provided in the manual. Experienced raters who developed the test had high correlations of 0.85 to 0.9 on four of the items, and 0.78 on word finding and 0.79 on paraphasia rating. The rating scale for grammatical form is difficult because only three levels are defined on the 7-point scale. The paraphasia rating in running speech is also quite subjective, with the middle range representing an only very slight deficit of "once per minute of conversation." The scoring of repetition does not allow for the occurrence of paraphasias, which is somewhat unrealistic. Sentence reading is also scored all or nothing. The comprehension of complex ideational material includes paragraph-length material and, therefore, is probably influenced by intellectual and memory factors. The auditory comrehension scale is converted from four auditory comprehension subtests and plotted on the Z score summary profile. The low interrater correlation, even among experienced raters on the word finding scale, is related to the difficulty in judging the categories of information proportional to fluency. This rating scale also differs from the others in that the normal score is in the middle. The rating of severity is accomplished by various definitions. The scoring of the other subtests is accomplished by converting to Z scores, which were standardized with a sample of 207 aphasics.

The revised score summary sheet was based on their new normative sample of 242 aphasics tested at the Boston V.A. Hospital between 1976 and 1982. The formulas for assigning patients into categories are contained in the manual. The type or category of aphasia is no longer determined by the patient's Z score profile but rather by their percentile ranking on the subtests. They designed a new subtest summary profile in the test booklet.

The length of the test, of course, is the price paid for its comprehensiveness. The administration, including all of the supplementary tests, may run upwards of six to eight hours, but it is rarely used in its entirety. Typical profiles of patients are provided in the manual, as well as ranges of speech profile ratings based on the clinical experience of the authors of the test. This requires the use of only the six rating scales and the auditory comprehension task. The profiles for conduction and transcortical aphasics are questionable, because they fail to include repetition. The manual indicates the difficulty in diagnosis and the complex interaction of severity with the typicality of the profiles. In fact, this leaves an estimated 60% of patients in the mixed or undiagnosed category. The test has also been standardized with an intercorrelation matrix on 111 patients and the Kuder-Richardson method of subtest reliability indicated good internal consistency within the subtests with respect to what the items were measuring. There is no test–retest reliability or interrater reliability information available except for the rating scales mentioned above. Normative data were published by Borod, Goodglass, and Kaplan (1980).[1] A comparison with the Western Aphasia Battery (WAB) classification system was done by Wertz (1983). The BDAE is published by Lea and Febiger (1972; revised version 1983) and includes the manual, stimulus cards, and examination booklet. The test has several translations.

The Western Aphasia Battery (WAB)

The WAB by Kertesz and Poole (1974) and the revised version by Kertesz (1982) follows the psycholinguistic principles of the BDAE and attempts to provide a comprehensive examination for both clinical and research use combined with practical length. Verbal sections of the test are (a) Fluency, grammaticality, and paraphasia rating of spontaneous speech, (b) Information content of spontaneous speech. Both (a) and (b) are elicited through standard interview questions and a picture description. The fluency, grammaticality, and paraphasia rating is a combined score of several dimensions with each point on a scale of 10 defined to fit a clinical category. These definitions are in each test booklet and further discussed in the manual. Information content measures functional communication and correlates well with the Aphasia Quotient (AQ; the total score). (c) Auditory verbal comprehension is assessed with three subtests. The first contains yes-no questions that are divided into sections of personal relevance, environmental orientation, and semantically simple common knowledge items and gram-

---

[1] A comparison of the test was made with the FCP, the CADL, and PICA (Holland, 1980).

matical items, using relational words. Next there is a pointing task for single word auditory discrimination using many categories, including body parts and fingers to detect the Gerstmann syndrome. Lastly, sentence comprehension is tested by commands of increasing length and grammatical complexity using sequential actions and relational words similar to the "Token Test." (d) Repetition is tested by single words, composite words, word–number combinations, and low and high probability sentences of increasing length and grammatical complexity. Additional items also test oral agility. Phonemic or word-order errors are penalized. (e) Naming on visual confrontation uses 20 high-frequency objects that are easily portable and commonly found in the home or the office. If visual stimulation fails, the patient is given the object to touch and manipulate and then, if required, a phonemic or semantic cue. These are scored less than uncued responses. For word fluency the patient must name as many animals as possible in one minute. Sentence completion and responsive speech are tests of word finding ability with conversational cues.

The reading section consists of sentence comprehension requiring completion, similar to the BDAE, but the items and foils are chosen to elicit visual and semantic confusions. Also the choice of items for completion are placed vertically to eliminate problems due to hemianopia. The scoring is graded to consider the length and complexity of the items. Reading commands aloud and comprehending are scored separately. If the combined score from these tasks is lower than a certain level then the following lower level tasks are used. These are word–object, word–picture, picture–word, spoken word–printed word matching, letter discrimination, spelling, and spelled word recognition. Writing is elicited by asking for the patient's name, address, for a picture description and by dictating a sentence. If this is above a certain cutoff level then the dictation of words, letters, numbers, and sentence copying is omitted, which shortens test administration time.

The remainder of the test deals with apraxia (20 items) including upper limb, intransitive, instrumental (transitive), bucco-facial and complex movements on command, and if necessary, imitation with and without object use. Constructional difficulties are tested by a drawing task (with scoring categories for quality, completeness, perspective, detail, and neglect), a short form of Koh's block design test adapted from the Wechsler Adult Intelligence Scale (WAIS), one and two-digit calculations, line bisection (to measure neglect), and Raven's Coloured Progressive Matrices (RCPM).

The first version of the WAB (Kertesz & Poole, 1974) was revised and extensively standardized (Kertesz, 1979; Shewan & Kertesz, 1980). The rationale and clinical and research experience with the test have been detailed in a monograph (Kertesz, 1979). Further revision of test items and the comparison of the two versions preceeded the most recent publication (Kertesz, 1982). The scoring system is simple, requiring no transformations, but described in sufficient detail in the manual to achieve good reliabilities, even in the scoring of spontaneous speech fluency (0.98), where judgmental scoring categories are used. High-intrarater correlations were noted on repeated scoring of the same test and there is good test–retest reliability with stable aphasics (Kertesz, 1979). The internal consistency of test items was high using Cronbach's alpha (0.90) and Bentler's theta (0.97) coefficients (Shewan & Kertesz, 1980).

The WAB has been compared with the NCCEA. Both tests were administered to 15 aphasics representing a wide range of difficulty levels and types, within two weeks. A high degree of correlation (0.96) was found for the total scores and 0.81–0.95 for the various comparable subtests. Criterion validity was measured by standardizing the WAB on various populations. Initially, 150 aphasics were clinically separated from nonaphasic brain-damaged individuals and the mean AQ of 93.8 of this population was taken as a cutoff point for aphasia. Subsequent standardizations include normal age-matched individuals, acute and chronic strokes, tumors, trauma, and recovered aphasics, altogether comprising 365 aphasics and 161 controls.

The classification system is based on the scores for spontaneous speech, fluency, comprehension, and naming. The unequivocal limits for each of the categories were established after assessment of the first 150 patients (Kertesz & Poole, 1974). Subsequently, numerical taxonomic studies showed the cohesiveness of 10 groups for acute and chronic populations separately (Kertesz, 1979). The initial clinical classification of Global, Broca's, Wernicke's, Conduction, Transcortical Motor, Transcortical Sensory, Isolation, and Anomic aphasias are retained in subsequent studies because they resembled the clusters on statistical analysis. The degree of clustering was high, supporting the rationale for such a classification (Kertesz & Phipps, 1977).

However, the taxonomy does not provide for all of the categories one might find desirable, such as verbal apraxia (Bartlett, 1983). Also, it is possible for a few borderline cases to change from one category to another due to a slight change in only one subtest score. Some clinicians feel it is better to retain a "mixed" classification for these borderline cases.

The Aphasia Quotient (AQ) is the sum of verbal subtest scores, expressed as a percentage of normal performance. The AQ can be easily calculated simply by adding the verbal subtest scores and then multiplying by two. For a Cortical Quotient (CQ) a simple addition of all subscores is used. Recently the Language Quotient (LQ), containing oral and written language scores, was used to follow recovery (Shewan & Kertesz, 1984).

Test administration is about 45 minutes for the oral language tasks and another 45 minutes to 60 minutes for reading, writing, and supplementary tasks. The examiner has to provide the 20 common test objects for naming. The test is published by Grune and Stratton, New York (Kertesz, 1982), and Igaku Shoin in Japan. It has been translated to German, Hungarian, Tamil, French, Portuguese, and Hebrew. It includes a manual with the classification system, a set of test cards, and 25 test booklets with score sheets.

TESTS OF COMMUNICATIVE FUNCTION

These tests attempt to rate the communicative abilities of aphasics on an informal basis without structural questions. Such tests observe language/behavior in more natural interactions, which occur in the patient's environment. Often the patient's relatives report that the patient comprehends a great deal and is able to let his wishes be known when formal tests reveal no scorable performance. Clinicians then form impressions about their patients' functional language based on these observations. These tests attempt, as objectively as possible, to provide an actual measure.

THE FUNCTIONAL COMMUNICATION PROFILE (FCP)

Taylor (1965) constructed this brief test consisting of 45 items that are informally rated on the basis of unstructured interviews and subjective scoring. The items are grouped into five categories covering movement, speaking, understanding, reading, and miscellaneous (includes writing and calculation). The scoring on a 9-point scale is measured as a proportion of estimated former ability. The emphasis is on language as used in common, everyday situations, such as the ability to handle questions, to indicate yes and no, to read headlines, and to handle money. Actual behavior should be evaluated, not just estimates of what the patient can do. The interview takes approximately 30 minutes.

Interrater reliability coefficients for each of the categories of the FCP have been reported (Sarno, 1969). This kind of subjective rating scale is most useful in the hands of skilled, experienced clinicians. The FCP was also compared with the NCCEA. Limited correlation was noted on measurements of improvement using follow-up tests. The FCP is not suitable for classifying patients into aphasic types but repeated examinations have been used to follow recovery (Sands, Sarno, & Shankweiler, 1969; Sarno & Levita, 1971). It is available from the Institute of Rehabilitation Medicine, New York University Medical Center, New York.

COMMUNICATIVE ABILITIES IN DAILY LIVING (CADL)

The CADL by Holland (1980) consists of 68 items in 10 categories. The examiner and patient participate in contextually enriched situations: using the telephone, going shopping, being inside of a car, going to the doctor, and an interview. Photographs, drawings, and objects are used as props to recreate these situations. False questions are also asked to assess the responses to misinformation. The scoring is only on a 3-point scale (correct, adequate, or wrong). The examiner is encouraged to interact in a friendly, participating way. The modality of response does not matter as long as correct gesture or writing is used. In fact, the patient need not speak to obtain a full score. The simplified scoring system eliminates the need for extensive training.

Standardization included high intertester and intrarater reliability for development staff, as well as outsiders instructed through the manual. The test correlated 0.87 with the FCP, 0.93 with PICA, and 0.84 with the BDAE in 80 patients. The CADL showed similar ranges of severity as did the aphasia types determined by the BDAE. Normative data were also collected on 130 normals grouped for age, sex, and institutionalization. The CADL was used to follow chronic patients with plateaued PICA scores whenever therapy aimed at functional communication. The scores increased significantly, indicating a dissociation of measures of functional communication and structural testing. The CADL is published by University Park Press, Baltimore.

SHORT SCREENING TESTS IN APHASIA

These tests have limited accuracy and their application is chiefly in a busy clinic where more comprehensive, in-depth assessment is impossible. There are many clinicians who use their own individual screening tests but few are standardized or available as a publication.

THE SKLAR APHASIA SCALE (SAS)

The SAS by Sklar (1973) examines four major language skills. Each of these skills have 5 items representing further areas. Auditory encoding includes identifying body parts, understanding questions, identifying objects in the environment, implements, and recalling the name of an object. Visual decoding matches printed words, words with pictures, sentence completion, arithmetic, and silent reading. Oral encoding scores functional speech, repetition of words, naming objects, reading aloud, describing actions in a picture, and retelling items remembered. Graphic encoding is writing name, address, copying words, written naming of objects, sentences from dictation, and picture description. Each item is scored on a 5-point scale: correct = 0, retarded = 1, assisted = 2, distorted = 3, no response = 4. A total impairment score (0–100) is calculated by adding the four subtest scores. A supplementary checklist of verbal behavior includes jargon, automatism, perseveration, dysarthria, dysphonia, oral apraxia, agrammatism, omissions, substitutions, distortions, retardation, word-finding difficulty, unintelligible responses, and impairment of self-monitoring. Normative data indicated that the items were easy for normal adults. Correlations between the SAS, Eisenson's test, Schuell's short examination, the Halstead-Wepman Aphasia screening test, the Wechsler-Bellevue, the Bender-Gestalt and the Goldstein Sheerer cube test were significant. The German version of the SAS (Cohen, et al., 1977) did not differentiate between types of aphasia but did separate aphasics from a variety of other controls (criterion validity). Test administration is less than 30 minutes. The SAS is available from Western Psychological Services, Los Angeles.

Aphasia Language Performance Scales (ALPS)

The ALPS by Keenan and Brassell (1975) contains four 10-item scales (listening, talking, reading, and writing). The scoring and administration are informal and the examiner may start with any item that is appropriate for the patient's level. This shortens the examination to about 30 minutes. Tests of test–retest reliability, internal consistency, and content validity (a comparison with the PICA) have been established. The ALPS is available from Pinnacle Press, Murfreesboro, Tennessee.

Aphasia Screening Test (AST)

The AST by Whurr (1974) samples a wide range of expressive and receptive functions with 30 relatively easy subtests of 5 items each. It is suited to a more severely affected population of aphasics. The scoring is done by adding the correct responses on the 30 subtests. No standardization data are available, although the test appears to have face validity for aphasia. The AST is available from the author in the Department of Speech Pathology, National Hospital, London, England.

APHASIA EXAMINATIONS FOR POLYGLOTS, AND IN OTHER LANGUAGES

Although many of the aphasia tests reviewed have been translated to several other languages, attempts have been made at designing functionally equivalent examinations in several languages, especially for polyglots or for the comparison of linguistic characteristics of various aphasics in different language areas. Benton originated this effort in English, French, German, Italian, and Spanish. The English version of the *Multilingual Aphasia Examination (MAE)* is published by Benton and Hamsher (1976). Paradis (1983), with the help of his graduate students from as many as 24 various countries, has also developed a linguistically oriented examination specifically for polyglots. The items are carefully constructed to have phonological, syntactic, and semantic equivalents across all languages whenever possible. At this time, the test is undergoing further standardization.

There are comprehensive aphasia tests developed in French, Italian, Dutch, Portuguese, and German centers. The recently published *Aachener Aphasic Test* (Huber, Poeck, Weniger, & Willmes, 1983) is well standardized and widely used in Germany. Italian and English versions are also in preparation.

MODALITY SPECIFIC TESTS OF APHASIA

THE TOKEN TEST (T.T.)

DeRenzi and Vignolo (1962) developed this test for comprehension disturbance that may occur in expressive syndromes. It consists of 61 commands of increasing length and complexity all concerning tokens of 5 colors, 2 shapes (circle and a square) and 2 sizes (large and small). The patient has to point, touch, pick up, and place the tokens according to a spoken command, (e.g., "touch all squares except the blue one"). Part 5 of the test is the most complex, utilizing prepositions, conjunctions, centering embedded sentences, and temporal sequences. The commands are designed to remain within the short-term memory capacity of patients. The test has been extensively standardized and modified (see review by Boller & Dennis, 1979). Years of education appears to influence performance and a correction factor is recommended (DeRenzi & Faglioni, 1978). It is a sensitive test for mild disturbances and reliably discriminates mild aphasia from normals. However, more severe aphasics do poorly on the test and cannot be distinguished, even if they have some residual comprehension. Nonaphasic and right- and left-hemisphere-damaged patients had a relatively high false positive rate on the test. Often it can be administered in usually less than 30 minutes, but it is boring and tiring to the patients.

DeRenzi (1979) published a shortened version. Mack and Boller (1979) created a Token Test that varied syntax, lexicon, and sentence length independently. This reduced the array from 20 to 8, minimized motor demands and revised the scoring to discriminate among aphasia types. The German version of the Token Test underwent modification and extensive study. Poeck, Kerschensteiner, & Hartje (1972) and Poeck, Orgass, Kerschensteiner, & Hartje (1974) showed no quantitative difference between motor, sensory, and anomic aphasics. However, they found that the test could highly discriminate between aphasics and nonaphasics, especially Part 5, and suggested that it may be used by itself. Part 5 of the test is clearly different from the other sections and this has recently been confirmed with probabilistic test models (Willmes, 1981).

THE REPORTER TEST (R.T.)

An expressive equivalent, the Reporter Test, was constructed by DeRenzi and Faglioni (1979). The patient verbally reports what the examiner does with the tokens. The Reporter Test is designed to be used following the Token Test. This test correlates with oral naming, but not with oral fluency.

AUDITORY COMPREHENSION TEST FOR SENTENCES (ACTS)

The ACTS by Shewan (1979), in its revised version, is a commercially published auditory sentence stimulus picture choice task, similar to many others used in psycholinguistic research. Forty-two sentences are varied in length, syntactic complexity, and lexical difficulty. A protocol sheet allows error analysis and weighted scoring, with speed of response considered. The manual contains the standardization of 150 aphasics and 30 normal controls. High internal reliability coefficients were established. It correlated highly with the WAB comprehension subtest (0.89), but not with the BDAE (0.52). The test takes approximately 15 minutes to administer. It is too difficult for more severe aphasics (similar to the Token Test) and has to be complemented with other tasks to detect lower levels of comprehension that are functionally quite significant. The ACTS is published by Biolinguistics Clinical Institute, Chicago.

BOSTON NAMING TEST (BNT)

The BNT by Kaplan, Goodglass, and Weintraub (1978) is an extended picture-naming task. Normative data for it, along with the BDAE, have been published by Borod et al. (1980). These show a considerable drop of scores for subjects over the age of 60. It is available from Lea and Febiger, Philadelphia, and it has been included in the latest edition of the BDAE.

CONCLUSION

This review attempted to describe the principles and practical aspects of aphasia testing with a descriptive evaluation of some historical and some current aphasia tests. Although the review has attempted to be objective and comprehensive, individual preferences and biases develop throughout one's training, theoretical orientation, and use of the tests. It is very difficult to recommend a single aphasia test as there is none that would be optimal for everyone. However, many readers will want some sort of selection. For a busy clinician, a short screening test such as the SAS or ALPS may suffice for the purpose of documentation. The most comprehensive and psycholinguistically sophisticated test is the BDAE, but its length and complex scoring deter many clinicians. Somewhere in between is the WAB with its practical scoring classification system, and length. Some clinicians will prefer the FCP or the CADL to assess functional capacity, whereas others may wish to use the Token Test as a measure of comprehension in the less affected patients. The PICA is most detailed in measuring the quality of response.

Aphasia testing is important to most clinicians because treating a patient without testing is like trying to navigate an uncharted sea. Not only the severity of impairment, but the pattern of deficits can be determined by the better tests, and both are helpful in formulating prognosis and treatment goals. Researchers need general aphasia testing for quantitation of change and definition of groups of patients with differing behaviour, in addition to the specifically constructed experimental tests. Few tests satisfy the needs for research as well as clinical practice and experience is required to determine which test is most suitable for what purpose.

REFERENCES

Appell, J., Kertesz, A., & Fisman, M. (1982). A study of language functioning in Alzheimer patients. *Brain and Language*, 17, 73–91.

Bartlett, C. L. (1983, November). *A critique of the WAB*. Paper presented at the meeting of the American Speech and Hearing Association, Cincinnati, OH.
Benson, D. F. (1967). Fluency in aphasia: Correlation with radioactive scan localization. *Cortex*, 3, 373–394.
Benton, A. L. (1967). Problems of test construction in the field of aphasia. *Cortex*, 3, 32–53.
Benton, A. L., & Hamsher, K. (1976). *Multilingual aphasia examination*. Iowa City: Department of Neurology, University Hospitals.
Blumstein, S. E., Baker, E., & Goodglass, H. (1977). Phonological factors in auditory comprehension in aphasia. *Neuropsychologia*, 15, 19–30.
Boller, F., & Dennis, M. (Eds.). (1979). *Auditory comprehension: Clinical and experimental studies with the Token Test*. New York: Academic Press.
Boller, F., & Green, E. (1972). Comprehension in severe aphasia. *Cortex*, 8, 382–394.
Boone, D. R. (1972). Porch index of communicative ability. In O. K. Buros (Ed.), *Seventh mental measurements yearbook: Vol. II*. Highland Park, NY: Gryphon Press.
Borod, S. C., Goodglass, H., & Kaplan, E. (1980). Normative data on the Boston Diagnostic Aphasia Examination, Parietal Lobe Battery and the Boston Naming Test. *Journal of Clinical Neuropsychology*, 2, 209–215.
Broca, P. (1861). Remarques sur le siège de la faculté du langage articulé, suivies d'une observation d'aphémie (perte de la parole). *Bulletin de la Société de l'Anatomie*, 36, 330–357.
Brookshire, R. H. (1979). The Language Modalities Test for aphasia. In F. L. Darley (Ed.), *Evaluation of appraisal techniques in speech and language pathology*. Reading, MA: Addison-Wesley.
Bub, D., & Kertesz, A. (1982). Deep agraphia. *Brain and Language*, 17, 146–165.
Cohen, R., Engel, D., Kelter, S., List, G., & Strohner, H. (1977). Validity of the Sklar Aphasia Scale. *Journal of Speech and Hearing Research*, 20, 146–154.
Crockett, D. J. (1977). A comparison of empirically derived groups of aphasic patients on the N.C.C.E.A. *Journal of Clinical Psychology*, 33, 194–198.
Dennis, M. (1976). Dissociated naming and locating of body parts after left anterior temporal lobe resection: An experimental case study. *Brain and Language*, 3, 147–163.
DeRenzi, E. (1979). A shortened version of the Token Test. In F. Boller & M. Dennis (Eds.), *Auditory comprehension: Clinical and experimental studies with the Token Test*. New York: Academic Press.
DeRenzi, E., & Faglioni, P. (1978). Normative data and screening power of a shortened version of the Token Test. *Cortex*, 14, 41–49.
DeRenzi, E., & Vignolo, L. (1962). The Token Test: A sensitive test to detect receptive disturbances in aphasics. *Brain*, 85, 665–678.
Eisenson, J. (1954). *Examining for aphasia: A manual for the examination of aphasia and related disturbances*. New York: Psychological Corporation.
Eisenson, J. (1972). *Aphasia in children*. New York: Harper & Row.
Emerick, L. L. (1971). *The appraisal of language disturbance. Manual*. Marquette: Northern Michigan University.
Gardner, H., Albert, M. L. & Weintraub, S. (1975). Comprehending a word: The influence of speed and redundancy on auditory comprehension in aphasia. *Cortex*, 11, 115–162.
Geschwind, N., & Fusillo, M. (1966). Color naming defects in association with alexia. *Archives of Neurology*, 15, 137–146.
Goodglass, H., & Kaplan, E. (1972). *Assessment of aphasia and related disorders*. Philadelphia: Lea & Febiger.

Goodglass, H., & Kaplan, E. (1983). *Assessment of aphasia and related disorders* (rev. ed.). Philadelphia: Lea & Febiger.
Goodglass, H., Klein, B., Carey, P. & Jones, K. (1966). Specific semantic word categories in aphasia. *Cortex, 2,* 74–89.
Goodglass, H., Quadfasel, F. A., & Timberlake, W. H. (1964). Phrase length and the type of severity of aphasia. *Cortex, 1,* 133–153.
Goodglass, H., Gleason, J., & Hyde, M. (1970). Some dimensions of auditory language comprehension in aphasia. *Journal of Speech and Hearing Research, 13,* 595–606.
Head, H. (1926). *Aphasia and kindred disorders of speech.* Cambridge: Cambridge University Press.
Hier, D. B., & Mohr, J. (1977). Incongruous oral and written naming. *Brain and Language, 4,* 115–126.
Holland, A. L. (1977). *Communicative abilities in daily living: Its measurement and observation.* Paper presented at the meeting of the Academy of Aphasia, Montreal.
Holland, A. L. (1980). *Communicative abilities in daily living. Manual.* Baltimore: University Park Press.
Howes, D., & Geschwind, N. (1964). Quantitative studies of aphasic language. *Association for Research in Neurology and Mental Disease, 42,* 229–244.
Huber, W., Poeck, K., Weniger, D., & Willmes, K. (1983). *Aachener-Aphasie Test.* Toronto: Verlag für Psychologie Göttingen.
Kaplan, E., Goodglass, H., & Weintraub, S. (1983). *The Boston Naming Test.* Philadelphia: Lea & Febiger.
Keenan, S. S., & Brassel, E. G. (1975). *Aphasia Language Performance Scales (ALPS).* Murfreesboro, TN: Pinnacle Press.
Kenin, M., & Swisher, L. P. (1972). A study of patterns of recovery in aphasia. *Cortex, 8,* 56–68.
Kerschensteiner, M., Poeck, K., & Brunner, E. (1972). The fluency–nonfluency dimension in the classification of aphasic speech. *Cortex, 8,* 233–247.
Kertesz, A. (1979). *Aphasia and associated disorders: Taxonomy, localization, and recovery.* New York: Grune & Stratton.
Kertesz, A. (1982). *The Western Aphasia Battery.* New York: Grune & Stratton.
Kertesz, A., & Phipps, J. (1977). Numerical taxonomy of aphasia. *Brain and Language, 4,* 1–10.
Kertesz, A., & Poole, E. (1974). The aphasia quotient: The taxonomic approach to measurement of aphasic disability. *Canadian Journal of Neurology, 1,* 7–16.
Liepmann, H. (1905). *Uber Störungen des Handelns bei Gehirnkranken.* Berlin: Karger.
Luria, A. R. (1966). *Higher cortical functions in man.* New York: Basic Books.
Mack, J. L., & Boller, F. (1979). Components of auditory comprehension: Analysis of errors in a revised Token Test. In F. Boller & M. Dennis (Eds.), *Auditory comprehension: Clinical and experimental studies with the Token Test.* New York: Academic Press.
Marie, P. (1906). Revision de la question de l'aphasie: La troisième circonvolution frontale gauche ne joue aucun rôle special dans la fonction du langage. *Séminaires de Médecin, 21,* 241–247.
Martin, A. D. (1977). Aphasia testing: A second look at the Porch Index of Communicative Ability. *Journal of Speech and Hearing Disorders, 42,* 547–561.
McNeil, M. R., Prescott T. E., & Chang, E. C. (1975). A measure of the PICA ordinality. In R. H. Brookshire (Ed.), *Clinical aphasiology conference proceedings.* Minneapolis, MN: BRK Publishers.
Moutier, F. (1908). *L'aphasie de Broca.* Paris: Steinheil.
Ombredane, A. (1951). *L'Aphasie et élaboration de la pensée explicite.* Paris: Presses Universitaire de France.

Paradis, M. (1983, November). *A Report on the Polyglot Aphasia Test*. Paper presented at the meeting of the American Speech and Hearing Association, Cincinnati, OH.
Poeck, K., Kerschensteiner, M., & Hartje, W. (1972). A quantitative study on language understanding in fluent and nonfluent aphasia. *Cortex, 8*, 299–302.
Poeck, K., Hartje, W., Kerschensteiner, M., & Orgass, B. (1973). Sprachverstandisstörungen bei aphasischen und nichtaphasischen Hirnkranken. *Deutsche Medizinische Wochenschrift, 98*, 139–147.
Poeck, K., Orgass, B., Kerschensteiner, M., & Hartje, W. (1974). A qualitative study on Token Test performance in aphasic and nonaphasic brain damaged patients. *Neuropsychologia, 12*, 49–54.
Porch, B. E. (1967). *Porch Index of Communicative Ability: Theory and development*. Palo Alto, CA: Consulting Psychologists Press.
Porch, B. E. (1971). *Porch Index of Communicative Ability: Administration, scoring and interpretation* (rev. ed.). Palo Alto, CA: Consulting Psychologists Press.
Porch, B. E., Collins, M., Wertz, R. T., & Friden, T. P. (1980). Statistical prediction of change in aphasia. *Journal of Speech and Hearing Research, 23*, 312–321.
Sands, E., Sarno, M. T., & Shankweiler, D. (1969). Long-term assessment of language function in aphasia due to stroke. *Archives of Physical Medicine and Rehabilitation, 50*, 202–207.
Sarno, M. T. (1969). *The Functional Communication Profile. Manual* (Rehabilitation Monographs No. 42). New York: New York University Medical Center, Institute of Rehabilitation Medicine.
Sarno, M. T., & Levita, E. (1971). Natural course of recovery in severe aphasia. *Archives of Physical Medicine and Rehabilitation, 52*, 175–179.
Schuell, H. (1957). A short examination for aphasia. *Neurology, 7*, 625–634.
Schuell, H. (1965). *The Minnesota test for differential diagnosis of aphasia*. Minneapolis, MN: University of Minnesota Press.
Schuell, H. (1974). Diagnosis and prognosis in aphasia. In L. F. Sies (Ed.), *Aphasia, theory and therapy*. Baltimore, MD: University Park Press.
Schuell, H., & Jenkins, J. J. (1961). Reduction of vocabulary in aphasia. *Brain, 84*, 243–261.
Shewan, C. M., & Kertesz, A. (1980). Reliability and validity characteristics of the Western Aphasia Battery (WAB). *Journal of Speech and Hearing Disorders, 45*, 308–324.
Shewan, C. M., & Kertesz, A. (1984). Effects of speech and language treatment on recovery from aphasia. *Brain and Language, 23*, 2:272–299.
Sklar, M. (1973). *Sklar Aphasia Scale: Protocol booklet*. Beverly Hills, CA: Western Psychological Services.
Spellacy, F. J., & Spreen, O. (1969). A short form of the Token Test. *Cortex, 5*, 390–397.
Spreen, O., & Benton, A. L. (1968). *Neurosensory Center Comprehensive Examination for Aphasia*. Victoria, BC: University of Victoria Press.
Spreen, O., & Benton, A. L. (1977). *Neurosensory Center Comprehensive Examination for Aphasia* (rev. ed.). Victoria, BC: Neuropsychology Laboratory, Department of Psychology, University of Victoria.
Taylor, M. L. (1965). A measurement of functional communication in aphasia. *Archives of Physical Medicine and Rehabilitation, 46*, 101–107.
Thompson, J., & Enderby, P. (1979). Is all your Schuell really necessary? *British Journal of Disorders of Communication, 14*, 195–201.
Wagenaar, E., Snow, C., & Prins, R. (1975). Spontaneous speech of aphasia patients: A psycholinguistic analysis. *Brain and Language, 2*, 281–303.
Weisenburg, T. H., & McBride, K. E. (1935). *Aphasia*. New York: Commonwealth Fund.
Wepman, J. M., & Jones, L. V. (1961). *Studies in aphasia: An approach to testing*. Chicago: Education-Industry Service.

Wertz, R. T. (1979). Word fluency measure. In F. L. Darley (Ed.), *Evaluation of appraisal techniques in speech and language pathology.* Reading, MA: Addison-Wesley.

Wertz, R. T. (1983, November). *A Comparison of the BDAE, PICA and the WAB.* Paper presented at a meeting of the American Speech and Hearing Association, Cincinnati, OH.

Whurr, R. (1974). *Aphasia screening test.* Unpublished manuscript, National Hospital, Queen's Square, Department of Speech Therapy, London, England.

Willmes, K. (1981). A new look at the Token Test using probabilistic test models. *Neuropsychologia, 19,* 631–645.

12

The Clinical Aspects of Memory Disorders

Contributions from Experimental Studies of Amnesia and Dementia

NELSON BUTTERS

It is well known to even fledgling clinical neuropsychologists that memory disorders are ubiquitous after brain damage and that standardized tests exist for documenting the presence and severity of these memory deficits. Severe impairments in the learning of new information and in the recall of public and personal events from the remote past occur after head trauma, long-term alcohol abuse, strokes, encephalitis, bilateral ECT, and as an early sign of progressive dementing illnesses. Tests such as the Wechsler Memory Scale and the Benton

---

Reprinted from the *Journal of Clinical Neuropsychology*, 6(1), pp. 17–36. Copyright 1984 by Swets & Zeitlinger, B.V. Reprinted by permission. The chapter was written while the author was affiliated with the Boston VA Medical Center and Boston University School of Medicine. A version of this chapter was delivered as the author's Presidential Address to Division 40 of the American Psychological Association in August, 1983, in Anaheim, CA.

NELSON BUTTERS • San Diego Veterans Administration Medical Center, 3350 La Jolla Village Drive, La Jolla, CA 92161. The studies reported in this Chapter were supported by funds from the Medical Research Service of the Veterans Administration, by NIAAA grant AA00187 to Boston University, by NINCDS grant NS16367 to Massachusetts General Hospital, and by NIA grant AG02269 to Beth Israel Hospital.

Visual Retention Test have proven valuable, but not perfect, tools for the assessment of these problems.

Despite this awareness of memory impairments, most neuropsychologists have been insensitive to the multidimensional nature of the symptoms. Usually implicitly, but sometimes explicitly, it has been assumed that all debilitating memory deficits, regardless of etiology, may be treated as a single symptom or cognitive problem. Amnesic patients with bilateral hippocampal lesions and patients with medial diencephalic damage have often been treated as two exemplars of a single underlying clinical entity. In recent years, this situation has changed as evidence has accumulated that the "amnesias" are as heterogeneous as the "aphasias" and the "apraxias" (for review, see Butters & Miliotis, 1985; Butters, Miliotis, Albert, & Sax, 1984; Squire, 1981, 1982). Based on studies from several laboratories it now appears that there are important qualitative differences among the anterograde and retrograde memory deficits of various neurological populations. Although failures in retention and recall following diencephalic and hippocampal lesions may have some superficial similarities, close scrutiny has shown these amnesic populations to have distinctive patterns of deficits when a broad range of memory capacities are assessed. Similarly, direct comparisons of the memory deficits of amnesic and demented patients have uncovered differences that may be of some importance in making prognostic and rehabilitative judgments.

The main purpose of this presentation will be to emphasize the differences between the memory disorders of amnesic (alcoholic Korsakoff patients) and demented patients (Huntington's Disease). We wish not only to convince the audience of the clinical importance of a thorough memory assessment but also to illuminate the ongoing symbiotic relationship between experimental and clinical neuropsychology. The discovery that memory tests are among the most sensitive psychometric instruments for distinguishing among various patient populations has evolved from "basic" research into brain–behavior relationships. Although it may be fashionable for some neuropsychologists to label themselves as "experimental" or "clinical," it is hoped that the data reviewed here will demonstrate that such divisions are both artificial and unrewarding.

COMPARISONS OF THE MEMORY DISORDERS OF PATIENTS WITH ALCOHOLIC KORSAKOFF'S SYNDROME AND PATIENTS WITH HUNTINGTON'S DISEASE

Severe memory disorders are not unique to amnesic patients. In fact, complaints about memory (anterograde and retrograde) are among the first symptoms of progressive dementing disorders (Miller, 1977). The major difference between pure amnesia and progressive dementia is that the memory loss of the demented patients is part of a broader intellectual decline. Although the amnesic patient's IQ usually remains within the normal range despite his severely impaired memory quotient, both the IQ and the MQ of the demented patient decline progressively as the illness advances. IQ and MQ scores in the 80s are common in the middle and late stages of the dementing process. Given the current evidence of the diversity of amnesic symptoms, it is of some interest to determine whether the severe memory disorders of amnesic and demented patients involve the same underlying processes.

Several studies (Albert, Butters, & Brandt, 1981a,b; Biber, Butters, Rosen, Gerstman, & Mattis, 1981; Butters & Grady, 1977; Butters, Tarlow, Cermak, & Sax, 1976; Meudell, Butters, & Montgomery, 1978; Oscar-Berman & Zola-Morgan, 1980a,b) comparing the memory disorders of amnesic patients with alcoholic Korsakoff's syndrome and of demented patients with Huntington's Disease (HD)[1] have indicated that the anterograde and retrograde memory deficits of these two patient populations are distinguishable. Because most of these previous investigations have been reviewed in detail elsewhere (Butters, Albert, & Sax, 1979; Butters & Cermak, 1980), we shall discuss them briefly and then concentrate more fully on some recent findings concerned with pictorial memory and with differences between the learning of declarative and procedural information. It should be evident from this presentation that the memory impairments of HD and Korsakoff patients involve different underlying processes and that our conception of the HD patients' deficits has undergone a natural evolution over the years. Although our initial studies led us to suggest that storage problems were the primary process involved in the HD patients' memory failures, the results of investigations completed during the past three years have demonstrated that retrieval deficits are perhaps the most important factors in the HD patients' learning and retentive difficulties.

In our initial studies, the abilities of HD and Korsakoff patients to retain information in short-term storage were compared using the Brown-Peterson distractor technique. On each trial the subjects were read or shown verbal stimuli (i.e., consonant trigrams like *JZD* or word

---

[1]Patients with Huntington's Disease have a genetically transmitted disorder that results in a progressive atrophy of the basal ganglia, especially the caudate nucleus. Their most common behavioral symptoms include involuntary choreic movements and a progressive dementia in which severe memory problems form an integral part of a broader intellectual decline (Caine, Ebert, & Weingartner, 1977; Weingartner, Caine, & Ebert, 1979).

triads such as *neck-chair-belt*) and then asked to count backwards by threes until the examiner said "stop." After 0, 3, 9, or 18 seconds of such counting (i.e., distraction), the examiner stopped the subjects and asked them to recall the stimulus material that had just been presented. It was found that, although the HD patients performed as poorly as did the alcoholic Korsakoff patients after 3, 9, and 18 second delays, only the Korsakoff patients' impaired recall was affected by proactive interference, rehearsal time, and encoding (Butters et al., 1976; Butters & Grady, 1977; Meudell et al., 1978). Manipulations of proactive interference (PI) and rehearsal time (i.e., time between the end of stimulus presentation and the beginning of the counting task) improved the Korsakoff patients' performance on distractor tasks, whereas such changes in experimental conditions had virtually no effect on the HD patients' ability to recall materials presented 9 or 18 seconds previously. Although the results of these investigations did not reveal the specific nature of the HD patients' anterograde memory deficits, they did suggest that these patients have difficulty storing new information (i.e., a deficit in consolidation). The HD patients' failure to improve with low PI conditions and with increased time for rehearsal suggested that these patients may lack some of the neuroanatomical structures necessary for consolidating and storing new information.

Oscar-Berman and Zola-Morgan (1980a,b) have compared alcoholic Korsakoff and HD patients on a series of visual and spatial discrimination tasks. In an initial experiment (Oscar-Berman & Zola-Morgan, 1980a), visual and spatial reversal learning tests were administered. Whereas the Korsakoff patients were impaired on both types of reversal problems, the HD patients had difficulty only with the visual reversals. An inspection of the types of errors compiled on the visual reversal tasks suggested that the HD and Korsakoff patients' deficiencies involved different learning, cognitive, and motivational mechanisms. In a second experiment (Oscar-Berman & Zola-Morgan, 1980b), the Korsakoff and HD patients learned a series of two-choice simultaneous and concurrent pattern discriminations. Again, both patient groups were impaired, but they differed in the nature of their deficits. The HD patients were equally impaired on simultaneous and concurrent discriminations whereas the Korsakoff patients encountered more difficulty with the concurrent than with the simultaneous tests. In summarizing their findings, Oscar-Berman and Zola-Morgan (1980b) suggested that, although both groups of patients were deficient in their ability to form stimulus–reinforcement associations, the Korsakoff, but not the HD, patients' deficiencies also involved an increased sensitivity to proactive interference and a lack of sensitivity to reinforcement contingencies.

To determine whether the HD patients also differ from the alcoholic Korsakoff patients in their ability to recall people and public events from the remote past, Albert, Butters, and Levin's (1979) remote memory battery was administered to alcoholic Korsakoff and HD patients (Albert et al., 1981a; Butters & Albert, 1982). As shown in Figure 1, the Korsakoff patients had a severe impairment in their ability to recall remote memories. Although this deficiency included all periods of their lives, it was characterized by a temporal gradient in which the Korsakoff patients' most remote memories (e.g., from the 1930s and 1940s) were relatively preserved. Like the alcoholic Korsakoff patients, the HD patients were severely impaired in their ability to identify famous people and to recall public events, but their retrograde amnesia was not characterized by a temporal gradient in which famous faces and public events from the 1930s and 1940s were relatively spared.

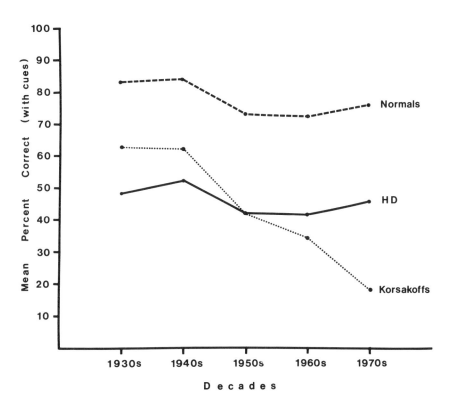

Figure 1. Performance of patients with alcoholic Korsakoff's Syndrome, patients with Huntington's Disease (HD), and normal controls on Famous Faces Test by Albert, Butters, and Levin (1979).

That is, the HD patients had as much difficulty identifying faces and events from the 1930s and 1940s as faces and events from the 1960s and 1970s. Wilson, Kaszniak, and Fox's (1981) report of very similar results for patients with senile dementia of the Alzheimer's type suggests that "flat" retrograde amnesias may be associated with a number of dementing illnesses.

In a second study, Albert et al. (1981b) compared the performances of advanced HD (diagnosed 3 to 7 years prior to testing) and recently diagnosed HD (less than 12 months prior to testing) patients on their remote memory battery. The recently diagnosed HD patients, who showed only a mild cognitive loss at this early stage of the disease, had a retrograde amnesia that was quantitatively less severe but qualitatively similar to that of the advanced patients. Like the advanced HD patients, the recently diagnosed HD patients' impairments in the identification of remote events and famous faces extended equally over all decades of their lives. On the basis of these results for the recently diagnosed HD patients, Albert et al. (1981b) concluded that the "flat" retrograde amnesia that seems to characterize all stages of the disease cannot be attributed to the dementing process per se. They also noted that the equal loss of remote memories over all time periods sampled was consistent with the thesis that these patients have a storage deficit affecting both the consolidation of new memory traces and the maintenance of memories formed prior to the onset of the disease.

Biber et al. (1981) used an orientation procedure in an attempt to improve the HD and Korsakoff patients' memory for faces. It had been reported (Bower & Karlin, 1974) that requiring normal subjects to make global-evaluative judgments (e.g., likeability) about a face prompts a thorough analysis of facial features, which, in turn, results in improved recognition. In contrast, requiring normal subjects to judge some isolated (i.e., piecemeal) facial features (e.g., straightness of hair) supposedly leads to a limited, inadequate analysis of facial features and ultimately to poor recognition performance. When Biber et al. (1981) administered these "high-level" and "low-level" orientation tasks to HD and alcoholic Korsakoff patients, they found that only the Korsakoff patients showed a significant improvement in face recognition after being required to judge the likeability of the to-be-remembered faces. The HD patients' recognition deficit was unaffected by the nature of the orientation task.

The results of Biber et al.'s 1981 investigation are consistent with the conclusions drawn from the previously cited comparisons of the memory disorders of alcoholic Korsakoff and HD patients. Once again the Korsakoff patients' impaired memory was aided by a procedure that promotes encoding, whereas the impairments of the HD patients remained impervious to the same experimental manipulation. This failure of the HD patients to improve significantly with the "high-level" orientation task, although not precluding other possible explanations, supported the notion that deficits in storage and consolidation play a vital role in the HD patients' memory problems.

I would now like to turn your attention to two recently completed studies that have forced us to alter our conception of the HD patients' memory impairment and also have uncovered new cognitive distinctions among patients with HD, Alzheimer's Disease (AD), and Korsakoff's syndrome. In the first of these two investigations, Butters, Albert, Sax, Miliotis, Nagode, and Sterste (1983) evaluated the beneficial effects verbal mediation and labeling might have on the amnesic and demented patients' ability to remember pictorial materials. In the other studies reviewed in this presentation, it was found that manipulations of experimental variables such as rehearsal time, intertrial rest intervals, and orientation instructions resulted in improved performance only for the alcoholic Korsakoff patients. However, because HD is a progressive dementia in which language abilities remain relatively intact until the terminal stages of the disease (Butters, Sax, Montgomery, & Tarlow, 1978), the possibility remained that providing these patients with verbal labels and mediators might reduce their severe difficulties in remembering pictorial stimuli. In view of numerous reports that alcoholic Korsakoff patients do not encode all of the semantic attributes of verbal material (for review, see Butters & Cermak, 1980), there was reason to believe that verbal mediators would have little impact on their memory problems. Likewise, the very prevalent and severe language impairments that usually accompany Alzheimer's Disease (Miller, 1977) could eliminate any beneficial consequences verbal mediators might have on an Alzheimer patient's ability to remember pictorial materials.

In this study Shneidman's (1952) Make-A-Picture-Story (MAPS) was modified to assess the pictorial memory of HD patients, patients with AD, alcoholic Korsakoff patients, patients with lesions restricted to the right hemisphere, and normal control subjects. Two conditions were employed: a no-story followed by a story condition. On the no-story condition, the subjects were shown pictures of six backgrounds (e.g., a raft floating on a large body of water, a living room) on which three cut-out human or animal figures had been placed. For example, a superman figure, a figure of an angry man, and a figure of a happy little boy were placed in the living room scene. The subjects were instructed to remember the identity and location of the specific figures in each scene and were allowed 30 seconds to study each of the six scenes. Five minutes following the presentation of the sixth scene the subjects were administered a forced-choice recognition test consisting of 15 pairs of figures. The subjects were required to indicate for each pair which of the two cut-out figures they had seen in one of the previously exposed scenes. For all 15 pairs, one of the figures had been exposed previously, and the other was a distractor item not previously seen by the subjects.

Ten minutes after the recognition test, a picture-context recognition test was administered. The 6 backgrounds were placed, one at a time, in front of the subjects, and 33 cut-out figures (18 targets and 15 distractors) were distributed symmetrically around the background. The subjects were asked to select from the 33 figures the ones that had been associated with the scene during the original exposure (i.e., learning) trial. The examiner recorded the identity of the figures selected, the location of the figures' positions on the backgrounds and the figures' spatial orientations.

After a 15-minute rest interval, the story condition was administered. Six different background scenes, each with three new cut-out figures, were shown to the subjects. The major difference in procedure was that during the 30-second study period provided for each picture the subjects were read a story about the events transpiring in the stimulus scene. Each story related not only what was occurring in the scene but also what had led to the depicted situation and how the situation would be resolved in the immediate future. As in the no-story condition, forced-choice and picture-context recognition tests followed the presentation of the sixth background.

The groups' performances on the forced-choice recognition tests are shown in Figure 2. Although for both story and no-story conditions the patients correctly recognized slightly fewer figures than did the normal control subjects, their performance was considerably better than chance (7.5 correct). Thus, it appeared that amnesic and demented patients could accurately discriminate familiar from unfamiliar figures.

Despite the patients' relatively intact performance on the forced-choice task, their recognition on the picture-context test, as seen in Figure 3, was severely impaired. A two-way analysis of variance yielded highly significant group, condition (story vs. no-story), and interaction (group × condition) effects. On the no-story condition, all four patient groups recognized significantly fewer figures than did the normal control group. Of the four patient groups, the alcoholic Korsakoff and the AD patients were the most impaired, although there was considerable variability within each patient group in terms of degree of impairment.

The results for the story condition (Fig. 3) clearly show that the four patient groups were aided differentially by the presentation of verbal mediators. The recognition scores of the HD and right hemisphere patients were significantly improved by the recitation of the

CLINICAL ASPECTS OF MEMORY DISORDERS 369

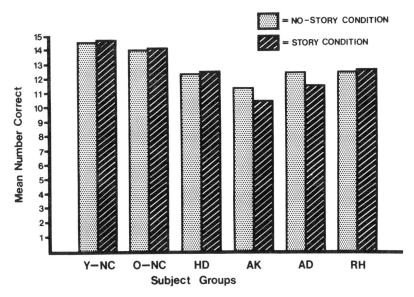

Figure 2. Forced-choice recognition on the picture memory (MAPS) test under the story and no-story conditions. AK = alcoholic Korsakoffs; HD = patients with Huntington's Disease; RH = patients with right hemisphere lesions; AD = patients with Alzheimer's Disease; Y-NC = young normal controls; O-NC = old normal controls.

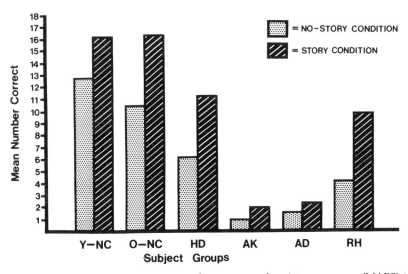

Figure 3. Picture-context recognition performance on the picture memory (MAPS) test under the story and no-story conditions. AK = alcoholic Korsakoffs; HD = patients with Huntington's Disease; RH = patients with right hemisphere lesions; AD = patients with Alzheimer's Disease; Y-NC = young normal controls; O-NC = old normal controls.

story, whereas the recognition scores of the Korsakoff and AD patients appeared insensitive to this mnemonic aid. It is important to note that the improvement of the HD and right hemisphere patients was not due to their superior performance on the no-story condition. When the amount of improvement between the story and no-story conditions was analyzed with a covariance design that statistically corrected for performance on the no-story condition, the HD and right hemisphere patients continued to demonstrate significantly more improvement than did the other two patient groups. Furthermore, no significant correlation was found between performance on the no-story condition and the amount of improvement induced by the story.

The findings of this investigation, especially those relating to the HD and Korsakoff patients, are of special importance because they provide additional legitimacy for other dissociations noted in this presentation. Although the studies of the patients' short-term memory, retrograde amnesia, and memory for faces suggested significant differences between HD and Korsakoff patients, none of them provided the elusive double dissocation needed to firmly establish our claim of qualitative differences in the memory disorders of the two groups. The fact that the Korsakoff, but not the HD, patients were affected by rehearsal time, intertrial rest intervals, and orientation procedures might have been a reflection of the totally debilitating effects of dementia rather than an indicator of qualitative differences in information processing. However, in the present study, the double dissociation between groups and tasks has been completed. For the first time in our comparative studies of Korsakoff and HD patients, an experimental manipulation (i.e., the introduction of verbal mediators) enhanced the learning performance of the HD patients more than that of the Korsakoff patients. As anticipated on the basis of their lack of aphasic symptoms, the HD patients were able to utilize language cues to facilitate associations between the cut-out figures and specific scenes and thereby improve their contextual memory. Conversely, the alcoholic Korsakoff patients, who are reputed to have deficits in verbal encoding (Butters & Cermak, 1980), were unable to utilize the stories in such a beneficial manner. It has been reported by Winocur and Kinsbourne (1978) that the contextual memory of alcoholic Korsakoff patients can be improved by increasing the saliency of cues present in the learning environment, but none of the procedures employed by these investigators involved the introduction of verbal mediators.

Although the present findings may support the thesis that HD and alcoholic Korsakoff patients have qualitatively distinct memory deficits, they offer little help in specifying the exact nature of the cognitive disorders. Clearly, our previous suggestion that HD patients lack the neuroanatomical structures necessary for storing information is untenable in light of this last study. If the HD patients had a storage problem, the introduction of verbal mediators should have been as unsuccessful as Biber et al.'s (1981) orientation task and Butters et al.'s (1979) use of increased rehearsal times and intertrial rest intervals. One point worth noting is that, unlike most of our previous studies that employed recall measures of retention, the investigation of pictorial memory required only recognition of the correct figure–background relationships. More compelling evidence that the HD patients' ability to recall previously presented stimuli is significantly more impaired than their capacity to recognize the same materials will be reviewed in the next investigation.

Like the HD patients, the patients with right-hemisphere lesions also benefited significantly from the introduction of the verbal mediators. The recitation of the stories apparently prompted a more complete analysis of the elements of the pictures and may also have provided valuable cues for linking the figures with specific contexts. Although there has been abundant documentation of the severe visuoperceptual deficits which accompany right-hemisphere damage (e.g., Benton, 1979; Milner, 1970), the feasibility of employing the linguistic capacities of the intact left hemisphere in rehabilitative efforts has not received adequate attention. Boller and DeRenzi (1967) have reported that right-hemisphere patients, as well as patients with left-hemisphere lesions and intact control subjects, find it easier to form associations between meaningful (i.e., verbalizable) than between meaningless (i.e., nonverbalizable) figures, but they did not evaluate whether their two patient groups would be differentially affected by imposing explicit verbal labels on the figures. The present findings suggest that it may be worthwhile to explore both the rehabilitative limits of verbal mediators and the mechanisms by which language can alter the perceptual and memory disorders of patients with right-hemisphere lesions.

The similarities and differences in the memory impairments of the HD and the AD patients should not go unmentioned. It is evident from the picture memory study that the dementias (like the amnesias) should not be treated as a single disorder. Whereas the HD patients can utilize language as a mnemonic aid for circumventing their pictorial memory problems, patients with Alzheimer's Disease may have lost this opportunity to employ linguistic mnemonics due to the aphasic symptoms usually associated with this disease process. The relevance of this difference for the demented patients' ability to remain in a noninstitutionalized setting should not escape an audience sophisticated in neuropsychology.

The second of our most recent investigations (Martone, Butters,

Payne, Becker, & Sax, 1984) emanated from a desire to examine what learning capacities were preserved in Huntington's Disease. For the past 15 years, research concerned with severe memory disorders has focused primarily on the patients' extensive anterograde and retrograde memory deficits (for review, see Butters, 1979; Butters & Cermak, 1980; Hirst, 1982; Squire, 1982). However, there has been a growing interest in those memory capacities that appear to be well preserved even in severely amnesic patients (Brooks & Baddeley, 1976; Cohen & Squire, 1980). There have been numerous demonstrations of the amnesics' ability to acquire and retain a variety of perceptual-motor skills on mirror-tracing, bimanual tracking, and pursuit-rotor tasks (Corkin, 1968; Cermak, Lewis, Butters, & Goodglass, 1973) despite the absence of any recollection by the patients of having performed the test previously. Good performances by amnesic patients have also been observed on tasks that are not primarily perceptual-motor in nature, such as rule-based paired-associate learning where word pairs are linked by a semantic or phonological rule (Winocur & Weiskrantz, 1976). Less formal demonstrations of preserved memory capacity have been reported by researchers who note that amnesic patients are often able to retain testing procedures across experimental sessions even when unable to recall the specific stimulus material (Corkin, 1968; Milner, Corkin, & Teuber, 1968). Such observations suggest that the acquisition and retention of at least some types of information are intact in patients with severe memory impairments.

A model to account for the pattern of preserved and impaired memory functions has recently been proposed (Cohen & Squire, 1980; Squire, 1982). It suggests that memory for information consisting of skills or procedures is spared in amnesia while memory for information which is data based or declarative in nature (e.g., specific facts) is impaired. According to this view, amnesic patients are able to learn and retain mirror-tracing and pursuit-rotor tasks because successful performance on these tests depends on the ability to learn and retain the procedures involved, but not on the ability to recall the specific content of the tasks. This proposed dissociation between two types of memory in amnesia was demonstrated experimentally by Cohen and Squire (1980) with a pattern-analyzing task that involved both skill learning (procedural knowledge) and verbal recognition (declarative knowledge). Subjects were required to read blocks of word triads that appeared as mirror images of themselves. Although half of the words were unique to each block, half were repeated on every block during the three test sessions. The results indicated that both the amnesic and normal control subjects showed significant and equivalent improvement at reading the unique, mirror-reflected triads over the three test days. Although the control subjects were much faster at reading the repeated than the unique words, the amnesic patients demonstrated only a slight improvement in reading speed for the repeated word triads. It seemed, then, that, although amnesic patients were able to learn and retain the general skills underlying mirror reading, they, unlike the normal control subjects, did not recognize that specific word triads had been presented on numerous trials. A verbal-recognition test administered following the skill-learning task confirmed that the Korsakoff patients could not identify the words employed on the mirror-reading task. In a recent report, Moscovitch (1984) replicated Cohen and Squire's (1980) findings with a task that requires patients with amnesic symptoms to read sentences written in transformed (i.e., rotated along the vertical axis) script.

The distinction between procedural and declarative information appears to characterize the memory defects of amnesic patients of numerous etiologies, but its generality to other populations of brain-damaged patients with memory disorders, such as progressive dementias, has yet to be fully evaluated. In the present study (Martone et al., 1984), the mirror reading task of Cohen and Squire (1980) was used to evaluate the skill learning and verbal recognition of HD, alcoholic Korsakoff, and normal control subjects. Because the mirror reading of unique words is a relatively pure indicator of skill learning, it was expected that normal acquisition would be reflected by a steady decrease in the time needed to read the unique, mirror-reflected word triads. Testing on the mirror-reading task was conducted on three successive days, with 60 trials (three blocks of 20 word triads) administered each day. The ability of the subjects to recognize verbal materials (i.e., declarative information) was assessed in two ways: (a) by any differences between the mirror reading of unique and repeated word triads; (b) by administering a verbal-recognition test following the last trial of the mirror-reading task. Normal subjects were expected to read the repeated word triads much faster than the unique triads and to be able to identify on the recognition test the words used on the mirror-reading task. Of these two measures, the word recognition test seemed to be the more uncontaminated indicator of declarative knowledge because the mirror reading of repeated words is affected by skill learning, priming effects (i.e., the recognition of one word elicits the recall of another) as well as by simple word recognition.

The major finding of this study (Martone et al., 1984) was a double dissociation between the HD and Korsakoff patients on rule learning and recognition memory. Figure 4 shows the mean latencies (log transformed) of the three subject groups for the unique word triads used on the mirror-reading task. Although both patient groups read the mirror-

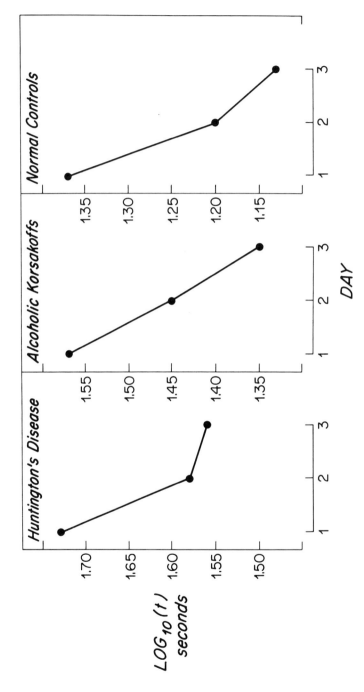

Figure 4. Performance of the three subject groups on the mirror reading of unique word triads. Mean time ($\log_{10}$ seconds) for the three test days (collapsing blocks) is shown.

reflected unique words more slowly than did the normal subjects, the alcoholic Korsakoff patients, but not the HD patients, demonstrated a normal rate of rule learning over the three test days. The Korsakoff patients and normal control subjects showed significant improvement from Day 1 to Day 2 and from Day 2 to Day 3, whereas the HD patients only improved from Day 1 to Day 2. The HD patients' mean latencies on Days 2 and 3 were essentially identical.

Figure 5 presents the performance of the three subject groups on both the unique and repeated triads over the three test days and the three blocks of 20 trials within each test day. The important result to note is the difference in latencies between the unique and repeated triads. This difference score (unique-repeated) is significantly greater for the normal controls and HD patients than for the alcoholic Korsakoff patients. That is, the HD patients and normal controls seem to recognize that certain triads are being repeated on each block of trials and subsequently identify these repeated mirror-reflected words in a short period of time. In contrast, the Korsakoff patients, who acquired the general rule in normal fashion, did not seem to recognize that specific word triads were being repeated on every block of 20 trials. This finding suggests that the Korsakoff patients are much more impaired in recognition memory than are the HD patients.

The latter conclusion was supported by the patients' performance on the recognition test that immediately followed administration of the mirror-reading task on Day 3 (Figure 6). This test required the subjects to identify from a 60-word list those words that had been used on the mirror-reading task. Thirty of the 60 words (15 unique words, 15 repeated words) had been presented on the rule learning task; the other 30 words were distractor (new) items. As shown in Figure 6, the HD patients could recognize (d') both unique and repeated words at a level comparable to the performance of the normal control subjects. However, the recognition scores of the Korsakoff patients for both unique and repeated words was significantly impaired in comparison to the performance of the HD and normal control groups.

The results of this investigation (Martone et al., 1984) suggest that the anterograde memory disorder of demented HD patients *cannot* be characterized by normal skill learning (i.e., procedural knowledge) paired with severely deficient recognition of specific verbal materials (i.e., declarative knowledge). In fact, the present results demonstrate a double dissociation between amnesic Korsakoff and demented HD patients on skill learning and verbal recognition. The Korsakoff patients, as had been reported previously by Cohen and Squire (1980), can acquire skills at a normal rate despite being severely impaired in their recognition memory. On the other hand, HD patients are impaired in

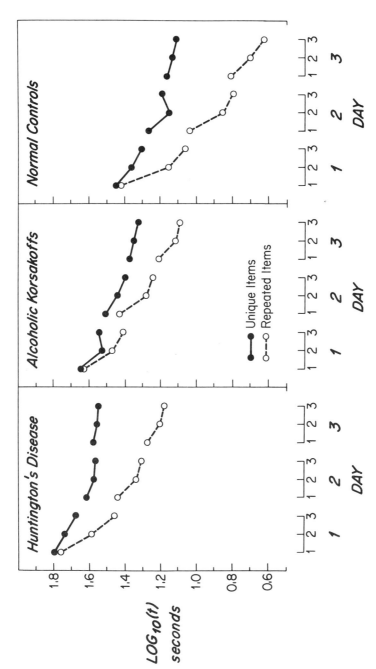

Figure 5. Performance of the three subject groups on the mirror reading of unique and repeated word triads. Mean time ($\log_{10}$ seconds) to read a word triad is shown for each block on all three test days.

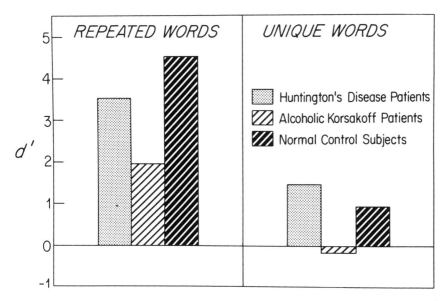

Figure 6. Mean recognition scores (d') of the three subject groups for both repeated and unique word triads.

their ability to acquire skills but retain their capacity to recognize previously presented verbal stimuli. The unsolicited casual remarks of the HD patients while performing the mirror-reading task are consistent with these quantitative results. Most of the HD patients and control subjects often uttered remarks like "I have seen these [words] before" while being shown the repeated words. Such acknowledgments were not offered by any of the alcoholic Korsakoff patients.

It is important to note that the HD patients' lack of skill learning cannot be attributed to the dysarthria that often accompanies their disorder. If dysarthria had limited the HD patients' mirror reading of unique word triads, it should have had the same debilitating effect for the repeated words. Given the large difference between HD patients' reading speed of the unique and repeated words, dysarthria does not appear to be a crucial factor in limiting the patients' acquisition of the mirror-reading skills. Furthermore, the HD patients who participated in this study were primarily in the middle stages of the disease process, and their dysarthria was relatively mild and was not considered a serious impediment to verbal communications.

In addition to emphasizing that important differences exist between the memory disorders of HD and alcoholic Korsakoff patients, the present study provides some clues as to the nature of the demented

HD patients' anterograde memory deficits. A review of our previous investigations concerned with the memory deficits of HD patients reveals that their memory performance seems most impaired when recall paradigms are employed. For example, when HD patients were asked to recall verbal materials after 3 to 18 seconds of distraction (i.e., the Brown-Peterson technique), their impaired performance was quantitatively similar to that of amnesic Korsakoff patients. However, in the present study, where a verbal recognition test (rather than a recall) paradigm was used, and on the previously discussed picture-context recognition task, the HD patients' performance was significantly superior to that of the Korsakoff patients. The alcoholic Korsakoff patients, like all amnesics, are impaired equally on recall and recognition tests of memory (for review, see Butters & Cermak, 1980; Hirst, 1982; Piercy, 1977), whereas the HD patients' impairments appear more prevalent and severe when recall rather than recognition is demanded. In retrospect, it appears likely that the very low MQs of the HD patients may be due to the fact that the Associative-Learning, Logical Memories, and Visual Reproduction tests of the unmodified Wechsler Memory Scale all require recall rather than recognition of previously presented materials.

If this proposed dissociation between the HD patients' recall and recognition abilities is substantiated by future studies, it will suggest that the HD patients' memory problems involve primarily an inability to search their recent and remote memories. Because recognition tests eliminate most of the patients' need to search their short- and long-term memories, they result in relatively intact performance in HD patients. The HD patients may adequately process and store verbal information but be unable to generate the search strategies needed to recover this material when a recall paradigm is employed. This interpretation is consistent with the HD patients' "flat" retrograde amnesia and with the previously discussed failures to improve recall from short-term memory by reducing proactive interference and by providing additional rehearsal time. Any patient with a general retrieval problem will have as much difficulty retrieving information from the remote as from the recent past.

It should be noted, however, that Biber et al.'s (1981) aforementioned findings on a facial recognition task are not consistent with this proposal concerning recall and recognition. Biber's investigation used a forced-choice recognition test to determine whether HD patients (as well as other patient populations) could identify photographs of faces they had viewed previously. The results showed that for all experimental conditions (baseline, high- and low-level orientation procedures) the HD patients were severely impaired in their recognition of faces. One possible explanation for the HD patients' poor recognition performance on this task involves the nature of the stimuli and the length of the exposure time employed. To assure that subjects could not rely on superficial cues (e.g., clothing, hair styles) to identify faces, Biber et al. selected photographs of cadets from the yearbook of a military academy. All of these cadets wore the same uniform and had short haircuts at the time their photographs were taken. Thus, to discriminate among the photographs, subjects had to focus their attention on the configurational features of the faces (e.g., the relationships among the eyes, nose, and mouth). Biber et al. further increased the difficulty of their test by limiting the exposure (i.e., study) time for each photograph to 7 seconds. Given the designed complexity of this task and the HD patients' inability to control their involuntary choreic movements, the HD patients may not have had sufficient time to analyze the critical configurational features of the faces. For example, due to their involuntary movements, the HD patients may have required 3 or 4 seconds to focus their attention on a given stimulus and subsequently had only 3 or 4 seconds to analyze the photograph of each cadet. If such were the case, it is not surprising then that the HD patients performed so poorly in comparison to normal control subjects and Korsakoff patients, both of whom had a full 7 seconds to analyze the critical features of each photograph.

In summary, the findings of the present and past studies suggest that the memory disorders of amnesic and of some demented populations have qualitatively distinct characteristics. Although amnesic patients with bilateral hippocampal or medical diencephalic lesions (e.g., alcoholic Korsakoff patients) are severely impaired in their recall and recognition of data-based verbal and nonverbal information, they retain the ability to learn and remember rule-based visuoperceptual, motor, and cognitive skills. In contrast, patients whose dementia results from atrophy or dysfunction of the basal ganglia (e.g., HD patients) appear to be retarded in their acquisition of general skills, whereas they continue to demonstrate a considerable ability to learn new facts when recognition tests of memory are employed. The possibility that limbic and basal ganglia structures may be involved in two distinct forms of memory is intriguing and deserving of further study with patients having other subcortical neurological disorders (e.g., patients with Parkinson's Disease, postencephalitic patients).

CONCLUSIONS

The results of these comparative studies of HD and Korsakoff patients reinforce conclusions based on a scrutiny of various forms of amnesia (for review, see Butters & Miliotis, 1985; Squire, 1982). These recent investigations of pictorial memory and of declarative and procedural knowledge, as well as the studies concerned with verbal learning, remote memory, and discrimination learning, suggest that HD, Korsakoff, and even Alzheimer patients fail to acquire and retrieve information for quite different reasons. The fact that all of the amnesic and demented groups discussed in this paper had low MQs revealed little about the nature of their impairments. Thus, reliance on a single quantitative measure of memory (e.g., the MQ) for the assessment of amnesic symptoms may have as many limitations as does the utilization of an isolated score on a naming or fluency test for the full description of aphasia.

Finally, I hope that this presentation has convinced you of the importance of experimental studies of memory for the clinical assessment of memory disorders. As I noted in my introduction, neither the practitioner of neuropsychology nor the experimentally oriented neuropsychologist can afford to ignore the fruits of each others' endeavors. Clinical practitioners should find ways of incorporating the reliable and valid findings of experimental studies of memory into their evaluation of patients, just as experimenters must always be aware of subtle clinical differences between and within the patient groups they study. I hope that clinical neuropsychology never suffers the malaise so common to academic psychology departments—namely, the total division of clinical and experimental programs. In my experience, clinical neuropsychology is a single specialty that requires and promotes the total interdependence of clinical practice and neurobehavioral research. This symbiotic relationship exists now and must continue to do so in the future to insure the continued growth and respect our specialty has enjoyed in the past 15 years.

REFERENCES

Albert, M. S., Butters, N., & Brandt, J. (1981a). Patterns of remote memory in amnesic and demented patients. *Archives of Neurology, 38,* 495–500.

Albert, M. S., Butters, N., & Brandt, J. (1981b). Development of remote memory loss in patients with Huntington's Disease. *Journal of Clinical Neuropsychology, 3,* 1–12.

Albert, M. S., Butters, N., & Levin, J. (1979). Temporal gradients in the retrograde amnesia of patients with alcoholic Korsakoff's disease. *Archives of Neurology, 36,* 211–216.

Benton, A. (1979). Visuoperceptive, visuospatial, and visuoconstructive disorders. In K. M. Heilman & E. Valenstein (Eds.), *Clinical neuropsychology* (pp. 186–232). New York: Oxford University Press.

Biber, C., Butters, N., Rosen, J., Gerstman, L., & Mattis, S. (1981). Encoding strategies and recognition of faces by alcoholic Korsakoff and other brain-damaged patients. *Journal of Clinical Neuropsychology, 3,* 315–330.

Boller, F., & DeRenzi, E. (1967). Relationship between visual memory defects and hemispheric locus of lesion. *Neurology, 17,* 1052–1058.

Bower, G. H., & Karlin, M. B. (1974). Depth of processing pictures of faces and recognition memory. *Journal of Experimental Psychology, 103,* 751–757.

Brooks, D. N., & Baddeley, A. D. (1976). What can amnesic patients learn? *Neuropsychologia, 14,* 111–122.

Butters, N. (1979). Amnesic disorders. In K. Heilman & E. Valenstein (Eds.), *Clinical neuropsychology* (pp. 439–474). New York: Oxford University Press.

Butters, N., & Albert, M. S. (1982). Processes underlying failures to recall remote events. In L. S. Cermak (Ed.), *Human memory and amnesia* (pp. 257–274). Hillsdale, N.J.: Lawrence Erlbaum Associates.

Butters, N., Albert, M. S., & Sax, D. (1979). Investigations of the memory disorders of patients with Huntington's Disease. In T. Chase, N. Wexler, & A. Barbeau (Eds.), *Advances in neurology, Volume 23: Huntington's Disease* (pp. 203–214). New York: Raven Press.

Butters, N., Albert, M. S., Sax, D. S., Miliotis, P., Nagode, J., & Sterste, A. (1983). The effect of verbal mediators on the pictorial memory of brain-damaged patients. *Neuropsychologia, 21,* 307–323.

Butters, N. & Cermak, L. S. (1980). *Alcoholic Korsakoff's Syndrome: An information-processing approach to amnesia,* New York: Academic Press.

Butters, N., & Grady, M. (1977). Effect of predistractor delay on the short-term memory performance of patients with Korsakoff's and Huntington's Disease. *Neuropsychologia, 13,* 701–705.

Butters, N., & Miliotis, P. (1985). Amnesic disorders. In K. Heilman & E. Valenstein (Eds.), *Clinical neuropsychology.* (2nd ed.) New York: Oxford University Press.

Butters, N., Miliotis, P., Albert, M. S., & Sax, D. S. (1984). Memory assessment: Evidence of the heterogeneity of amnesic symptoms. In G. Goldstein (Ed.), *Advances in clinical neuropsychology, Vol. 1,* New York: Plenum Press.

Butters, N., Tarlow, S., Cermak, L. S., & Sax, D. (1976). A comparison of the information processing deficits of patients with Huntington's Chorea and Korsakoff's syndrome. *Cortex, 12,* 134–144.

Butters, N., Sax, D., Montgomery, K., & Tarlow, S. (1978). Comparison of the neuropsychological deficits associated with early and advanced Huntington's Disease. *Archives of Neurology, 35,* 585–589.

Caine, E. D., Ebert, M. H., & Weingartner, H. (1978). An outline for the analysis of dementia: The memory disorder of Huntington's Disease. *Neurology, 27,* 1087–1092.

Cermak, L. S., Lewis, R., Butters, N., & Goodglass, H. (1973). Role of verbal mediation in performance of motor tasks by Korsakoff patients. *Perceptual and Motor Skills, 37,* 259–262.

Cohen, N., & Squire, L. R. (1980). Preserved learning and retention of pattern analyzing skills in amnesia: Dissociation of knowing how and knowing that. *Science, 210,* 207–210.

Corkin, S. (1968). Acquisition of motor skill after bilateral medial temporal-lobe excision. *Neuropsychologia, 6,* 255–265.

Hirst, W. (1982). The amnesic syndrome: Descriptions and explanations. *Psychological Bulletin, 91,* 435–460.

Martone, M., Butters, N., Payne, M., Becker, J., & Sax, D. S. (1984). Dissociations between skill learning and verbal recognition in amnesic and dementia. *Archives of Neurology, 41,* 965–970.

Meudell, P., Butters, N., & Montgomery, K. (1978). Role of rehearsal in the short-term memory performance of patients with Korsakoff's and Huntington's Disease. *Neuropsychologia, 16,* 507–510.

Miller, E. (1977). *Abnormal aging: The psychology of senile and presenile dementia.* London: Wiley.

Milner, B. (1970). Interhemispheric differences in the localization of psychological processes in man. *British Medical Bulletin, 27,* 272–275.

Milner, B., Corkin, S., & Teuber, H. L. (1968). Further analysis of the hippocampal amnesic syndrome: 14-year follow-up study of H. M. *Neuropsychologica, 6,* 215–234.

Moscovitch, M. (1984). The sufficient conditions for demonstrating preserved memory in amnesia. In L. Squire & N. Butters (Eds.), *The neuropsychology of memory.* New York: Guilford Press.

Oscar-Berman, M., & Zola-Morgan, S. M. (1980a). Comparative neuropsychology and Korsakoff's Syndrome. I. Spatial and visual reversal learning. *Neuropsychologia, 18,* 499–512.

Oscar-Berman, M., & Zola-Morgan, S. M. (1980b). Comparative neuropsychology and Korsakoff's Syndrome. II. Two-choice visual discrimination learning. *Neuropsychologia, 18,* 513–525.

Piercy, M. (1977). Experimental studies of the organic amnesic syndrome. In C. W. M. Whitty & O. L. Zangwill (Eds.), *Amnesia,* (2nd ed., pp. 1–51). London: Butterworths.

Shneidman, E. S. (1952). *Make A Picture Story Test.* The Psychological Corporation, New York.

Squire, L. R. (1981). Two forms of human amnesia: An analysis of forgetting. *The Journal of Neuroscience, 1,* 635–640.

Squire, L. (1982). The neuropsychology of memory. *Annual Review of Neurosciences, 5,* 241–273.

Weingartner, H., Caine, E., & Ebert, M. H. (1979). Imagery, encoding, and retrieval of information from memory: Some specific encoding-retrieval changes in Huntington's Disease. *Journal of Abnormal Psychology, 88,* 52–58.

Wilson, R. S., Kaszniak, A. W., & Fox, J. H. (1981). Remote memory in senile dementia. *Cortex, 17,* 41–48.

Winocur, G., & Kinsbourne, M. (1978). Contextual cueing as an aid to Korsakoff amnesics. *Neuropsychologia, 16,* 671–682.

Winocur, G., & Weiskrantz, L. (1976). An investigation of paired-associate learning in amnesic patients. *Neuropsychologia, 14,* 97–110.

13
Visual-Spatial Disabilities

NILS R. VARNEY and ABIGAIL B. SIVAN

INTRODUCTION

The cognitive disturbances to be discussed in this chapter have as their most significant feature an inability to appreciate the spatial aspects of visual experience. Like the aphasic disorders, visual-spatial deficits can be quite heterogeneous, both in the constellation of deficits shown by individual patients and in qualitative features of their abnormal performances. Unlike aphasic disorders, however, visual-spatial deficits often go unnoticed by the affected patient and his relatives and come to light only when directly evaluated with neuropsychological tests. Thus, without formal assessment, many of the disabilities to be discussed here would escape the notice of neuropsychologists, neurologists, or rehabilitation personnel.

There are a number of theoretical issues, some nearly a century old, that remain open to question today. In some instances, the theoretical controversy has little relevance to clinical assessment. However, in other areas, such as constructional apraxia, unresolved theoretical issues make clinical assessment more complex and/or inconclusive. For a more complete discussion of these theoretical issues, the reader should consult Benton (1979, 1982), DeRenzi (1982) and Ratcliff (1982).

NILS R. VARNEY • Veterans Administration Medical Center, and Department of Psychiatry, University of Iowa College of Medicine, Iowa City, IA 52240.   ABIGAIL B. SIVAN • Child Development Clinic, Department of Pediatrics, University of Iowa, Iowa City, IA 52242.

CONSTRUCTIONAL APRAXIA

Constructional praxis is a broad concept that applies to any type of performance in which parts must be joined or assembled into a single entity or object. Following the designation of Kleist (1912), its pathological counterpart is called "constructional apraxia." Kleist defined constructional apraxia as a "disturbance in formative activities such as assembling, building, and drawing in which the spatial form of the product proves unsuccessful" for reasons other than paralysis, ataxia, or apraxia of single movements (cited by Strauss, 1924). In other words, the constructionally apraxic patient cannot put things together in an organized manner despite being able to execute each of the individual movements required (e.g., draw lines, but not form a diamond; pick up blocks, but not build a pyramid; see Figure 1).

Assessment of constructional praxis is appropriate with a wide variety of clinical case material, but it is essential in the evaluation of patients with known or suspected right-hemisphere disease. Constructional apraxia is one of the most frequent behavioral sequalae of right-brain damage, and visual-constructional performance is an important means of detecting right-brain disease and of following its progression or remission. However, without formal specialized testing most constructional deficits are likely to go undiagnosed. Few patients are aware of being impaired, or offer any evidence of the disorder in daily behavior.

Kleist introduced the concept of constructional apraxia as a distinct variety of what was then termed "optic apraxia," a general label for all disturbances in visually guided behavior. He and Strauss (1924) also intended constructional apraxia to denote a specific type of visuoconstructional disability; a "perceptuo-motor" deficit in which intact visual perceptions could not be translated into appropriate action, with the responsible lesion being in the posterior-parietal area of the dominant hemisphere. Subsequent clinical observations confirmed that constructional apraxia was a distinct disability, but suggested that other aspects of the original Kleist-Strauss formulation were too narrow. Specifically, constructional deficits were observed among patients with right-brain lesions, and were often associated with a variety of other cognitive disturbances, such as the Gerstmann syndrome, visual-perceptive impairments, and general intellectual loss. As a result, constructional apraxia has come to be used in reference to most types of visual-constructional disability.

The fact that constructional apraxia is so broadly defined had resulted in the development of a number of specific spatial-organizational tasks to detect its presence. Test currently in clinical and investi-

VISUAL-SPATIAL DISABILITIES   385

gative use include freehand drawing, simple design copying, Rey's complex figure, stick arrangement, vertical-block construction, three-dimensional block construction, Kohs block design, and the Wechsler Adult Intelligence Scale (WAIS) Object Assembly (c.f., Lezak, 1976 for illustrative examples). Clearly, these diverse tasks are not equivalent in

Figure 1. Examples of abnormal constructional praxis.

their demands on sustained attention, capacity for deliberation, perceptual judgment, motor skills, higher intellectual functioning, or visual-perceptive ability. Nevertheless, all are measures of constructional praxis.

As constructional praxis tasks vary, so too do the performances of brain-injured patients. The most important distinction in this regard is between graphomotor performance (e.g., drawing, design copying) and assembling performance (e.g., block construction, stick arrangement). Benton (1967) and Dee (1970) found dissociation in the level of performance on these two general types of tasks to be relatively common, with a significant minority of constructionally apraxic patients being impaired on only one task or the other. However, even on more closely related performances, such as block construction and block design, there are significant differences in individual performance, frequency of failure, and differential sensitivity to right- and left-sided lesions (c.f., Benton, 1969, and Table 1).

For the most part, studies concerned with the relationship between constructional apraxia and locus of lesion have indicated that right-brain-injured patients are likely to show more frequent and more severe constructional deficits than those with left-sided lesions (e.g., Arrigoni & DeRenzi, 1964; Benton, 1969, 1967, 1968; Benton & Fogel, 1962; Piercy, Hécaen, & de Ajuriaguerra, 1960; Piercy & Smyth, 1962). However, the between-group differences observed here were on the order of two or three to one, too small a diference to suggest right-hemisphere dominance for constructional activities. Similarly, there is a tendency for constructional deficits to be more frequent in association with posterior lesions, but only relatively so, there being occasional cases resulting from anterior lesions as well (Benson & Barton, 1970; Benton, 1973;

Table 1. *Performances of Patients with Right and Left Sided Brain Lesions on Tests of Constructional Praxis (Benton, 1969b)*

Tests	Right	Left
Design copying	29%	14%
3-D construction	54%	23%
Stick construction	34%	26%
Block designs	34%	30%

Note. From "Constructional Apraxia" by A. L. Benton in *Contributions to Clinical Neuropsychology*, edited by A. L. Benton. Copyright 1969 by Aldine Publishing Company. Reprinted by permission.

Black & Strub, 1976). In practical terms, therefore, visuoconstructional deficits have no specific implications for side or site of cerebral lesion.

Given the many different cognitive demands involved in any constructional task, it could be suspected that each hemisphere makes a special contribution to visual-constructional performance. The idea most frequently mentioned in this regard, first advanced by Duensing (1953), is that right-sided lesions result in "spatio-agnostic" or perceptual deficits and that left-sided lesions result in "ideo-motor apractic" or executive deficits, the latter being similar to that of the Kleist-Strauss definition of constructional apraxia. Some selected case material has been offered in support of this view (c.f., Warrington, 1969). In addition, some constructional tasks have been found to be more differentially sensitive to right-brain lesions than others (Benton, 1967). However, Piercy and Smyth (1962) and Dee (1970) found that most visual-constructional deficits resulting from lesions of either hemisphere are closely associated with visual-perceptive deficits. A small minority of cases appeared to have executive constructional impairments, but these patients had right lesions as often as left lesions. Thus, although there is reason to suspect that each hemisphere makes a relatively different contribution to constructional performance, these differences are not sufficiently great to provide meaningful clinical diagnostic distinctions.

A more important correlate of constructional performance related to the side of the lesion is receptive language impairment. Benton (1973) found that most left brain-damaged patients who failed in three-dimensional block construction were aphasics with impaired Token Test performances. Aphasics with the poorest Token Test scores were the most frequently and severely impaired in block construction, and as Token Test scores improved, constructional deficits became increasingly less common. Those with normal Token Test sscores, whether aphasic or nonaphasic, usually showed intact block construction. However, some with the most severe receptive aphasias also performed normally in block construction, so constructional apraxia did not appear to be an inevitable correlate of sensory aphasia.

With constructional apraxia being a frequent correlate of unilateral lesions in either hemisphere, it would be surprising if visual-constructional impairments were not also common in association with general mental impairment. Indeed, brain-injured patients with estimated WAIS 1Q losses of 20 to 49 points are far more likely to show constructional deficits than those with more nearly normal intellectual functioning (Benton, 1969b). However, other aspects of the relationship between dementia and constructional praxis are somewhat less predictable. For example, the intellectually demanding WAIS Block Design subtest was not failed more frequently by deteriorated patients than a more basic block construction task, and the latter was failed more frequently than Block Design by less demented patients. In addition, constructional praxis was often intact among patients with significant intellectual loss, and was sometimes impaired among those with 1Q declines of less than 20 points. Thus, there would appear to be considerable room for individual variability in the relation between constructional performance and intellectual decline.

RECOGNITION AND DISCRIMINATION OF FACES

PROSOPAGNOSIA

Prosopagnosia is a quite uncommon disorder in which affected patients lose their ability to visually identify or recognize familiar faces, including those of their immediate family. Such patients are still able to identify familiar persons by their voice, and may succeed in visual recognition as the result of distinctive clothing, hair style, gait, or posture. They are also usually able to offer verbal descriptions of faces, and are typically able to name objects on sight.

Given the fact that prosopagnosia is quite uncommon, it has taken some time to identify its neuropathologic correlates. For a while, it was suspected that posterior right-hemisphere lesions were responsible. However, this was based on clinical rather than post-mortem studies (c.f., Hécaen & Angelergues, 1963). By contrast, autopsy data has consistently indicated bilateral posterior involvement, typically with lesions involving the mesial occipital-temporal regions bilaterally (cf., Benton, 1979). Similar results were reported by Damasio, Damasio, and Van Hoesen (1982) based on CT scan data. The findings of Damasio et al., (1982) are also important in that they indicated that unilateral lesions to either the right or left mesial occipital-temporal region did not result in prosopagnosia. Thus, current evidence strongly suggests that bilateral, symmetric lesions are essential for the development of the disorder.

Clinical evaluation of the suspected prosopagnostic will usually require a certain amount of improvisation. In addition to having the patient identify familiar individuals first hand, it is often of value to have the patient's family provide photos of familiar family members for identification. This may reveal impairments that are not so apparent in first hand clinical evaluation, in which it is difficult to control for nonrelevant stimuli (e.g., familiar clothing or hair styles). It is also usually appropriate to employ a set of photographs of prominant public figures (e.g., Time or Newsweek covers).

Facial Discrimination

Because prosopagnosia appeared to be such a rare disorder, and because there was, in the early 1960s, a lingering suspicion that the disorder could result from only right posterior lesions, a number of neuropsychological laboratories developed evaluation procedures for the assessment of "subclinical" prosopagnosia. The suspicion of these investigators was that the use of objective tests of facial recognition would reveal many more subtle disturbances in facial recognition. The tests developed in various laboratories were similar in that they required patients to identify a photograph of an unfamiliar person shown from various angles and in different lighting conditions. An example is shown in Figure 2.

In accord with initial assumptions, disturbances in unfamiliar facial recognition proved to be far more common than prosopagnosia, and were considerably more common in association with right- as opposed to left-sided lesions. However, even those cases with grossly defective unfamiliar facial recognition performance were free of prosopagnosia, and prosopagnostic patients were able to perform normally in unfamiliar facial recognition. Thus, rather than identifying subtle varieties of prosopagnosia, these tests identified a completely different disturbance in the perception or discrimination of faces.

The anatomic correlates of impaired facial discrimination are considerably less specific than in prosopagnosia. One major difference is that defects in facial discrimination can occur from unilateral lesions of either the right or left hemisphere. Among patients with right-posterior lesions, impaired performance will be quite frequent (i.e., 53%). This is significantly greater than the proportion of failure observed among right anterior (26%), left posterior (27%), or left anterior (14%) patients (Benton, Hamsher, Varney, & Spreen, 1983, p. 41). In addition, virtually all instances of impaired facial discrimination observed in association with left-sided lesions were also associated with receptive language impairment. Thus, although an impairment in facial discrimination has no specific implications for side or locus of lesion, it indicates right-sided disease when the failure is not associated with receptive aphasia. Given the frequency with which right-posterior patients fail, performances of this type are particularly useful in the detection of such lesions.

DRESSING APRAXIA

As its name suggests, dressing apraxia is an inability to clothe oneself. The affected patient will usually show a disorganized pattern

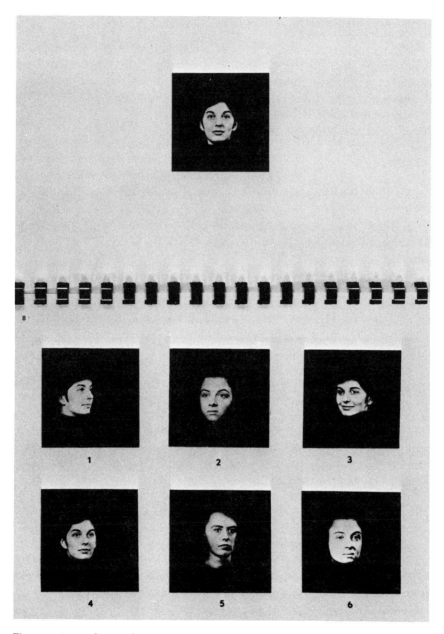

Figure 2. A sample item from a test of facial recognition. From *Contributions to Neuropsychological Assessment* by A. L. Benton, K. Hamsher, N. Varney, and O. Spreen. Copyright 1983 by Oxford University Press. Reprinted by permission.

Table 2. Frequency of Defective Performance in Facial Recognition

Group	Percentage defect
Normal subjects (n=286)	3.5
Right anterior (n=23)	26
Right posterior (n=36)	53
Left anterior nonaphasic (n=15)	0
Left posterior nonaphasic (n=14)	0
Left anterior, aphasic without comprehension defect (n=5)	0
Left posterior, aphasic without comprehension defect (n=8)	0
Left anterior, aphasic with comprehension defect (n=17)	29
Left posterior, aphasic with comprehension defect (n=27)	44

Note. From Contributions to Neuropsychological Assessment by A. L. Benton, K. Hamsher, N. Varney, and O. Spreen. Copyright 1983 by Oxford University Press. Reprinted by permission.

of orienting clothes to his or her body, and as often as not, will simply make stabbing motions at the clothes with the arms or legs.

As a clinical entity, dressing apraxia is easy to evaluate. One need only present a bathrobe or lab coat folded in a disorganized manner and ask the patient to put the garment on. Tying shoe laces is another suitable task. It is important to keep in mind that many dressing apraxics are mortally embarrassed by their deficit, and frequently mistake the disorder as positive proof that they are senile or in "a second childhood." As a result, some brief counseling will usually be in order.

Dressing apraxia is usually seen in association with lesions of the right hemisphere, and is often observed in association with visual-constructional disability. However, the disorder is only one-fourth as common as constructional apraxia. Dressing apraxia is very rare among left-brain-injured patients. Specific aspects of the relationship between dressing apraxia and locus of lesion within the right hemisphere remain open to question.

There has been relatively little in the way of systematic investigation as to the nature of dressing apraxia, and some deny its existence as an independent symptom (cf., Poeck, 1969). Clearly, patients with unilateral neglect will have difficulty orienting clothing and their own bodies for independent dressing. Similarly, given the spatially amorphous nature of clothes, one would expect patients with significant visual-spatial disorientation to also have problems with dressing. At the same time, no investigator has established a consistent relationship between dressing apraxia and other more "primary" cognitive impairments. Thus, at present, it seems appropriate to regard dressing apraxia as an independent symptom.

OPTIC ATAXIA

Optic ataxia refers to a specific disability that is evidenced by inaccurate reaching for objects and localizing stimuli in space. As a consequence, patients with optic ataxia often show impairment in activities of daily living, such as lighting a cigarette, dressing, or cutting food, while at the same time often showing adequate performances on tests of perception not requiring a pointing response. Thus, the problem in optic ataxia is not in the process of perception but in the mechanism of localization or visually guided movements. In other words, the problem is not purely a perceptual one but a perceptual-motor one concerned with the connection or disconnection between the perception and the motor movement.

Optic ataxia was first noted as a part of a syndrome (Balint, 1909) that included, in addition to the defect in visually guided movements, paralysis of gaze with an inability to shift the eyes and inattention to objects and events in other parts of the visual field. Early reports of patients with the syndrome indicated that these patients suffered injuries in the occipital and posterior parietal-temporal areas. More recent evidence suggests that the defect is more specifically localized in Brodmann's area 7, which contains two types of neural cells particularly relevant to visual orientation (Ratcliff, 1982).

The diagnosis of optic ataxia may be made on the basis of tests that document the patient's inability to perform coordinated perceptual-motor tasks when these are performed under visual control. This can be achieved by showing that the patient is capable of doing "on-body" tasks while at the same time not being able to do the same task "off-body." Pointing to parts of the body can be executed by the patient on his own body, whereas pointing to the same parts of the examiner cannot. Well-rehearsed movements such as buttoning or bringing a cigarette to the patient's mouth are easily performed until the patient is asked to observe what he is doing and perhaps light the examiner's cigarette. Then it appears that "visual guidance actually seemed to disrupt the performance" (Damasio & Benton, 1979, p. 172).

Whole-body movements, such as walking from room to room, are unimpaired. In contrast, movements that require differentiation of the upper extremities, such as finger localization and finger naming, are defective. Damasio and Benton (1979) even suggest that the recognition defect extends beyond the hand to the wrist, elbow and shoulder.

There are many characteristics which are sometimes associated with the defect, but none have been found to be highly correlated. Included in these are those in the Balint syndrome (ocular fixation and inattention to other parts of the visual field), visual-field defects, abnormal eye movements, difficulties in sound localization, topographical disorientation, and bilaterality of lesions.

TOPOGRAPHICAL DISORIENTATION

Another type of spatial disorientation is the failure to locate familiar places (cities or buildings) on a map of one's native country or one's hometown. Associated with this disability is the failure to describe the spatial arrangement of even more familiar surroundings such as rooms in one's house or streets in one's neighborhood.

There appear to be two levels on which this deficit occurs. One level is represented by a failure in perception, such as visual-field defect in which one part of the map or representation is systematically distorted or ignored. The second level is defined by a failure on the part of the patient to make a mental representation of a particular spatial configuration that has been repeatedly experienced in the past.

At one time, it was thought that this deficit was produced by bilateral disease involving the occipital and posterior-parietal and temporal areas (Benton, 1982). More recent evidence, however, suggests that right hemisphere lesions in the retro-rolandic areas can produce the deficit (Benton, Levin, & Van Allen, 1974; Hécaen & Angelergues, 1963).

Topographical orientation or disorientation has been assessed by a number of tasks. Most popular are map-reading or path finding. Perhaps most interesting are the tasks used by Benton, Levin, and Van Allen (1974). These involve two verbal tests as well as one nonverbal test. The first test required patients to state in what state various cities were located (e.g., Chicago, Little Rock). The second test consisted of 15 questions asking the patient to state the direction he would travel in going from one city to another. The third, nonverbal test consisted of a large map of the United States on which the boundaries of the states were indicated and on which the location of Chicago was shown. The patient was asked to mark the location of 10 cities and states on the map (see Figure 3).

These tests were differentially associated with the presence of brain injury. Performance on the Verbal Association test did not discriminate between the control and brain-diseased groups. In contrast, map localization elicited defective performance in the brain-diseased patients irrespective of the side of the lesion. Only the vector score on the map localization task discriminated significantly between brain-damaged groups. Selected patients with brain disease showed shifts in localization indicative of neglect of the half space contralateral to the

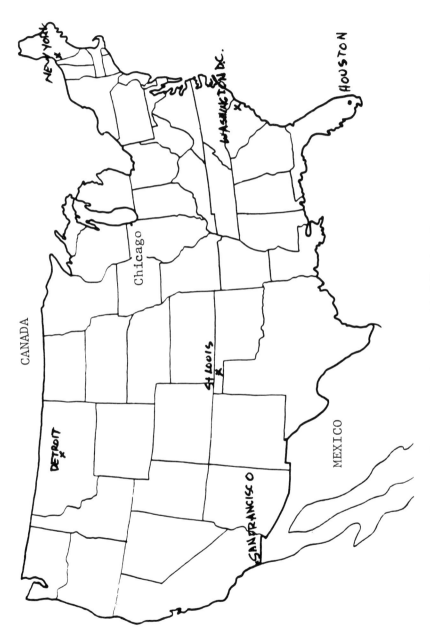

Figure 3. A test for topographical disorientation.

side of the lesion. These patients, however, did not include all those with visual-field defects. It is noteworthy that the verbal tests make no contribution to the diagnosis of this spatial deficit.

NEGLECT

In unilateral neglect, patients fail to respond to stimuli on one side of the body (i.e., hemispace). In its most clinically florid form, neglect can be quite obvious. Affected patients will run into door frames, read only half of printed material, eat only from half of their plate, and generally fail to acknowledge any stimuli from the affected hemispace. Most cases with left-unilateral neglect suffer right parietal lesions. However, right-sided neglect can also be observed occasionally, and parietal-lobe lesions are not always involved.

One of the easiest procedures for demonstrating neglect is to have patients point to the midpoint of a ruler or line. The patient with left-sided neglect (right-hemisphere lesion) will locate the midpoint well to the right of center. That is, if the patient does not realize that there is a left half of space, the localization will err away from the neglected side.

A second, and more sensitive procedure for demonstrating neglect, first developed by Albert (1973), involves a line cancellation task such as that shown in Figure 4. The patient's task is simply to mark each

Figure 4. A line cancellation task for assessment of neglect.

line. In profound neglect, all or nearly all lines to one side will be omitted (Figure 5). In less severe neglect or inattention disorders, a significant proportion of lines will be omitted (Figure 6). The usual procedure is to score the left, right, and middle thirds of the task separately.

Lateral neglect may also be demonstrated on drawing or copying tasks. Some sample items are shown in Figures 7a and 7b. Similarly, neglect may also be shown on tests of constructional praxis or reading (i.e., hemi-alexia). Although lateral neglect is often observed in association with visual-field defects, many hemianopic patients do not neglect, and some neglecting patients are not hemianoptic.

Assessment of tactile neglect can be accomplished in two ways. One involves a tactile line bisection procedure similar to the visual line bisection task described above. Another described by Heilman (1979) involves touching lateral parts of the patient. The patient with tactile neglect will point to the correct body part, but on the incorrect side. For example, if one were to touch the left knee, the neglecting patient would point to the right knee. To use this procedure, one must first determine that primary somesthesis is normal bilaterally, at least in the dermal areas being assessed. Auditory neglect is somewhat more difficult to assess without specialized equipment. However, one potentially useful procedure is to stand to the patient's left or right side, speak into his left or right ear somewhat softly, and have him perform an aurally

Figure 5. A line cancellation task with an example of neglect.

VISUAL-SPATIAL DISABILITIES

Figure 6. A line cancellation task with an example of inattention.

mediated task such as the Token Test or WAIS Arithmetic. Performance on the neglecting side will be poorer than on the normal side.

Unlike patients with only visual-field defects, it is usually difficult to train neglecting patients to compensate for their impairment. For most affected patients, neglect may result in total vocational disability. For obvious reasons, visual neglect is a prima facie grounds for revocation of a driver's license.

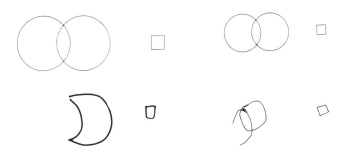

Figure 7. An example of neglect from a copying task.

PERCEPTION OF DIRECTION

There is a considerable body of literature that suggests that perception of direction is more strongly mediated in the right hemisphere, both for visual and tactile performance in normal and brain-injured populations. Some of these tasks, such as those developed by Carmon and Benton (1969) involve rather complex instrumentation that is not commercially available. However, there remain a number of different tasks that are either commercially available or that can be constructed at modest expense.

The most easily accessible is the Judgement of Line Orientation test of Benton et al. (1983), a sample item of which is shown in Figure 7. The task requires patients to identify the directional orientation of 2 cm lines from an array of 11 cm lines that are presented simultaneously. Impaired performance on the task was observed in only 10% of patients with left brain lesions as compared to 46% of those with right-sided lesions. Among right brain-damaged patients with posterior lesions, impaired judgment of line orientation was observed in 75% of cases. By contrast, only 13% of cases with perirolandic lesions and 0% of those with prefrontal lesions were impaired. Thus, impaired performance is quite infrequent except among right-brain-damaged patients with posterior lesions. A tachistoscopic analogue of this test has also been developed (Benton, Hannay, & Varney, 1975) in which stimuli are exposed for 3 seconds, and are followed 1 second later with the multiple choice response array in which the orientation and placement of the response alternatives are in exactly the same location visually as the stimuli. Although this task would appear relatively easier than the booklet task, this has not proved to be the case. Using the same patient population, Benton, Varney, and Hamsher (1977) found that failure of the tachistoscopic version was considerably more frequent than failure of the booklet version, with the difference being explained by the fact that patients with more anterior right-sided lesions failed the tachistoscopic version. No cases have been observed in which failure of the booklet version was associated with normal performance on the tachistoscopic version.

DeRenzi, Faglioni, & Scotti (1970) have developed a three-dimensional task that may be used tactually or visually. Basically, this task involves a single rod that is on a stand, and that can be oriented in any direction in three-dimensional space. Subjects are required to match the orientation of the stimulus with a similar piece of equipment. Error is determined in reference to the end point of the response rod in reference to three-dimensional space coordinates (i.e., latitude and longitude).

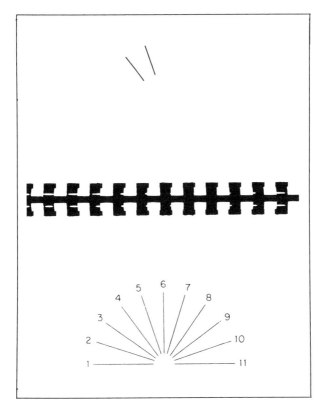

Figure 8. A sample item from a test of perception of direction. From *Contributions to Neuropsychological Assessment* by A. L. Benton, K. Hamsher, N. Varney, and O. Spreen. Copyright 1983 by Oxford University Press. Reprinted by permission.

Benton et al. (1978) have developed a two-dimensional tactile perception of direction task that is closely similar to the visually mediated task of Benton, Hannay, and Varney (1975). Here, stimuli are metal rods imbedded in a wooden board.

CONCLUSION

At one time, this chapter might have been titled "Cognitive Sequalae of Right-Hemisphere Disease." Up until the early 1970s, most of the specific cognitive deficits falling into the general class of visual-spatial impairments were thought to be mediated primarily in the right hemisphere, with the right hemisphere being as dominant for these skills as the left hemisphere is for language. Unfortunately, or at least inconveniently, it has been repeatedly demonstrated that visual-spatial defects associated with right-brain lesions also occur regularly in association with left-sided lesions. Although some visual-spatial disorders occur relatively more frequently in association with right-sided lesions, the ratio is usually on the order of 2 to 1, far lower than the 20 to 1 ratio associated with left-hemisphere dominance for language. Thus, although visual-spatial disorders represent symptoms of primary interest in right-hemisphere disease, they are not in and of themselves diagnostic of right- versus left-sided lesions. The clinical fallout from this situation is two-fold. First, there are few, if any, tasks that by themselves conclusively demonstrate right-hemisphere damage. Second, there are no visual-spatial tasks which can be used within the context of receptive aphasia to reliably prove additional right-sided damage.

REFERENCES

Albert, M. (1973). A simple test of visual neglect. *Neurology, 23,* 658–664.

Arrigoni, G., & DeRenzi, E. (1964). Constructional apraxia and hemispheric locus of lesion. *Cortex, 1,* 180–197.

Balint, R. (1909). Die seelenlahmung des "Schauens." Cites in Heilman and Valenstein (Eds) *Clinical Neuropsychology.* Oxford University Press, New York, 1979.

Benson, F. and Barton, M. (1970). Disturbances in constructional ability. *Cortex, 6,* 29–49.

Benton, A. L. (1969). The visual retention test as a constructional praxis task. *Confinia Neurologica, 22,* 141–155.

Benton, A. L. (1967). Construction apraxia and the minor hemisphere. *Confinia Neurologica, 29,* 1–16.

Benton, A. L. (1968). Differential effects of frontal lobe disease. *Neuropsychologia, 6,* 53–60.

Benton, A. L. (1969a). Disorders of spatial orientation. In P. Vinken & G. Bruyn (Eds.), *Handbook of Clinical Neurology: Vol. III* (pp. 212–228). Amsterdam: North-Holland.

Benton, A. L. (1969b). Constructional apraxia. In A. L. Benton (Ed.), *Contributions to Clinical Neuropsychology,* (pp. 129–142). Chicago: Aldine.

Benton, A. L. (1973). Visuoconstructional disability in patients with cerebral disease. *Documenta Ophthalmologica, 34,* 67–76.

Benton, A. L. (1979). Visuoperceptive, visuospatial, and visuoconstructive disorders. In K. Heilman & E. Valenstein (Eds.), *Clinical Neuropsychology* (pp. 186–232). New York: Oxford University Press.

Benton, A. L. (1982). Spatial thinking in neurological patients: Historical aspects. In M. Potegal (Ed.), *Spatial Abilities.* New York: Academic Press.

Benton, A. L., & Fogel, M. (1962). Three-dimensional constructional praxis. *Archives of Neurology, 7,* 347–359.

Benton, A. L., Hamsher, K., Varney, N., & Spreen, O. (1983). *Contributions to neuropsychological assessment.* New York: Oxford University Press.

Benton, A. L., Levin, H., & Van Allen, M. (1974). Geographic orientation in patients with unilateral cerebral disease. *Neuropsychologia, 12,* 183–191.

Benton, A. L., Varney, N., & Hamsher, K. (1978). Visuospatial judgement. *Archives of Neurology, 35,* 364–367.

Black, F. W., & Strub, R. L. (1976). Constructional apraxia in patients with discrete missle wounds of the brain. *Brain, 12,* 212–220.

Carmon, A., & Benton, A. L. (1969). Tactile perception of direction and number in patients with unilateral cerebral disease. *Neurology, 19,* 525–532.

Damasio, A. R., & Benton, A. L. (1979). Impairment of hand movements under visual guidance. *Neurology, 29,* 170–178.

Damasio, A. R., Damasio, H., & Van Hoesen, G. (1982). Prosopagnosia: Anatomical basis and behavioral mechanisms. *Neurology, 32,* 331–341.

Dee, H. L. (1970). Visuoconstructional and visuoperceptive deficits in patients with unilateral cerebral lesions. *Neuropsychologia, 8,* 305–314.

DeRenzi, E. (1982). *Disorders of space exploration and cognition.* New York: John Wiley & Sons.

DeRenzi, E., Faglioni, P., & Scotti, G. (1970). Hemispheric contribution to exploration of space through the visual and tactile modalities. *Cortex, 6,* 191–203.

Duensing, F. (1953). Raumagnostische and ideatorisch-apraxische Stoerung des gestalten Handelns. *Deutsche A. f. Nervenheilkunde, 170,* 72–94.

Hécaen, H. & Angelergues, R. (1963). *La cecite psychique.* Paris: Masson.

Heilman, L. (1979). Neglect and related disorders. In K. Heilman & E. Valenstein (Eds.), *Clinical neuropsychology* (pp. 268–302). New York: Oxford University Press.

Kleist, K. (1934). Kriegverletzungen des Gehirns in ihrer Bedeutung fuer die Hirnlokalisation und Hirnpathologie. In O. von Schjerning (Ed.), *Handbuch der aerztlichen Erfahrung im Weltkriege* (Vol. 4, pp. 343–390). Leipzig: Barth.

Lezak, M. (1976). *Neuropsychological assessment.* New York: Oxford University Press.

Piercy, M., & Smyth, V. (1962). Right hemisphere dominance for certain nonverbal intellectual skills. *Brain, 85,* 775–790.

Piercy, M., Hécaen, H., & de Ajuriaguerra, J. (1960). Constructional apraxia associated with cerebral lesions: Left and right cases compared. *Brain, 83,* 225–242.

Poeck, K. (1969). Modern trends in clinical neuropsychology. In A. L. Benton (Ed.), *Contributions to clinical neuropsychology* (pp. 1–39). Chicago: Aldine.

Ratcliff, G. (1982). Disturbances of spatial orientation associated with cerebral lesions. In M. Potegal (Ed.), *Spatial abilities* (pp. 301–311). New York: Academic Press.

Strauss, H. (1924). Ueber konstructive Apraxie. *Monatsschrift fuer Psychiatric und Neurologic, 63,* 739–748.

Warrington, E. (1969). Constructional apraxia. In Vinken and Bruyn (eds) *Handbook of Clinical Neurology,* Vol. 4, North Holland: Amsterdam.

Index

Academic achievement
  Halstead-Reitan Battery, 169
  Wide Range Achievement Test, 93–94
Activities of Daily Living (ADL) concept, 61–62
Affect, 148–149
Alcoholic Korsakoff's syndrome, See Korsakoff's syndrome
Alcoholism, 20–21
Alzheimer's disease, 22, 23, 24
American Psychological Association, 244
Amnesia, 362–363
  See also Memory
Anxiety, 51
Aphasia, 329–360
  Aphasia Language Performance Scales, 353
  Aphasia Screening Test, 353
  Appraisal of Language Disturbance, 344–345
  Auditory Comprehension Test for Sentences, 355
  Boston Diagnostic Aphasia Examination, 345–348
  Boston Naming Test, 356
  Communicative Abilities in Daily Living, 352
  Eisenson's Examination for, 338–339
  flexible battery, 132
  Functional Communication Profile, 351–352
  history of assessment in, 329–330
  Language Modalities Test, 339
  Minnesota Test for Differential Diagnosis, 340–341

Aphasia (Cont.)
  Neurosensory Center Comprehensive Examination, 341–342
  polyglots and, 354
  Porch Index of Communicative Abilities, 342–344
  principles of testing in, 331–338
  Reporter Test, 355
  Sklar Aphasia Scale, 352–353
  Token Test, 354–355
  Western Aphasia Battery, 348–351
Aphasia Language Performance Scales, 353
Aphasia Screening Test, 353
Appraisal of Language Disturbance, 344–345
Apraxia
  constructional, 384–388
  dressing apraxia, 389–391
  flexible battery and, 132–133
Arithmetic skills. See Calculational skills
Assessment
  Huntington's disease, 305–309
  neuropsychology and, 76–80
  normative criterion assessment, 151
  screening versus, 107–108
  test batteries and, 142–146
  See also Diagnosis; Tests and test batteries; entries under names of specific tests
Ataxia (optic), 392–393
Attention
  Halstead-Reitan Battery, 170
  measures of, 125
Attitude (of patient), 47, 50–51

403

Auditory Comprehension Test for Sentences, 355
Auditory Trail Making, 125

Background Interference Procedure (BIP), 97–99
Behavior
  brain damage and, 79
  Halstead-Reitan Battery and, 168
  test batteries and, 136–137
  test interpretation and, 147
Behavioral environment. See Environment
Behavioral neurology, 7–8
Bender-Gestalt Test, 94–99
Benton Line Orientation Test, 131
Bias, 138–142
Boston Diagnostic Aphasia Examination, 345–348
Boston Naming Test, 356
Brain damage/dysfunction
  behavior and, 79
  Bender-Gestalt Test and, 96
  intelligence and, 82–83
  Luria-Nebraska Battery and, 205–207
  Minnesota Multiphasic Personality Inventory and, 100–101
  organicity and, 78–79
  schizophrenia and, 31–32, 85–86, 101
  treatment and, 54
  Wechsler Memory Scale and, 90–91
Brief Psychiatric Rating Scale, 60

Calculational skills
  Halstead-Reitan Battery, 171
  Luria-Nebraska Battery, 195, 211
Cerebral blood flow, 24
Cerebrovascular disease, 21–25
Children, 149–150
Clinical interview, 45–73, 123
  assessment methods, 56–69
  complaint/presenting problem, 48
  diagnostic data, 49–50
  evaluation framework, 4–7
  history and, 63–65
  level of consciousness and, 48–49
  medical records and, 65–69
  patient's attitude and, 50–51
  premorbid intelligence and, 49
  purposes of, 46–53
  rating scales and, 60–63

Clinical interview (Cont.)
  severity of impairment, 51–52
  test selection and, 52–53
  treatment planning and, 53–56
Clinical neuropsychology. See Neuropsychology
Cognition
  focal cognitive deficits, 128–134
  Halstead-Reitan Battery, 172–173
  nonfocal cognitive deficits, 124–128
Cognitive rehabilitation, 18
Cognitive tests, 80–99
  Bender-Gestalt test, 94–99
  Wechsler Intelligence Scales, 81–89
  Wechsler Memory Scale, 89–93
  Wide Range Achievement Test, 93–94
Communication skills, 170–171
Communicative Abilities in Daily Living, 352
Complaint, 48
Computer, 12–19
  administration/scoring and, 12–14
  bias and, 141–142
  cognitive rehabilitation and, 18
  data analysis/interpretation and, 14–18
  geriatrics and, 18–19
Concentration, 170
Confidentiality, 12
Constructional apraxia, 384–388
Continuous Performance Test, 125
Cooperation (patient's), 48–49

Data identification, 164
Deficits
  focal cognitive deficits, 128–134
  nonfocal cognitive deficits, 124–128
  vocabulary in, 122
Demography, 84–85
Depression
  clinical interview and, 51
  neuropsychology and, 30–31
  specific interviews and, 59
Diagnosis
  brain damage and, 78
  clinical interview and, 47, 49–50
  disagreement in, 57
  neuropsychology and, 77
  standard battery and, 80–107
  See also Assessment
Digit Symbol Substitution Test, 125

INDEX   405

Direction, 398–399
Dressing apraxia, 389–391
DSM-III, 57–59

Eisenson's Examination for Aphasia, 338–339
Emotion. See Affect
Environment, 55–56
Epilepsy, 102
Ethics, 12

Face recognition tests, 131–132
Facial discrimination, 389
Family, 54–55
Family history, 50
Feedback, 7
Fixed batteries
  Halstead-Reitan Battery/Luria-Nebraska Battery compared, 245
  See also Halstead-Reitan Battery; Luria-Nebraska Battery
Flexible batteries, 9–11, 121–134
  arguments for use of, 133
  bias and, 141–142
  clinical interview and, 45, 47, 52–53
  focal cognitive deficits, 128–134
  nonfocal cognitive deficits, 124–128
  See also entries under names of specific tests
Focal cognitive deficits, 128–134
Focal tests, 140–141
Forensic neuropsychology, 28
Frontal lobe damage, 125–126
Functional assessment, 61–63
Functional Communication Profile, 351–353

General interview, 57–59
General rating scales, 60–61
Geriatrics, 27–28
  cerebrovascular disease, 21–25
  computer and, 18–19
  neuropsychology and, 2
Glasgow Coma Scale, 60

Halstead, W. C., 236–243
Halstead-Reitan Battery, 155–192
  bias and, 139
  clinical application of, 163–174
  development of, 156–163
  differences with Luria-Nebraska Battery, 252–261

Halstead-Reitan Battery (Cont.)
  dominance of, 277
  future developments of, 184–189
  history of, 236–243
  Huntington's disease and, 305–307
  intelligence comparisons, 294–295
  limitations of, 278
  Luria-Nebraska Battery compared, 261–268, 309–311, 317–318
  rater comparisons, 279–281
  report examples, 174–184
  schizophrenia and, 312–314
  similarities with Luria-Nebraska Battery, 243–252
  statistical comparisons, 281–294
Hamilton Depression Scale, 59
Handedness, 147
Head injury
  clinical interview and, 50, 63–65
  neuropsychology and, 19–20
Hemispheric specialization
  focal cognitive deficits, 128–134
  intelligence scales and, 87–88
  Luria-Nebraska Battery, 196
  memory tests, 91
  personality tests, 102
History (patient)
  clinical interview and, 49–50
  evaluation framework, 3–7
  purposes of, 46–53
Huntington's disease
  Halstead-Reitan Battery, 305–307
  Luria-Nebraska Battery, 307–309
  memory disorders in, 362–379

Idiographic testing, 146
Individual-centered normative approach, 9
Individual differences
  relevance of, 76
  testing and, 77
Individualized test batteries, 47
Infarction (thalamic), 318–324
Intelligence
  Halstead-Reitan Battery, 169
  Halstead-Reitan Battery/Luria-Nebraska Battery compared, 294–295
  Huntington's disease and, 305–306
  Luria-Nebraska Battery, 195, 211–212
  memory disorders and, 363
  origins of testing for, 76

Intelligence (Cont.)
    thalamic infarction and, 322
    See also Premorbid intelligence
Interview. See Clinical interview
Intramodal tests, 144–146

Korsakoff's syndrome, 127
    memory disorders, 362–379
    See also Alcoholism

Language
    flexible battery and, 132
    Halstead-Reitan Battery and, 170–171
    Huntington's disease and, 307
    See also Aphasia
Language Modalities Test, 339
Law, 28
Level of consciousness, 46, 48–49
Level of deficit. See Severity of impairment
Luria, Alexander, 236–243
Luria-Nebraska Battery, 193–233
    advantages of, 278
    compared to Halstead-Reitan Battery, 261–268, 309–311, 317–318
    differences with Halstead-Reitan Battery, 252–261
    history of, 236–243
    Huntington's disease and, 307–309
    intelligence comparisons, 294–295
    interpretation of, 204–227
    item pattern and qualitative analysis, 226–227
    localization and, 219–223
    memory assessment and, 295–298
    pattern analysis, 213–219, 223–226
    rater comparisons, 279–281
    reliability of, 197–198
    scales of, 194–196
    schizophrenia and, 314–317
    similarities with Halstead-Reitan Battery, 243–252
    statistical comparisons, 281–294
    thalamic infarction and, 318–324
    validity of, 198–203

Make-A-Picture-Story (MAPS), 367–370
Malingering, 28
Medical neuropsychology, 25–26
Medical records
    clinical interview and, 65–69
    evaluation framework, 4

Memory disorders, 361–382
    cognitive tests, 89–93
    flexible battery in, 127–128
    Halstead-Reitan Battery and, 170
    Halstead-Reitan Battery/Luria-Nebraska Battery compared, 246
    Korsakoff's syndrome/Huntington's disease compared, 362–379
    Luria-Nebraska Battery and, 195, 211, 295–298
    thalamic infarction and, 319, 321, 323–324
Mental retardation, 76
Michigan Alcoholism Screening Test (MAST), 59–60
Minnesota Multiphasic Personality Inventory (MMPI), 58, 99–103
Minnesota Test for Differential Diagnosis of Aphasia, 340–341
Motor disorders
    flexible battery, 132–133
    Luria-Nebraska Battery, 194, 208–209

Naming
    aphasia tests, 335–336
    Boston Naming Test, 356
Neglect, 395–397
Neurodiagnosis. See Assessment; Diagnosis
Neurological examination, 1
Neuropsychological batteries. See Tests and test batteries
Neuropsychology
    assessment role and, 76–80
    computers in, 12–19
    credentialing in, 79
    depression and, 30–31
    evaluation approaches, 7–11
    evaluation framework in, 3–7
    forensic, 28
    future of, 108–110
    geriatric, 27–28
    medical, 25–26
    new developments in, 19–25
    rehabilitation and, 28–30
    role in, 1–3
    schizophrenia and, 31–32
    single case investigations, 26–27
Neurosensory Center Comprehensive Examination for Aphasia, 341–342
Nonfocal cognitive deficit, 124–128

Optic ataxia, 392–393
Organicity, 77–79
  Bender-Gestalt Test, 95–96
  Halstead-Reitan/Luria-Nebraska compared, 280
  Rorschach test, 103–104

Pathognomonic scale, 196, 212
Patient-centered testing. See Flexible batteries
Patient's attitude, 47, 50–51
Pattern analysis, 85–86
Pediatrics, 149–150
Perception of direction, 398–399
Personality tests, 99–107
  Halstead-Reitan Battery, 173
  Minnesota Multiphasic Personality Inventory, 99–103
  Rorschach Test, 103–106
  Thematic Apperception Test, 107
Pharmacology, 311–312
Physical therapy, 61–62
Pittsburgh Initial Neuropsychological Test System (PINTS), 144
Polyglots, 354
Porch Index of Communicative Abilities, 342–344
Premorbid intelligence
  clinical interview and, 46–47, 49
  intelligence scales and, 83
  See also Intelligence
Presenting problem, 48
Problem solving, 172, 173
Prosopagnosia, 131–132, 388
Psychology, 76–77
Psychometric testing, 76–77
  See also Tests and test batteries; entries under names of specific tests

Quality of life, 55

Rating scales, 60–63
Reading skills
  aphasia and, 336–337
  Luria-Nebraska Battery and, 211
Referral, 166–167
Rehabilitation
  functional assessment and, 61–62
  Halstead-Reitan Battery and, 187–188
  neuropsychology and, 28–30
Reporter Test, 355

Rey Auditory Verbal Learning Test, 128
Rhythm, 194, 209
Rorschach Test, 103–106

Schizophrenia
  brain damage contrasted, 85–86, 101
  Halstead-Reitan Battery and, 312–314
  Luria-Nebraska Battery, 314–317
  neuropsychology and, 31–32
  personality tests and, 100
Screening
  assessment versus, 107–108
  Halstead-Reitan Battery/Luria-Nebraska Battery compared, 246
  Luria-Nebraska Battery in, 318–324
  test batteries and, 142–144
Sensory-perceptual/motor functioning, 169
Severity of impairment, 51–52
Sex differences
  alcoholism, 20
  cognitive tests, 88
Significant others, 4
Single case investigations, 26–27
Sklar Aphasia Scale, 352–353
Specific interviews, 59–60
Specific rating scales, 61
Speech, 195, 210
Standard batteries. See Halstead-Reitan Battery; Luria-Nebraska Battery
Stroke
  neuropsychology and, 21–25
  patient's attitude and, 50, 51
Symbol Digit Modalities Test, 125–126

Tactile stimuli, 194–195, 209–210
Tests and test batteries
  affect and, 148–149
  aphasia and, 331–338
  characteristics of, 136–137
  childhood and, 149–150
  clinical interview and, 45–46, 47–48, 53–56, 58
  computer and, 12–19
  evaluation approaches, 9–11
  evolving nature of, 123
  flexible battery and, 121–134
  future of, 108–110
  history of, 77–79
  idiographic testing, 146
  interpretation of results, 146–148
  intramodal tests, 144–146

Tests and test batteries (Cont.)
  level of consciousness and, 48–49
  level of impairment and, 52
  normative/criterion assessment, 151
  role of, 75–119
  screening batteries, 142–144
  selection of, 52–53
  specific interviews and, 59
  targeted tests, 122
  theoretical bias in development of, 138–142
  types of, 137–138
  value of, 79–80
  See also Cognitive tests; Personality tests; entries under names of specific tests
Thalamic infarction, 318–324
Thematic Apperception Test, 107
Token Test, 354–355
Topographical disorientation, 393–395
Treatment. See Rehabilitation

Uniform batteries, 141–142
  See also Halstead-Reitan Battery; Luria-Nebraska Battery

Visual-spatial deficits, 383–401
  constructional apraxia, 384–388
  dressing apraxia, 389–391
  facial discrimination, 389
  focal cognitive deficits, 128
  Halstead-Reitan Battery, 171
  Luria-Nebraska Battery, 195, 210
  memory disorders and, 364
  neglect, 395–397
  optic ataxia, 392–393
  perception of direction, 398–399
  prosopagnosia, 388
  tests of, 122, 130–131
  topographical disorientation, 393–395

Wechsler Block Design, 130
Wechsler Intelligence Scale, 81–89, 122
Wechsler Memory Scale, 89–93, 122, 127–128, 296–297
Western Aphasia Battery, 348–351
Wide Range Achievement Test, 93–94
Wisconsin Card Sorting Test, 126
Word Fluency Test, 126
Wordlist generation, 126
Writing
  aphasia and, 336–337
  Luria-Nebraska Battery and, 195, 210–211

Youth. See Children